Methods for NEURAL ENSEMBLE RECORDINGS

METHODS & NEW FRONTIERS IN NEUROSCIENCE

Series Editors
Sidney A. Simon, Ph.D.
Miguel A.L. Nicolelis, M.D., Ph.D.

Published Titles

Apoptosis in Neurobiology
Yusuf A. Hannun, M.D., Professor of Biomedical Research and Chairman/Department of Biochemistry and Molecular Biology, Medical University of South Carolina
Rose-Mary Boustany, M.D., tenured Associate Professor of Pediatrics and Neurobiology, Duke University Medical Center

Methods for Neural Ensemble Recordings
Miguel A.L. Nicolelis, M.D., Ph.D., Professor of Neurobiology and Biomedical Engineering, Duke University Medical Center

Methods of Behavioral Analysis in Neuroscience
Jerry J. Buccafusco, Ph.D., Alzheimer's Research Center, Professor of Pharmacology and Toxicology, Professor of Psychiatry and Health Behavior, Medical College of Georgia

Neural Prostheses for Restoration of Sensory and Motor Function
John K. Chapin, Ph.D., Professor of Physiology and Pharmacology, State University of New York Health Science Center
Karen A. Moxon, Ph.D., Assistant Professor/School of Biomedical Engineering, Science, and Health Systems, Drexel University

Computational Neuroscience: Realistic Modeling for Experimentalists
Eric DeSchutter, M.D., Ph.D., Professor/Department of Medicine, University of Antwerp

Methods in Pain Research
Lawrence Kruger, Ph.D., Professor or Neurobiology (Emeritus), UCLA School of Medicine and Brain Research Institute

Motor Neurobiology of the Spinal Cord
Timothy C. Cope, Ph.D., Professor of Physiology, Emory University School of Medicine

Nicotinic Receptors in the Nervous System
Edward D. Levin, Ph.D., Associate Professor/Department of Psychiatry and Pharmacology and Molecular Cancer Biology and Department of Psychiatry and Behavioral Sciences, Duke University School of Medicine

Methods in Genomic Neuroscience
Helmin R. Chin, Ph.D., Genetics Research Branch, NIMH, NIH
Steven O. Moldin, Ph.D, Genetics Research Branch, NIMH, NIH

Methods in Chemosensory Research
Sidney A. Simon, Ph.D., Professor of Neurobiology, Biomedical Engineering, and Anesthesiology, Duke University
Miguel A.L. Nicolelis, M.D., Ph.D., Professor of Neurobiology and Biomedical Engineering, Duke University

The Somatosensory System: Deciphering the Brain's Own Body Image,
Randall J. Nelson, Ph.D., Professor of Anatomy and Neurobiology,
University of Tennessee Health Sciences Center

New Concepts in Cerebral Ischemia
Rick C. S. Lin, Ph.D., Professor of Anatomy, University of Mississippi Medical Center

DNA Arrays: Technologies and Experimental Strategies
Elena Grigorenko, Ph.D., Technology Development Group, Millennium Pharmaceuticals

Methods for Alcohol-Related Neuroscience Research
Yuan Liu, Ph.D., National Institute of Neurological Disorders and Stroke, National Institutes of Health
David M. Lovinger, Ph.D., Laboratory of Integrative Neuroscience, NIAAA

Methods for NEURAL ENSEMBLE RECORDINGS

Edited by Miguel A.L. Nicolelis

CRC PRESS

Boca Raton London New York Washington, D.C.

Library of Congress Cataloging-in-Publication Data

Methods for neural ensemble recordings / edited by Miguel A.L. Nicolelis.
p. cm. — (CRC Press methods in the life sciences. Methods & new frontiers in neuroscience)
Includes bibliographical references and index.
ISBN 0-8493-3351-2 (alk. paper)
1. Electroencephalography. 2. Microelectrodes. 3. Neurons. I. Nicolelis, Miguel A. L. II. Series.
[DNLM: 1. Neurons—physiology. 2. Brain—physiology. 3. Monitoring, Physiologic—methods. 4. Electrodes. WL 102.5M592 1998]
QP376.5.M47 1998
573.8′53619′028--dc21
DNLM/DLC
for Library of Congress

98-38795
CIP

This book contains information obtained from authentic and highly regarded sources. Reprinted material is quoted with permission, and sources are indicated. A wide variety of references are listed. Reasonable efforts have been made to publish reliable data and information, but the authors and the publisher cannot assume responsibility for the validity of all materials or for the consequences of their use.

Direct all inquiries to CRC Press LLC, 2000 N.W. Corporate Blvd., Boca Raton, Florida 33431.

International Standard Book Number 0-8493-3351-2
Library of Congress Card Number 98-38795
Printed in the United States of America 3 4 5 6 7 8 9 0
Printed on acid-free paper

The Editor

Miguel A. L. Nicolelis was born in Sao Paulo, Brazil, in 1961. He received an M.D. degree from the University of Sao Paulo Medical School in 1984 and a Ph.D. in physiology in 1988 from the Department of Physiology in the Institute of Biomedical Science at the University of Sao Paulo. From 1989 to 1992, Dr. Nicolelis was a postdoctoral fellow in the Department of Physiology at Hahnemann University. In 1994, Dr. Nicolelis joined the faculty of the Department of Neurobiology at Duke University, where he is currently an associate professor.

Dr. Nicolelis' research focuses on the investigation of the basic physiological mechanism through which populations of cortical and subcortical neurons encode tactile information. His published work includes studies on the development, plasticity, and normal function of the mammalian somatosensory system. For his research efforts, Dr. Nicolelis has received the Oswaldo Cruz Award for excellence in biomedical research (1984), the Whitehead Scholar Award (1994), a fellowship from the McDonnel Pew Foundation (1994), the Whitehall Young Investigator Fellowship (1994), and the Klingenstein Fellowship Award (1996).

Contributors

Luiz Antonio Baccalá, Ph.D.
Escola Politécnica
Universidade de São Paulo
São Paulo, Brazil

Amy Brisben, Ph.D.
Department of Neurobiology
Duke University Medical Center
Durham, North Carolina

John K. Chapin, Ph.D.
Department of Neurobiology and Anatomy
Allegheny University of the Health Sciences
Philadelphia, Pennsylvania

Amish S. Dave, B.S.
Department of Organismal Biology and Anatomy
The University of Chicago
Chicago, Illinois

Sam A. Deadwyler, Ph.D.
Department of Physiology and Pharmacology and Neuroscience Program
Wake Forest University School of Medicine
Winston-Salem, North Carolina

George L. Gerstein, Ph.D.
Department of Neuroscience
University of Pennsylvania
Philadelphia, Pennsylvania

John J. Gilpin
Department of Organismal Biology and Anatomy
The University of Chicago
Chicago, Illinois

Robert E. Hampson, Ph.D.
Department of Physiology and Pharmacology and Neuroscience Program
Wake Forest University School of Medicine
Winston-Salem, North Carolina

Mark Laubach, Ph.D.
Department of Neurobiology
Duke University Medical Center
Durham, North Carolina

Daniel Margoliash, Ph.D.
Department of Organismal Biology and Anatomy and Committee on Neurobiology
The University of Chicago
Chicago, Illinois

Karen A. Moxon, Ph.D.
Department of Neurobiology
and Anatomy
Allegheny University of the Health
Sciences
Philadelphia, Pennsylvania

Miguel A. L. Nicolelis, M.D., Ph.D.
Department of Neurobiology
Duke University Medical Center
Durham, North Carolina

Koichi Sameshima, M.D., Ph.D.
Faculdade de Medicina
Universidade de São Paulo
São Paulo, Brazil

Edward M. Schmidt, Ph.D.
Laboratory of Neural Control
National Institute of Neurological
Disorders and Stroke
National Institutes of Health
Bethesda, Maryland

Cornelius Schwarz, M.D., Ph.D.
Department of Physiology and
Neuroscience
New York University Medical Center
New York, New York

Christopher R. Stambaugh, B.S.
Department of Neurobiology
Duke University Medical Center
Durham, North Carolina

John P. Welsh, Ph.D.
Department of Physiology and
Neuroscience
New York University Medical Center
New York, New York

Bruce C. Wheeler, Ph.D.
Electrical and Computer Engineering
Department and
Beckman Institute
University of Illinois
at Urbana-Champaign
Urbana, Illinois

Albert C. Yu, M.D.
Committee on Neurobiology
The University of Chicago
Chicago, Illinois

Contents

Foreword

Technological advances often have great impact on the subsequent progress of a research field. A good example is the introduction of the oscilloscope for measuring weak, brief, electrical brain events. The oscilloscope, in conjunction with the design of microelectrodes and procedures for amplifying weak signals, led to a golden age of neurophysiological discoveries from the 1930s to 1950s when basic principles of brain function now described in textbooks were first revealed. Recently, a potentially comparable advance has been made with the introduction of noninvasive methods for imaging regional differences in neuronal activity in the human brain.

In this volume, the authors consider another technological breakthrough that has the potential to complement, equal, and even exceed the other two. A major problem in understanding how the central nervous system works is that information processing in any brain depends on the coordinated activity of large neuronal populations. Although valuable information can be gathered either by measuring the overall responsiveness of entire brain regions (e.g., in field potential and EEG recordings) or by serially recording the activity of individual neurons, one at a time, neither of these approaches provides direct windows to the investigation of the principles that underlie the collective and interactive responses of the many neurons involved in processing a given event. Thus, until now we have had only provocative glimpses of what the "processing machine" of the central nervous system is really about. Nonetheless, we have made much progress from these glimpses. We now know that information processing of a particular event occurs in many brain regions at once, and it almost certainly recruits nearby neurons differently. Consequently, the only effective way to study these distributed neural processes is to simultaneously sample the firing profiles of large numbers of single neurons within and across many brain structures.

The great advantage of simultaneously recording from many neurons has long been recognized, but practical applications of this approach have been extremely difficult to implement. Advances in the design and production of parallel recording systems, microelectrodes, and data collection, display, and analysis were all needed. Over the last decade, many technical and conceptual problems have been addressed, with clear examples of great success. These goals are being realized, and a growing number of researchers are on the forefront of applying these methods to a host of waiting problems.

The value of this volume is to introduce and explain in detail the progress that has been made in techniques and procedures for simultaneous many-neuron recordings. The authors are the leaders of this advance. There is much to learn from their efforts, but more important, these studies illustrate the changes that have taken place. The future is here, and we should start designing experiments, collecting data, and altering textbooks.

Jon H. Kaas
Centennial Professor of Psychology
Vanderbilt University
Nashville, Tennessee

Introduction

During the last decade, more neuroscientists have become interested in investigating the properties of populations of single neurons in a variety of species and preparations. As a result of these efforts, a large number of new electrophysiological methods and multivariate data analysis procedures have been added to the arsenal of extracellular neurophysiology. However, most of these methods have only been described in original papers in which little space can be devoted to a detailed description of new techniques. The main goal of this book, therefore, is to provide an in-depth description of the large spectrum of new techniques for neural ensemble recordings for those who want to employ these methods in their research but have not been able to find a single reference text to guide their initial efforts.

In this book, the fundamental methodological challenges involved in neural ensemble recordings are covered in eleven chapters. The first of these challenges involves choosing the type of multielectrode design that best fits the research question, the neural circuit, and the animal species chosen for a project. To help the reader explore the multiple options available today, a large variety of multielectrode designs are reviewed in Chapter 1. To complement this discussion, Chapter 2 introduces the many issues that have to be addressed when one opts to design one's own multielectrode array. Next, Chapter 3 describes the current trends in hardware and software used to sample the extracellular activity of hundreds of neurons simultaneously. Chapter 4 introduces the fundamentals of single neuron isolation and provides an overview of how these methods can be used for on-line, automatic discrimination of action potentials in extracellular recordings. Then, Chapters 5 through 7 illustrate particular experimental preparations which have taken advantage of current methods for neural ensemble recordings. Chapter 5 illustrates the use of such an approach for studying the properties of the cerebellar cortex. Chapter 6 introduces the reader to a new approach for investigating information coding in the singing bird. This section of the book concludes in Chapter 7 with the description of methods for multisite neural ensemble recordings in behaving primates.

The final four chapters of the book deal with another great challenge of this area of research—methods for neural ensemble data analysis. Chapter 8 introduces the reader to classical and new correlation-based techniques to analyze population recordings. Chapter 9 describes how directed coherence analysis, a method formerly employed to analyze EEG recordings, can be applied to the analysis of data derived from populations of single neurons located in different brain areas. In Chapter 10, the reader is introduced to a data analysis routine based on several classical multivariate statistical methods, such as principal component analysis, factor analysis, and discriminant analysis. The book concludes (Chapter 11) by providing the reader with a brief comparison of several data analysis methods described in this volume.

I could not let this opportunity pass without thanking some of the many people who made this book a reality. First, I would like to thank my good friend and colleague at Duke, Dr. Sidney Simon, the editor of the CRC Methods in Neuroscience Series, for giving me the opportunity to put this book together. Without his enthusiasm and continuous support, this project would never have been done. I would also like to thank the authors for their enthusiasm, patience, and dedication, and for sticking with a "rookie" editor all the way through. Finally, I thank Barbara Norwitz, a CRC senior editor, for her support and for helping us at critical stages of this project.

Miguel A.L. Nicolelis
Duke University
Durham, North Carolina
November 1998

Chapter

Electrodes for Many Single Neuron Recordings

Edward M. Schmidt

Contents

1.1 Introduction

For many years, neurophysiologists have investigated the activity of the brain by examining the firing patterns of single neurons one at a time while stimuli were presented or the animal performed trained tasks. Depending upon the size of the electrode employed, the activity of several neurons was sometimes recorded simultaneously and the resultant record separated into single-unit activity by hardware devices or computer programs. Although many scientists would have liked to record from many neurons simultaneously, the electrode technology and computer processing were not readily available. A number of electrode technologies have been and

0-8493-3351-2/99/$0.00+$.50

are being developed (see reference 1 for an earlier review), and low-cost, high-performance computers are now available. The area of multiple simultaneous single neuron recording will be the wave of the future. In this first chapter, we review the microelectrodes and drives used for *in vivo* recording from the central nervous system.

1.2 Wire Electrodes

One of the simplest techniques for recording many single neurons is the use of small-diameter insulated wires that are simply cut to expose the cross-sectional surface area of the end of the wire. This is also one of the oldest techniques employed for chronic recording of many neurons. Strumwasser[2] individually implanted four to six 80-µm-diameter stainless-steel wires in the brains of ground squirrels and was able to record discharges of single neurons for periods of a week or more. Olds[3,4] modified the technique and used 67-µm-diameter nichrome wires and implanted nine of them in the brain of each rat.

The majority of investigators who have used the wire electrode technique have simply cut off the end of the wire. However, a few investigators have etched fine wires to a few-micron tip diameter and then reinsulated the wire and exposed only the sharpened tip.[5,6] An alternate technique is to grind preinsulated wires to a conical tip.[7-9]

Table 1.1 summarizes the different materials and wire sizes that have been used for the construction of wire electrodes for chronic recording.

1.2.1 Electrode Insertion

A number of techniques have been described for the insertion of wire electrodes or wire bundles into the brain for long-term recording (see Chapters 5–7 for examples). Recordings from cortical structures have been obtained by inserting the individual wires,[10,13] or a preformed group of wires[20] into the cortex. Burns et al.[6] glued a 30-µm-diameter platinum wire with hot sucrose to a 150-µm-diameter tungsten needle that had been etched to a fine point. The assembly was inserted at a 45° angle to the surface of the cortex for a distance of approximately 2 mm. After the sucrose dissolved, the tungsten needle was removed leaving the fine wire in the cortex.

Wire bundles are usually used for recording from deep brain structures. As the diameter of the wire decreases, they are no longer stiff enough to be inserted directly into the brain. Marg and Adams[5] used a group of five free wires bound together at their tips by a fine plastic tube. The tube, placed on the surface of the brain, held the wires together and kept them from buckling as they were inserted into the brain. O'Keefe and Bouma[14] made up electrode arrays and passed the free end of the wires through the barrel of a hypodermic needle until they protruded. The needle was then inserted into the cortex at the desired location. The free end of each individual wire was then pulled, cutting it off within the brain. After all wires were cut off, the

TABLE 1.1
Materials and Wire Sizes that Have Been Employed for Long-Term Chronic Recording

Wire type	Diameter (μm)	Preparation	Reference
Iridium	37.5	Etched	10
Platinum	30	Etched, platinum black	6
Platinum–iridium	25	Cut off	11
Platinum–iridium	25	Cut off	12
Platinum–iridium	25	Cut off	13
Nichrome	67	Cut off	3
Nichrome	65	Cut off	14
Nichrome	62	Cut off	15
Nichrome	60	Cut off	16
Nichrome	25	Cut off	14
Stainless Steel	80	Cut off	2
Stainless Steel	70	Cut off	17
Stainless Steel	50	Cut off	18
Stainless Steel	25	Cut off	19
Stainless Steel	18	Cut off	19
Tungsten	50	Etched	5
Tungsten	50	Cut off	20
Tungsten	25	Ground	9
Tungsten	13	Cut off	21

needle was removed. In a later technique, the ends of eight 25-μm-diameter wires were cut level with each other and attached to a thin probe by dipping them into melted carbowax. The assembly was inserted into the brain and after half a minute, the wax dissolved and the probe was withdrawn. Palmer[11] employed a similar technique using polyethylene glycol to temporarily glue the wires to a sharpened 150-μm-diameter tungsten rod.

Hypodermic tubing has been used to strengthen wire bundle arrays for placement in deep structures. Chorover[12] passed the wires through the hypo tubing and then bound the free wire ends together with dextrose, which dissolved in the brain after insertion. Four to ten strands of 13-μm-diameter tungsten wire have been glued together with cyanoacrylate and then glued inside a 30 gauge hypo tube.[21] The electrode assembly was then lowered into the basal ganglia of awake behaving primates with a microdrive. Kubie[22] devised a method for constructing a head-mounted drive assembly for a bundle of ten 30-μm-diameter nichrome wires inserted into a 26 gauge hypo tube. The three adjusting screws in the head-mounted assembly advanced the wire bundle 454 μm per turn. Diana et al.[19] described a single screw, head-mounted microdrive that housed four 18-μm- or 25-μm-diameter stainless steel wires within a 24 gauge hypo tube. The drive advanced the wire bundle 300 μm per turn of the adjusting screw.

By using 50-μm-diameter stainless steel wire, sucrose can be used to glue the wires together during insertion without the aid of a stiffening device.

1.2.2 Recording Characteristics

Little information is available about the recording characteristics of the different materials and wire sizes that have been used for wire electrodes. Better single unit isolation was reported using 13-μm-diameter tungsten wire as compared to 25-μm-diameter tungsten or stainless steel wire.[21] Diana et al.[19] reported that 18-μm-diameter stainless steel wires provided more single units than 25-μm-diameter wires. The type of insulation used was also a factor. With some of the insulation materials, the electrode impedances were quite low, indicating that a much larger surface area was present than just the cut-off end of the wire. They also compared the recording ability of wires to sharpened tungsten electrodes and found similar yields with the 18-μm-diameter stainless steel electrodes. Kaltenbach and Gerstein[9] reported that sharpened 25-μm-diameter tungsten wires resulted in better single unit isolation than cut-off wires. However, Nicolelis et al.[20] reported that 50-μm-diameter blunt wires proved to be more suitable for long-term recordings than fine wire tips.

The speed of insertion of the wires or wire bundles may be an important factor in the success of an implant. The only insertion speed reported in the literature was ~100 μm/min.[20] This slow speed may have minimized neuronal damage during insertion.

The wire type and insulation material used to make wire electrodes may have a profound effect on their long-term recording properties. Different metals have markedly different corrosion properties in a saline environment. Detailed studies on tissue reaction to the different metals are essential for selecting the "best" metal for long-term recording. The insulation on the wire is another factor that can affect long-term recordings. The first concern with insulating materials is tissue reaction, while the second is long-term adhesion of the insulation to the electrode. Teflon[20] and Parylene-C®[23,24] have demonstrated encouraging results for long-term implants.

Because it is very difficult to conduct satisfactory comparison studies covering wire size and tip shape, much of the work on obtaining long-term recordings will remain art rather than science.

1.2.3 Floating Electrodes

Wire electrodes or wire bundles work well for recording from deep structures or from cortical areas of small animals such as rats where little relative movement exists between skull and brain. For larger animals, such as cats and primates, sufficient brain movement occurs during free head movements to render stiff wire electrodes fixed to the skull useless for cortical recording. Burns et al.[6] were the first to recognize this problem and implanted individual 30-μm-diameter platinum wires, with a harpoon, at a 45° angle to the cortex. Salcman and Bak[25] etched the tip of a

Teflon-coated, 25-μm-diameter platinum/iridium wire and inserted it into a 1.5- to 2.0-mm-long, tapered micropipette to form a flexible electrode. The glass-insulated and strengthened tip of the wire was inserted into the cortex through a craniotomy. The free end of the wire was attached to a skull-mounted connector and a chamber of methyl–methacrylate covered the skull defect. The chamber was filled with saline to reestablish CSF pressure. A later development of this "floating" microelectrode incorporated a 25-μm-diameter gold wire that was welded to an etched 37.5-μm-diameter iridium electrode.[10] The electrode and gold wire were insulated with Parylene-C. The flexibility of the gold wire allowed long-term recording of cortical neurons in primates.[23] An attempt to simplify the construction of floating electrodes was described by Legendy et al.[13] using 25-μm-diameter platinum/iridium wires that were cut at a bevel at the time of implantation. A loop was formed in the wire to provide strain relief. Since the platinum/iridium wire was much stiffer than the gold wire, this type of electrode was more easily dislodged by rapid head movements of the animal.

A novel chronically implantable floating electrode has been described by Kennedy.[26] The electrode consists of an insulated gold wire (or wires), with end exposed, fixed inside a hollow glass cone. A piece of sciatic nerve is placed inside the cone prior to implantation in the cortex. Cortical neurites grow into the cone from surrounding neurons, and their activity is recorded via the wire (or wires) in the cone. Thus the electrode combines standard wire recording techniques with neural regeneration.

1.2.4 Improved Single-Unit Separation

Wire electrodes usually record the activity of several neurons in the vicinity of the tip. Differences in wave shape can usually be used to separate the records into single unit activity.[27-29] At times, different neurons can produce similar wave shapes, and separation of the record into single unit activity is not possible. Two special electrodes have been designed to help overcome this problem. The "stereotrode"[30] was constructed by twisting two 25-μm-diameter platinum/iridium wires together and cutting the ends flush. Each electrode records a neuron's activity from a slightly different position, providing a pair of signals for discriminating single-unit activity. The technique has been extended to a tetrode electrode[31] that consists of four 12-μm-diameter nichrome wires that are twisted together and inserted into a 30 gauge guide tube. The tip of the resultant electrode was beveled on four sides at a 30° angle. The stereotrode and tetrode provide better isolation of single-unit activity but are more difficult to manufacture than straight wire electrodes.

Currently under development by Uwe Thomas Recording (Marburg, Germany) are two multicore quartz fiber electrodes. The first electrode consists of a 26-μm-diameter platinum/10% tungsten central core surrounded by three, 14-μm-diameter platinum (90%)/tungsten (10%) wires encased in a 96-μm-diameter quartz sheath as shown in Figure 1.1 (upper). Grinding the tip at different angles changes the displacement between the central core and the three surrounding wires. Using the same manufacturing technology, an electrode has been produced with

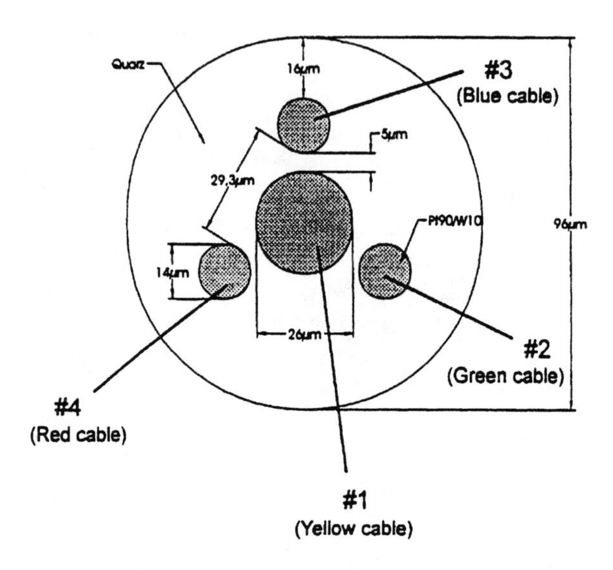

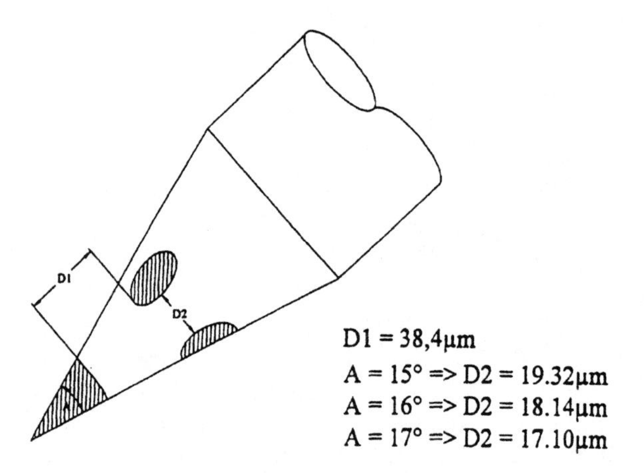

FIGURE 1.1
A four-core multifiber electrode made with platinum (90%)/tungsten (10%) wire encased in quartz. A cross section of the fiber is shown in the upper part of the figure, while the lower part of the figure shows the wire exposures and positions when the tip is ground to different angles. The electrodes are available from Uwe Thomas Recording (Marburg, Germany).

seven, 7 to 8-µm-diameter platinum (90%)/tungsten (10%) wires encased in a 100- to 110-µm-diameter quartz sheath as shown in Figure 1.1 (lower). Six wires are concentrically spaced around a central core. By conical grinding of the tip, an electrode arrangement similar to the four-core electrode is produced, but by beveling the tip, an ellipsoidal arrangement of recording sites is produced. These multifiber electrodes promise to provide an exciting new tool for exploring the nervous system.

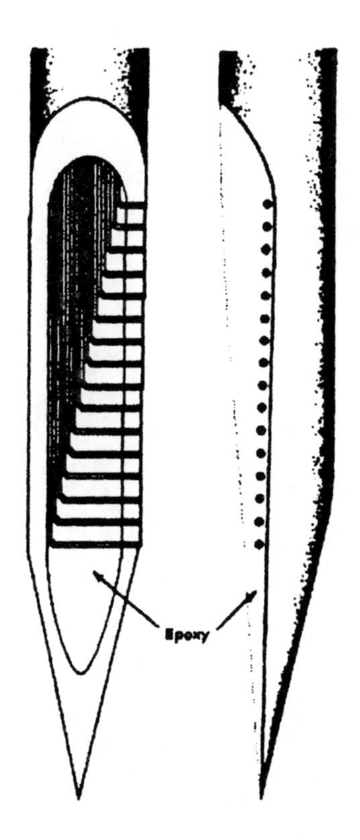

FIGURE 1.2
Sixteen 25-μm-diameter wires are threaded through a hypodermic tube that has been cut away at the tip. The wires are bent at right angles and embedded in epoxy. The two views are rotated 90°. (From Barna, J. S., Arezzo, J. C., and Vaugh, H. J. J., A new microelectrode array for the simultaneous recording of field potentials and unit activity, *Electroenceph. Clin. Neurophysiol.*, 52, 494, 1981. With kind permission from Elsevier Science Ireland Ltd., Bay 15K, Shannon Industrial Estate, Co. Clare, Ireland.)

1.2.5 Linear and Parallel Electrode Arrays

Linear arrays of wire electrodes have been described by several investigators. Barna et al.[32] removed half of the wall of the lower end of a 30 gauge hypodermic needle and threaded 16 insulated, 25-μm wires through the needle. The wires were bent at right angles at equal spacings from the needle tip and fixed with a drop of epoxy. The wires were cut flush with the edge of the needle as shown in Figure 1.2. Jellema and Weijnen[33] fabricated a similar linear array of 12 recording electrodes from 25-μm insulated nichrome wires. They split lengthwise a Teflon tube (300-μm inner diameter) to form a mold. Holes were pierced in the tube with a fine needle at the desired inter-electrode distance. The wires were inserted through the holes, bent at right angles to lie within the tube, and embedded in epoxylite resin as shown in Figure 1.3. The cured electrode assembly was removed from the tube and the wires cut flush with the surface. This technique provided a more precise localization of the tips of

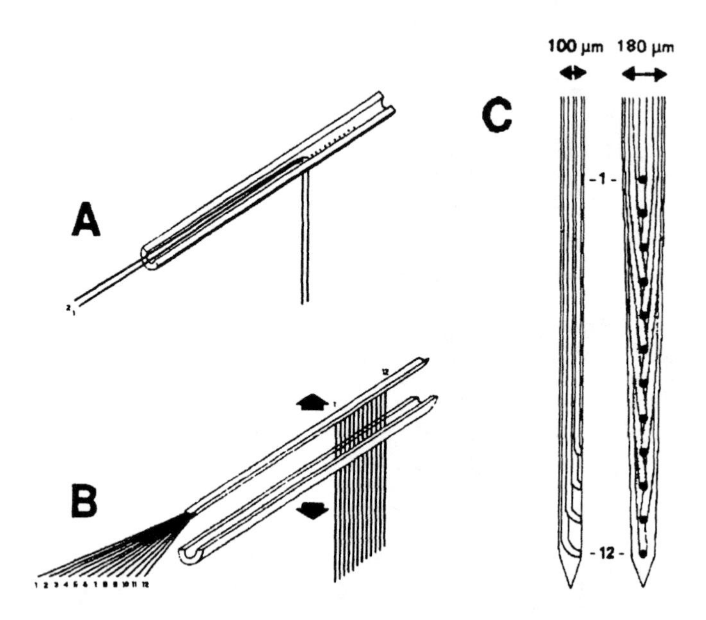

FIGURE 1.3
Parts A and B show two steps in the construction of a 12-channel multiwire electrode. In part A, two insulated nichrome wires have been inserted through punched holes in a Teflon® mold and bent at right angles. In part B, the wires that have been embedded in epoxy are removed from the mold. Two views of the finished microelectrode are shown in part C. (From Jellema, T. and Weijnen, J. A. W. M., A slim needle-shaped multiwire microelectrode for intracerebral recording, *J. Neurosci. Meth.*, 40, 203, 1991. With kind permission from Elsevier Science-NL, Sara Burgerhartstraat 25, 1055 KV Amsterdam, The Netherlands.)

the electrodes than that described by Barna et al.[32] and also removed the equipotential surface of the metal needle from the recording area.

One of the disadvantages of the previously described linear arrays is that the recording surfaces are in close proximity to the area of tissue that is damaged by the insertion of the electrode. To circumvent this problem, Kruger[34] inserted 12 platinum/iridium wires (5-μm-diameter wire, coated with quartz-glass for a total diameter of 30 μm) through folded electron microscope carrier grids as shown in Figure 1.4. The grids maintain the parallel alignment of the electrodes with tips in a line perpendicular to the first grid. The electrodes were inserted into the cortex at a 45° angle to the surface. The tips then lie in a line perpendicular to the cortical surface. The area from which activity was recorded was not severely damaged by the passage of the electrodes.

Linear arrays have been constructed from 13-μm-diameter tungsten wires that have been cut to different lengths and glued together with Super Glue.[21] A scanning electron micrograph of a linear bundle is shown in Figure 1.5. Gundjian et al.[35] glued six borosilicate glass capillary tubes together and hot drew them to form 40- to 50-μm-diameter tubes and further drew them to a sharp sealed tip. A CO_2 laser was used to cut side holes of 30- to 40-μm-diameter at intervals of 200 μm in respective tubes. The tubes were then filled with 9.9 *M* NaCl for recording electrophysiological data.

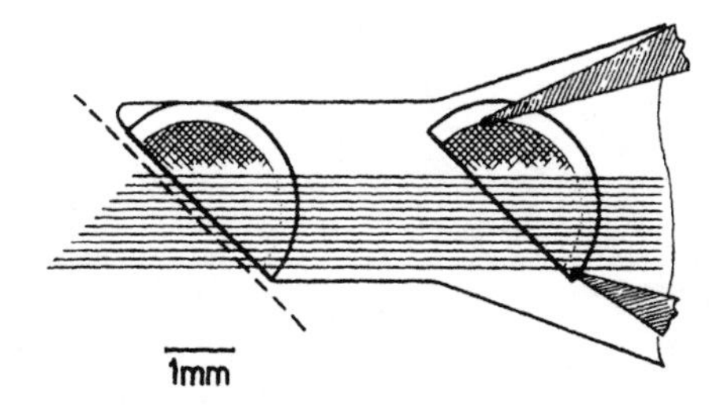

FIGURE 1.4
A 12-fold planar microelectrode array. The semicircular structures are folded circular electron microscope object-holder grids that hold the microelectrodes in a parallel arrangement. The dashed line represents the cortical surface. The tips of the microelectrodes lie in a line normal to the cortical surface. (From Kruger, J., A 12-fold microelectrode for recording from aligned cortical neurones, *J. Neurosci. Meth.*, 6, 347, 1982. With kind permission from Elsevier Science-NL, Sara Burgerhartstraat 25, 1055 KV Amsterdam, The Netherlands.)

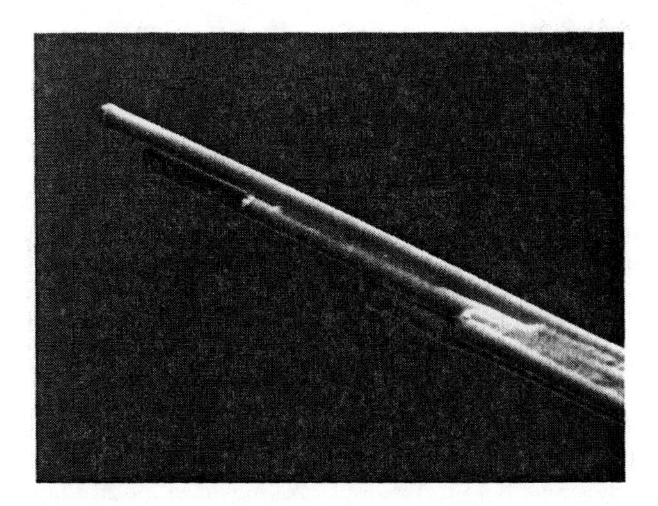

FIGURE 1.5
Scanning electron micrograph of five strands of insulated 13-μm-diameter tungsten wire that have been cut off at different lengths and glued together with Super Glue® to increase their stiffness and prevent divergence of the individual wires as they are advanced through the brain. (From Jaeger, D., Gilman, S., and Aldridge, J. W., A multiwire microelectrode for single unit recording in deep brain structures, *J. Neurosci. Meth.*, 32, 143, 1990. With kind permission from Elsevier Science-NL, Sara Burgerhartstraat 25, 1055 KV Amsterdam, The Netherlands.)

Verloop and Holsheimer[36] describe a simple method for constructing an array of parallel wire electrodes. The concept of positioning a number of wires in parallel at equal distances is based on the use of two springs, made from wire with a diameter that equals the desired spacing between the electrodes. The springs, kept in position by parallel bars, hold the electrode wires between their coils, as shown in Figure 1.6. A mold in the center of the jig is filled with epoxy to hold the wires in position. The wires are cut to the desired length on one side of the mold and attached to a connector on the other side.

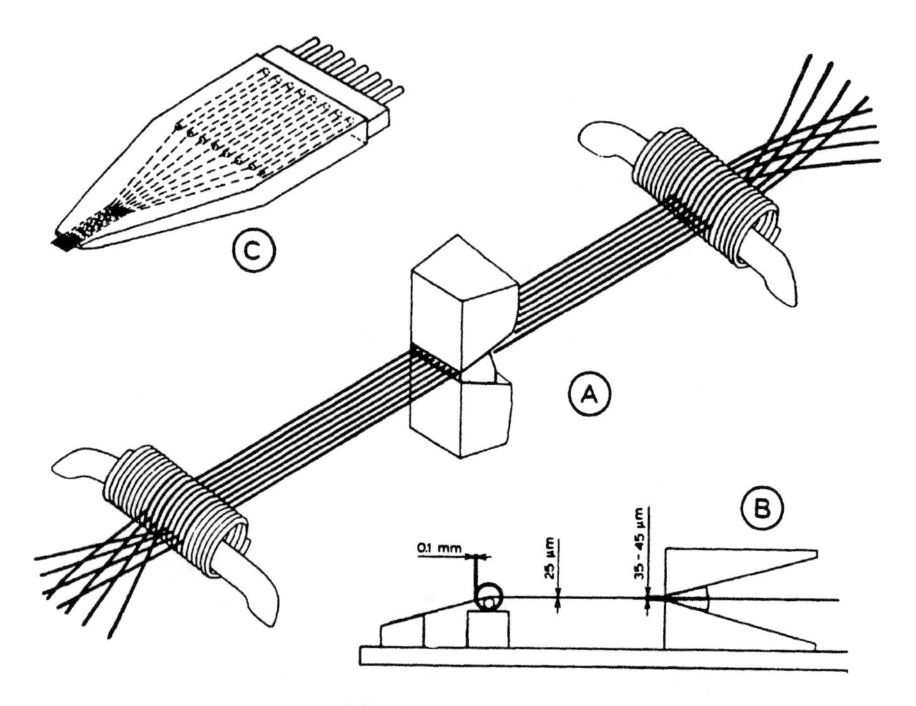

FIGURE 1.6
Simple technique for assembling arrays of parallel microwires. The wires are held parallel to each other by the springs shown in Part A. A mold is placed between the two springs and filled with epoxy. A side view of the jig is shown in Part B, while the completed parallel wire array is shown in Part C. (From Verloop, A. J. and Holsheimer, J., A simple method for the construction of electrode arrays, *J. Neurosci. Meth.*, 11, 173, 1984. With kind permission from Elsevier Science-NL, Sara Burgerhartstraat 25, 1055 KV Amsterdam, The Netherlands.)

One-dimensional parallel electrode arrays have been expanded into two dimensions by adding additional rows of parallel electrodes. Kruger and Bach[7] used electron microscope grids to maintain parallel alignment of 30 sharpened, glass-coated platinum/iridium wires (5-µm-diameter wire, 30-µm-diameter with glass). An example of a 5×6 array with 160 µm distance between microelectrodes is shown in Figure 1.7. To insert this array into the cortex, the dura was opened and the pia was held by a tangentially inserted three-pronged fork made from fine metal microelectrodes. The prongs were located between the rows of microelectrodes during insertion to minimize cortical depression. Following insertion, the fork was removed.

Several parallel electrode arrays are available from a number of commercial vendors. A drawing of several different arrays that are available from FHC (Bowdoinham, Maine) is shown in Figure 1.8. NB Labs (Denison, Texas) fabricates parallel wire arrays and also wire bundles, examples of which are shown in Plate 1. Custom fabricated electrode arrays are also available from Microprobe, Inc. (Potomac, Maryland).

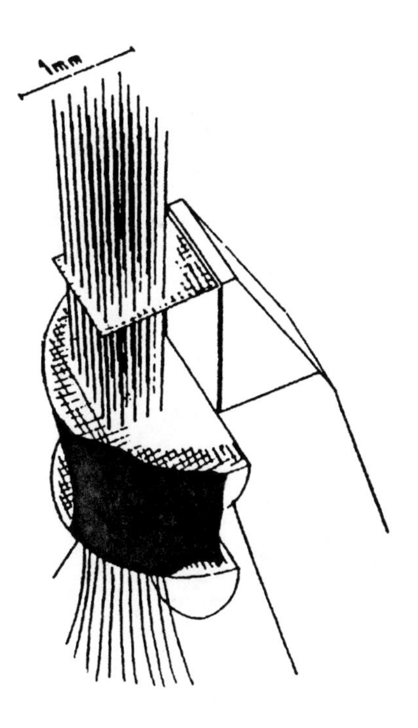

FIGURE 1.7
A 5 × 6 array of wire microelectrodes. Parallel alignment of the microelectrodes is accomplished by passing them through three electron microscope object-holder grids. The microelectrodes are securely held between the lower two grids with pitch (black patch). (From Kruger, J. and Bach, M., Simultaneous recording with 30 microelectrodes in monkey visual cortex, *Exp. Brain Res.*, 41, 191, 1981. With kind permission from Springer-Verlag, New York.)

1.3 Printed Circuit and Semiconductor Microelectrodes

The need to record simultaneously from many neurons to obtain a better understanding of the functions of the central nervous system seemed to be met with the ability to mass produce microelectrodes utilizing the developments of integrated-circuit manufacturing technology. The reader is referred to two earlier reviews that cover some of the developments of printed circuit microelectrodes.[1,37]

Wise et al.[38] described the first photoengraved microelectrodes for single-unit recording. Recording neural activity with these electrodes, with geometric recording sites ranging in size from 15 to 75 μm^2, they found that the smaller area recording sites produced larger spikes.[39] They also noted that unless the recording sites were within 20 μm of the probe tip, the microelectrodes were not able to record spike activity. This was probably due to damage produced by the acute insertion of the probe, resulting in a low impedance shunt path along the shaft of the microelectrode.

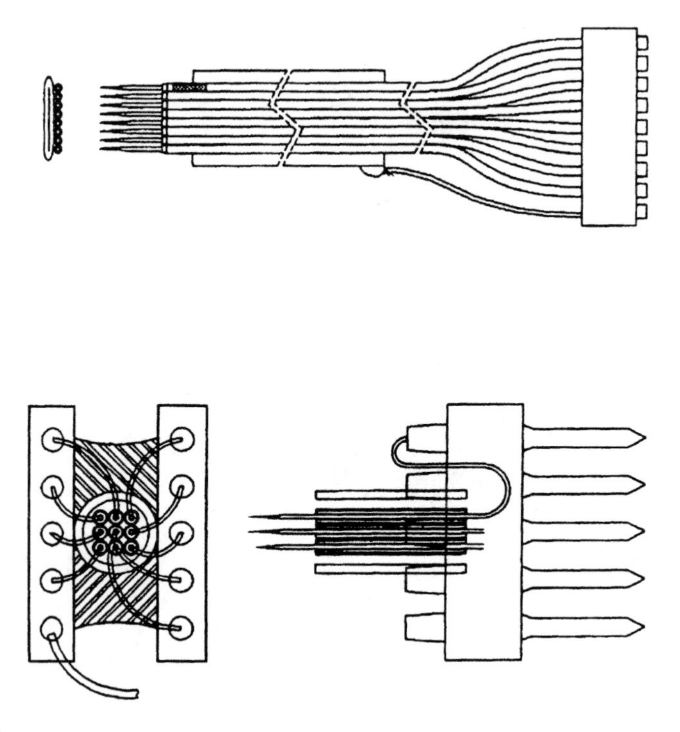

FIGURE 1.8
Examples of two custom needle microelectrode arrays available from FHC, Bowdoinham, Maine. The upper diagram shows a parallel array of epoxy insulated tungsten microelectrodes, while the lower diagram shows a three-by-three array of microelectrodes.

With multisite microprobes, Drake et al.[40] were able to simultaneously record from four sites spaced at 100-μm intervals along the shank of the probe. Thus, multisite probes can record from sites other than the tip if the surrounding tissue is not too badly damaged. This was also confirmed with chronically implanted multisite microprobes.[41,42]

Fabrication of hybrid buffer amplifiers on the carrier shank of probes[43] was the first step in the integration of microelectronics on probe structures. An active microelectrode was described by Jobling et al.[44] where nine recording sites and associated buffer amplifiers were fabricated with standard integrated circuit processing techniques. With these microelectrodes, electrostatic noise pick-up, or cross talk, from stimulators was significantly reduced.

Prohaska et al.[45] used thin-film technology to fabricate an eight-site recording probe with 50 × 50 μm contacts spaced at 300-μm intervals along a glass substrate. After processing, the glass substrate was cut and polished to a fine needle. The shank of the needle was rather large (300 μm) compared to those produced by other fabrication techniques.

Electron-beam lithography has been employed to define several thin-film metal electrodes on the outer surface of glass micropipettes.[46] Recording sites were 5-μm-wide rings spaced 100 μm apart along the length of the probe. A recording

FIGURE 1.9
A group of passive silicon microprobes, displayed on the surface of a penny, that were manufactured at the University of Michigan. The single-tine microelectrode has five recording sites distributed along the shaft, while the multi-tined microelectrodes have either one or two recording sites on each shaft. (Photo courtesy of the Center for Neural Communications Technology at the University of Michigan, Ann Arbor.)

site located 250 μm from the tip recorded neural activity in the marginal gyrus of the cat. Unit activity was observed during insertion and also during withdrawal of the electrode, suggesting minimal tissue damage and that neurons close to the shank can survive.

Photolithography has made it possible to produce a microelectrode with 24 recording sites.[47] The probes were fabricated on a 17-μm-thick molybdenum substrate with either a 1 × 24 linear array along one edge or a 2 × 12 array along both edges. Blum et al.[48] fabricated four- and six-site planar probes by sandwiching gold electrodes and leads between two polyimide dielectric layers that were bonded to a 15-μm-thick molybdenum structural support. Peckerar et al.[49] have described a procedure that is fully compatible with integrated circuit fabrication for constructing either a 20- or 40-channel microelectrode on a 50-μm silicon substrate. After processing, the substrate is thinned to less than 20 μm. All of these probe designs were for acute recording experiments.

Currently, the Center for Neural Communication Technology (University of Michigan, Ann Arbor) has the largest selection of passive and active silicon microprobes. Figure 1.9 shows a group of passive microprobes displayed on the surface of a penny. The single-tine microelectrode has five recording sites distributed along the shaft, while the multi-tined microelectrodes have either one or two recording sites on each shaft. A unique step in the manufacture of the University of Michigan microprobes is the use of a deep boron diffusion process to form an etch stop and define the shape of the probe. Microelectronics can be integrated on the back of the probe.

Stanford University (Palo Alto, California) has also produced a number of different silicon microprobes. Their latest designs have been developed in conjunction with Dr. James Bower at the California Institute of Technology.[50] Figure 1.10 is a scanning

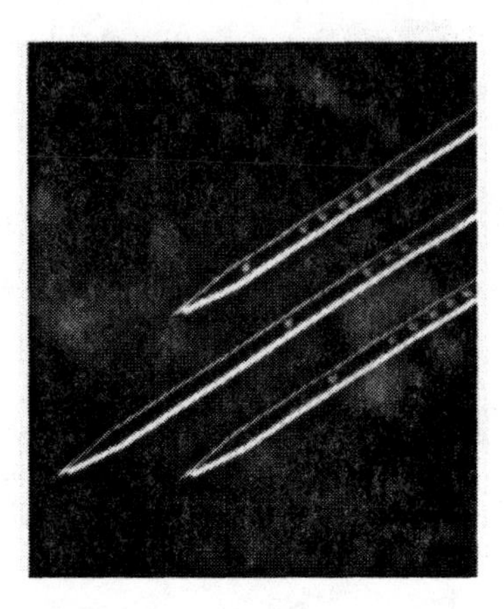

FIGURE 1.10
A three-shank silicon microprobe developed at Stanford University. Each shank has multiple iridium recording sites. Note the extremely sharp tips to minimize tissue compression during insertion. (From Kewley, D. T., Hills, M. D., Borkholder, D. A., Opris, I. E., Maluf, N. I., Storment, C. W., Bower, J. M., and Kovacs, G. T. A., Plasma-etched neural probes, *Sensors and Actuators*, A 58, 27, 1997. With permission from Elsevier Science, The Boulevard, Langford Lane, Kidlington 0X5 1GB, UK.)

electron micrograph of a three-tined microprobe having 16 iridium recording sites near the tips of the shafts. Note the extremely sharp tips of the individual shafts.

A novel approach to chronic recording from multiple sites along the shaft of a microprobe is being developed by Dr. Jerry Pine at the California Institute of Technology.[51] The probe consists of multiple wells into which cultured neurons are placed. At the bottom of each well is an electrode for recording or stimulating the neuron that is trapped in the well. The electrode has an integrated lead wire for external connection. When implanted, it is hoped that neurons in the probe will synaptically integrate with the host nervous system.

An acute recording probe is under development at the University of Arkansas.[52] They use sputter-deposited, thin-film carbon for the recording sites. The University of Michigan has also sputter-deposited carbon on the recording sites of some of their multichannel microelectrodes.[53] One of the advantages of carbon electrode sites is that they can be coated with materials such as Nafion for *in vivo* electrochemical and single-unit electrophysiological recordings.

1.3.1 Two- and Three-Dimensional Silicon Microelectrode Arrays

Two-dimensional electrode arrays have been constructed from individual wires or microelectrodes bonded together to form a desired configuration. The complexity of three-dimensional arrays lends itself to integrated circuit fabrication techniques. Normann and colleagues[54,55] described techniques for making 100 needle-like silicon electrodes in a 10×10 array. The array was micromachined from a monocystalline block of silicon using a diamond dicing saw followed by chemical etching. A

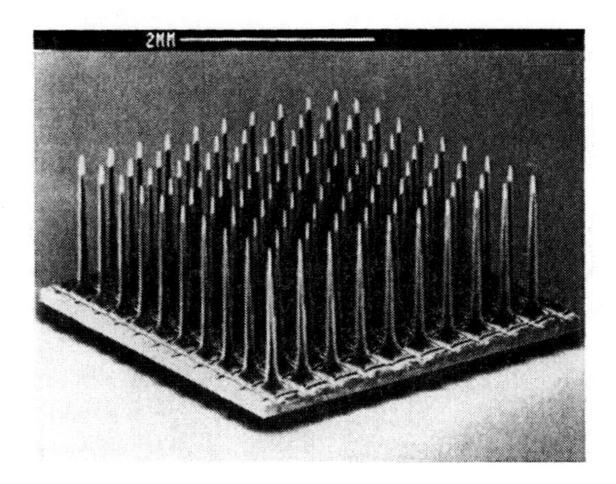

FIGURE 1.11
Scanning electron micrograph of a 100-electrode University of Utah silicon needle array. The array is insulated with silicon nitride, and the tip of each electrode is metallized with platinum. (Photo courtesy of Dr. Richard Normann.)

scanning electron micrograph of a 100-electrode array is shown in Figure 1.11. The array is insulated with silicon nitride, and the tip of each electrode is metallized with platinum. With this array, there is a single recording site at the tip of each silicon needle. Although the silicon needles are quite sharp, an array of this magnitude cannot be inserted into the cortex without excessive dimpling. To overcome this problem, a pneumatic insertion device was developed to rapidly drive the array into the cortex.[56]

Another approach for constructing three-dimensional microelectrode arrays was developed at the University of Michigan. Figure 1.12 shows eight planar multi-shank silicon microprobes, with each microprobe composed of 16 shanks and multiple recording sites along each shank, assembled into a micromachined silicon alignment platform to form a three-dimensional recording array. The microprobe fabrication techniques are compatible with integrated circuit fabrication, and thus microcircuits can be constructed on the silicon platform. These circuits can select recording sites that have neuronal activity, amplify the signals, and multiplex the desired signals onto a single data line. If the lengths of the individual probes are different, the array might be inserted through the pia into the cortex without excessive dimpling. However, if excessive dimpling occurs, then a rapid insertion tool similar to that described by Rousche and Normann[56] might be required.

1.4 Microdrives for Independently Positioning Multiple Microelectrodes

Over 30 years ago, investigators were becoming interested in controlled recording from more than one microelectrode at a time to observe the complex interactions of neuronal populations. Blum and Feldman[57] devised a single micrometer that could

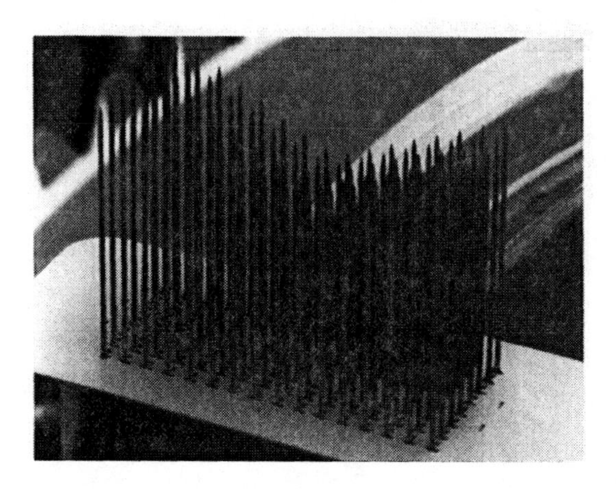

FIGURE 1.12
A three-dimensional multielectrode system developed at the University of Michigan. The probe consists of eight, 16-shank planar microelectrode arrays assembled in a silicon alignment platform. (Photo courtesy of the Center for Neural Communications Technology at the University of Michigan, Ann Arbor.)

move four microelectrodes as a group or move each microelectrode independently. When the microelectrodes were driven simultaneously, unit activity was rarely picked up by more than one microelectrode. However, being able to manipulate each microelectrode independently made it possible to obtain simultaneous unit records from each of the four microelectrodes. This particular microelectrode drive, because of size, can only be used with acute preparations, or where an animal's head is restrained.

Reitbock and Werner[58] developed a system for independently driving up to seven microelectrodes. They used a series of piezoelectric clutches and brakes to either hold the microelectrode in a fixed position, or to couple it to a platform that was moved by a stepping motor. A later version was constructed for independently positioning 19 microelectrodes[59] using the same piezoelectric clutch and brake arrangement. Both the seven- and 19-microelectrode drive systems are currently manufactured by Uwe Thomas Recording (Marburg, Germany). A photograph of a seven microelectrode drive system is shown in Plate 2.

Another multiple microelectrode positioning system, manufactured by Uwe Thomas Recording, is shown in Figure 1.13. The basic drive is for seven microelectrodes, but it is also made for 16 channels. The driving force is supplied by a stretched rubber tube that surrounds and grasps each microelectrode. When the tension in the rubber tube is reduced, the microelectrode is driven into the brain. The tension in the tube is varied by a computer-controlled DC motor equipped with a position sensor that enables a resolution of 0.27 μm.

Frederick Haer and Co. (Bowdoinham, Maine) has developed a four- or eight-microelectrode drive system. Stepper motors drive hydraulic couplers to eliminate microelectrode vibration. Each microelectrode can be advanced 15 mm in steps as small as 0.25 μm. A main drive cylinder can drive all of the electrodes together an additional 15 mm. A drawing of the eight-microelectrode drive system is shown in Figure 1.14.

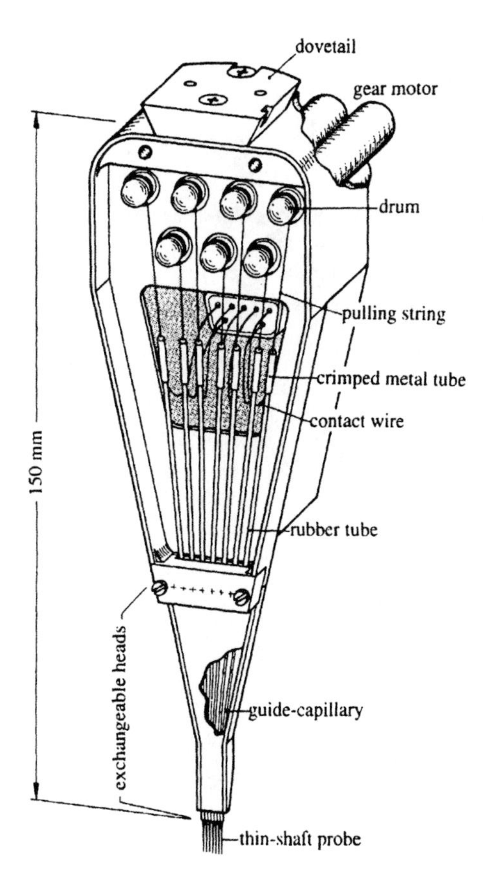

FIGURE 1.13
Prototype of a rubber tube drive with seven independently moveable probes. The rubber tubes, which grasp the microelectrodes, are stretched by strings that are wound on a drum of a computer-controlled gear motor. Changing the tension in the rubber tube positions the microelectrodes. (Photo courtesy of Uwe Thomas Recording, Marburg, Germany.)

Alpha Omega Engineering (Nazareth, Israel) has also developed a modular (four drives per module) microelectrode positioning system. The drive module contains the motors and feedback elements needed to control the resolution of the movements of the microelectrodes to 1 μm through flexible shafts. Microelectrode position control is achieved with an interface card installed in a PC. The interface card supports two modules, for a total of eight microelectrodes. Additional cards can be installed in the PC for additional microelectrodes. A sketch of the drive system is shown in Figure 1.15.

The preceding microelectrode drive systems, like the one described by Blum and Feldman,[57] because of their size can only be used when an animal's head is restrained.

Humphrey[60] devised a chronically implantable multiple microelectrode drive system that could be used to independently control the position of five microelectrodes. Each microelectrode was positioned with a screw that had 36 threads/cm

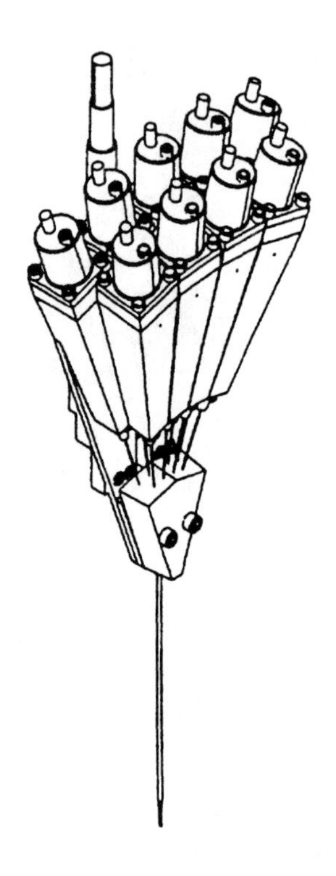

FIGURE 1.14
System for independently driving eight microelectrodes through hydraulic coupled stepper motors. The movement of individual microelectrodes is as small as 0.25 μm with a total travel of 15 mm. All microelectrodes can be moved together over a distance of 15 mm. (FHC, Bowdoinham, Maine.)

(277 μm/rev.) making movements of 5 to 10 μm easily obtainable. A modern version of a chronically implantable multiple microelectrode drive system has been developed by David Kopf Instruments (Tujunga, California). This device consists of 14 independently adjustable micropositioners intended for driving tetrode electrodes, but could be used for stereotrodes or more conventional wire electrodes. A drawing of the microdrive is shown in Figure 1.16. Each micropositioner provides adjustments of 5 to 10 μm (312 μm/rev) with a total electrode movement of 7.5 mm.

1.5 Concluding Remarks

A number of promising microelectrodes have or are currently being developed for recording from many neurons simultaneously. Also, a number of microelectrode drive systems have recently been introduced. The microelectrode technology is now at hand for simultaneously recording from large numbers of neurons. The dream of many scientists to monitor simultaneously large numbers of individual neurons is becoming a reality.

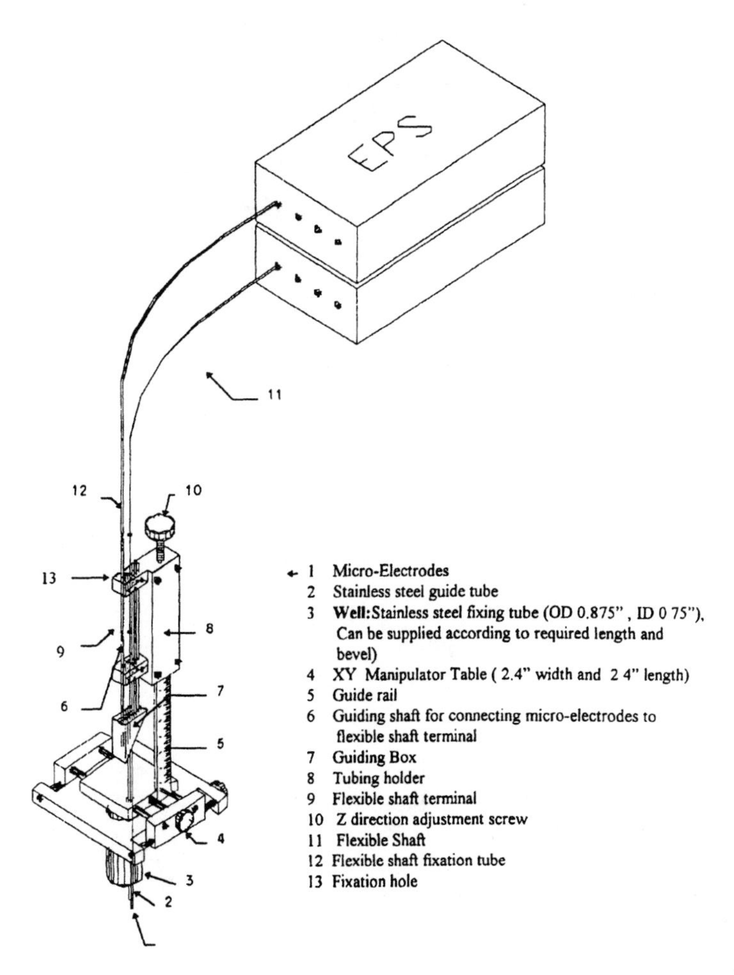

FIGURE 1.15
Sketch of a modular multiple microelectrode positioning system. Each Electrode Positioning System (EPS) module can independently position four microelectrodes with motors coupled to the microelectrodes by flexible cables. Positioning resolution is 1 μm. (Alpha Omega Engineering, Nazareth, Israel.)

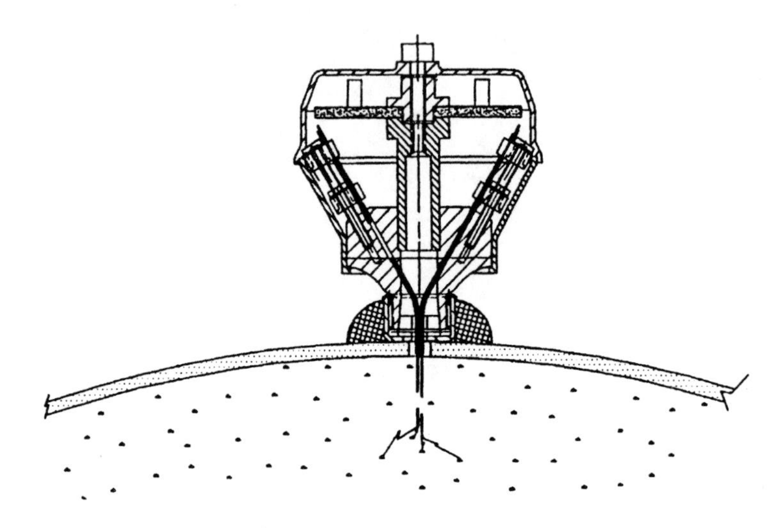

FIGURE 1.16
Line drawing of a chronically implantable microelectrode drive system that is small enough for rodents. There are 14 screw drives arranged in a circular fashion. The drive is designed for tetrode wire electrodes, but stereotrodes or individual wire electrodes can be driven by the device. (David Kopf Instruments, Tujunga, California.)

References

1. Kruger, J., Simultaneous individual recordings from many cerebral neurons: Techniques and results, *Rev. Physiol. Biochem. Pharmacol.*, 98, 177, 1983.
2. Strumwasser, F., Long-term recording from single neurons in brain of unrestrained mammals, *Science*, 127, 469, 1958.
3. Olds, J., Operant conditioning of single unit responses, *Proc. XXIII Int. Congr. Physiol. Sci. Tokyo*, 372, 1965.
4. Olds, J., Disterhoft, J. F., Segal, M., Kornblith, C. L., and Hirsh, R., Learning centers of rat brain mapped by measuring latencies of conditioned unit responses, *J. Neurophysiol.*, 35, 202, 1972.
5. Marg, E. and Adams, J. E., Indwelling multiple micro-electrodes in the brain, *Electroenceph. Clin. Neurophysiol.*, 23, 277, 1967.
6. Burns, B. D., Stean, J. P. B., and Webb, A. C., Recording for several days from single cortical neurons in completely unrestrained cats, *Electroenceph. Clin. Neurophysiol.*, 36, 314, 1974.
7. Kruger, J. and Bach, M., Simultaneous recording with 30 microelectrodes in monkey visual cortex, *Exp. Brain Res.*, 41, 191, 1981.
8. Reitboeck, H. J., Fiber microelectrodes for electrophysiological recordings, *J Neurosci. Meth.*, 8, 249, 1983.
9. Kaltenbach, J. A. and Gerstein, G. L., A rapid method for production of sharp tips on preinsulated microwires, *J. Neurosci. Meth.*, 16, 283, 1986.
10. Salcman, M. and Bak, M. J., A new chronic recording intracortical microelectrode, *Medical and Biological Engineering*, 14, 42, 1976.

11. Palmer, C., A microwire technique for recording single neurons in unrestrained animals, *Brain Res. Bull.*, 3, 285, 1978.
12. Chorover, S. L. and Deluca, A. M., A sweet new multiple electrode for chronic single unit recording in moving animals, *Physiol. Behav.*, 9, 671, 1972.
13. Legendy, C. R., Salcman, M., and Brennan, N., A multiple floating microelectrode for chronic implantation and long-term single unit recording in the cat, *Electroenceph. Clin. Neurophysiol.*, 58, 285, 1984.
14. O'Keefe, J. and Bouma, H., Complex sensory properties of certain amygdala units in the freely moving cat, *Experimental Neurology*, 23, 384, 1969.
15. Linseman, M. A. and Corrigall, W. A., Neurophysiological evidence of movement of chronically-implanted fine wire electrodes in recordings of field potentials in hippocampus, *Physiol. Behav.*, 26, 729, 1981.
16. Best, P. J., Knowles, W. D., and Phillips, M. I., Chronic brain unit recording for pharmacological applications, *J. Pharmacol. Meth.*, 1, 161, 1978.
17. Rasmussen, D. D. and Malvern, P. V., Chronic recording of multiple-unit activity from the brain of conscious sheep, *Brain Res. Bull.*, 7, 163, 1981.
18. Kosobud, A. E. K., Harris, G. C., and Chapin, J. K., Behavioral associations of neuronal activity in the ventral tegmental area of the rat, *J. Neurosci.*, 14, 7117, 1994.
19. Diana, M., Garcia-Munoz, M., and Freed, C. R., Wire electrodes for chronic single unit recording of dopamine cells in substantia nigra pars compacta of awake rats, *J. Neurosci. Meth.*, 21, 71, 1987.
20. Nicolelis, M. A. L., Ghazanfar, A. A., Faggin, B. M., Votaw, S., and Oliveira, L. M. O., Reconstructing the engram: Simultaneous, multisite, many single neuron recordings, *Neuron*, 18, 529, 1997.
21. Jaeger, D., Gilman, S., and Aldridge, J. W., A multiwire microelectrode for single unit recording in deep brain structures, *J. Neurosci. Meth.*, 32, 143, 1990.
22. Kubie, J. L., A driveable bundle of microwires for collecting single-unit data from freely-moving rats, *Physiol. Behav.*, 32, 115, 1984.
23. Schmidt, E. M., Bak, J. M., and McIntosh, J. M., Long-term chronic recording from cortical neurons, *Experimental Neurology*, 52, 496, 1976.
24. Schmidt, E. M., Parylene as an electrode insulator: a review, *JEPT*, 10, 19, 1983.
25. Salcman, M. and Bak, M. J., Design, fabrication, and *in vivo* behavior of chronic recording intracortical microelectrodes, *IEEE Trans. Biomed. Eng.*, BME-20, 253, 1973.
26. Kennedy, P. R., The cone electrode: a long-term electrode that records from neurites grown onto its recording surface, *J. Neurosci. Meth.*, 29, 181, 1989.
27. Wheeler, B. C. and Heetderks, W. J., A comparison of techniques for classification of multiple neural signals, *IEEE Trans. Biomed. Eng.*, 12, 752, 1982.
28. Schmidt, E. M., Computer separation of multi-unit neuroelectric data: a review, *J. Neurosci. Meth.*, 12, 95, 1984.
29. Schmidt, E. M., Instruments for sorting neuroelectric data: a review, *J. Neurosci. Meth.*, 12, 1, 1984.
30. McNaughton, B. L., O'Keefe, J., and Barnes, C. A., The stereotrode: A new technique for simultaneous isolation of several single units in the central nervous system from multiple unit records, *J. Neurosci. Meth.*, 8, 391, 1983.

31. Gray, C. M., Maldonado, P. E., Wilson, M., and McNaughton, B., Tetrodes markedly improve the reliability and yield of multiple single-unit isolation from multi-unit recordings in cat striate cortex, *J. Neurosci. Meth.*, 63, 43, 1995.
32. Barna, J. S., Arezzo, J. C., and Vaugh, H. J. J., A new microelectrode array for the simultaneous recording of field potentials and unit activity, *Electroenceph. Clin. Neurophysiol.*, 52, 494, 1981.
33. Jellema, T. and Weijnen, J. A. W. M., A slim needle-shaped multiwire microelectrode for intracerebral recording, *J. Neurosci. Meth.*, 40, 203, 1991.
34. Kruger, J., A 12-fold microelectrode for recording from aligned cortical neurones, *J. Neurosci. Meth.*, 6, 347, 1982.
35. Gundjian, A., Ibisoglu, H., and Kostopoulos, G., Laser-drilled multichannel multidepth extracellular microelectrodes, *Medical & Biological Engineering & Computing*, 24, 420, 1986.
36. Verloop, A. J. and Holsheimer, J., A simple method for the construction of electrode arrays, *J. Neurosci. Meth.*, 11, 173, 1984.
37. Pickard, R. S., A review of printed circuit microelectrodes and their production, *J. Neurosci. Meth.*, 1, 301, 1979.
38. Wise, K. D., Angell, J. B., and Starr, A., An integrated-circuit approach to extracellular microelectrodes, *IEEE Trans. Biomed. Eng.*, BME-17, 238, 1970.
39. Starr, A., Wise, K. D., and Csongradi, J., An evaluation of photoengraved microelectrodes for extracellular single-unit recording, *IEEE Trans. Biomed. Eng.*, 291, 1973.
40. Drake, K. L., Wise, K. D., Farraye, J., Anderson, D. J., and BeMent, S. L., Performance of planar multisite microprobes in recording extracellular single-unit intracortical activity, *IEEE Trans. Biomed. Eng.*, 35, 719, 1988.
41. Carter, R. R. and Houk, J. C., Multiple single-unit recordings from the CNS using thin-film electrode arrays, *IEEE Trans. Rehab. Eng.*, 1, 175, 1993.
42. Heetderks, W. J. and Schmidt, E. M., Chronic, multi-unit recording of neural activity with micromachined silicon microelectrodes, *Proc. RESNA 95*, 659, 1995.
43. Wise, K. D. and Angell, J. B., A low capacitance multielectrode probe for use in extracellular neurophysiolgy, *IEEE Trans. Biomed. Eng.*, BME-22, 212, 1975.
44. Jobling, D. T., Smith, J. G., and Wheal, H. V., Active microelectrode array to record from the mammalian central nervous system *in vitro*, *Medical & Biological Engineering & Computing*, 19, 553, 1981.
45. Prohaska, O., Olcaytug, F., Womastek, K., and Petsche, H., A multielectrode for intracortical recordings produced by thin-film technology, *Electroenceph. Clin. Neurophysiol.*, 42, 421, 1977.
46. Pochay, P., Wise, K. D., Allard, L. F., and Rutledge, L. T., A multichannel depth probe fabricated using electron-beam lithography, *IEEE Trans. Biomed. Eng.*, BME-26, 199, 1979.
47. Kuperstein, M. and Whittington, D. A., A practical 24 channel microelectrode for neural recording *in vivo*, *IEEE Trans. Biomed. Eng.*, BME-28, 288, 1981.
48. Blum, N. A., Carkuff, B. G., Charles, H. K. J., Edwards, R. L., and Meyer, R. A., Multisite microprobes for neural recordings, *IEEE Trans. Biomed. Eng.*, 38, 68, 1991.
49. Peckerar, M., Shamma, S. A., Rebbert, M., Kosakowski, J., and Isaacson, P., Passive microelectrode arrays for recording of neural signals: A simplified fabrication process, *Review of Scientific Instruments*, 62, 2276, 1991.

50. Kewley, D. T., Hills, M. D., Borkholder, D. A., Opris, I. E., Maluf, N. I., Storment, C. W., Bower, J. M., and Kovacs, G. T. A., Plasma-etched neural probes, *Sensors and Actuators*, A 58, 27, 1997.
51. Pine, J., *Cultured Neuron Probe,* California Institute of Technology, Pasadena, NIH Neural Prosthesis Contract No. N01-NS-3-2393 1-15, 1993–1997.
52. Sreenivas, G., Ang, S. S., Fritsch, I., Brown, D. B., Gerhardt, G. A., and Woodward, D. J., Fabrication and characterization of sputtered-carbon microelectrode arrays, *Anal. Chem.*, in press, 1997.
53. van Horne, C. G., BeMent, S., Hoffer, B. J., and Gerhardt, G. A., Multichannel semiconductor-based electrodes for *in vivo* electrochemical and electrophysiological studies in rat CNS, *Neuroscience Letters*, 120, 249, 1990.
54. Campbell, P. K., Jones, K. E., Huber, R. J., Horch, K. W., and Normann, R. A., A silicon-based, three dimensional-neural interface: Manufacturing processes for an intracortical electrode array, *IEEE Trans. Biomed. Eng.*, 38, 758, 1991.
55. Jones, K. E., Campbell, P. K., and Normann, R. A., A glass/silicon composite intracortical electrode array, *Ann. Biomed. Eng.*, 20, 423, 1992.
56. Rousche, P. J. and Normann, R. A., A method of pneumatically inserting an array of penetrating electrodes into cortical tissue, *Ann. Biomed. Eng.*, 20, 413, 1992.
57. Blum, B. and Feldman, B., A microdrive for the independent manipulation of four microelectrodes, *IEEE Trans. Biomed. Eng.*, 12, 121, 1965.
58. Reitbock, H. J. and Werner, G., Multi-electrode recording system for the study of spatio-temporal activity patterns in the central nervous system, *Experientia*, 39, 339, 1983.
59. Reitboeck, H. J., Adamczak, W., Eckhorn, R., Muth, P., Thielmann, R., and Thomas, U., Multiple singleunit recording: design and test of a 19-channel micro-manipulator and appropriate fiber electrodes, *Neuroscience Letters*, 7, S148, 1981.
60. Humphrey, D. R., A chronically implantable multiple micro-electrode system with independent control of electrode positions, *Electroenceph. Clin. Neurophysiol.*, 29, 616, 1970.

Chapter 2

Multichannel Electrode Design: Considerations for Different Applications

Karen A. Moxon

Contents

2.1 Introduction

There are several different types of multiple-site recording electrodes designed to effectively increase the yield of recorded neurons (Table 2.1; see also Chapter 1). In most instances these electrodes are easy to use. However, they are designed to work optimally under certain conditions, and this must be considered when selecting the appropriate electrode for the application. Moreover, successful recordings are dependent on proper handling of the electrode. Generally, special care must be used when

0-8493-3351-2/99/$0.00+$.50

TABLE 2.1
Summary of the Different Types of Experimental Preparations that Use Multiple Single-Unit Electrodes and the Different Types of Electrodes

Experimental Preparation	
Chronic	**Acute**
Multiple, single-unit recording	Multiple, single-unit recording
Cochlear implants	Peripheral nerves
Visual prosthesis	Muscle fibers
Motor prosthesis	Cardiac tissue
	Brain slice preparations

Types of Electrodes	
Multiple Wire Electrodes	**Printed Circuit Electrodes**
Tungsten	Probe-type arrays
Stainless steel	Surface-mount electrodes
Carbon fiber	Sieve-type macroelectrodes

the electrodes are implanted. Successful use of these electrodes requires skill acquired through practice.

Multiple-site recording electrodes have been developed for recording from peripheral nerve tissue,[1-4] cultured nerve tissue,[5,6] and from central nervous tissue *in vivo.*[7-9] Currently there are ideal, high-impedance multiple-site electrodes that can be used for acute recordings as well as low-impedance multiple-site electrodes that can be used for chronic recordings in awake, freely moving animals. In addition to different electrical characteristics, there are also elegant architectural styles that have been developed to allow the user access to novel nerve preparations. The application of different shapes and sizes can lead to increased understanding of neural information processing.

Since the late 1960s, many investigators have followed earlier examples from single-unit recording, improved on these early attempts, and developed completely novel electrode systems. Chapter 1 of this book provides a comprehensive list. Here, our objective is to give the reader a broad overview, and list commercially available electrodes. While it is possible to make many of these multiple-site electrodes in the lab, it is time consuming, and the commercially available electrodes are not prohibitively expensive. In addition, most electrode manufacturers will accommodate novel designs requested by users. So the investigator must weigh the "cost" of their time to produce these electrodes against any compromises they must make to use commercially available electrodes.

2.1.1 History

Investigators have conducted successful multiple single-unit recordings since the sixties.[10,11] Microwires have been the dominate choice. Microwires are generally made of metal, with a diameter ranging from 12 to 50 microns, and coated with a thin insulating material. Initially, bundles of microwires implanted in the cortex showed promise for recording several cells simultaneously. The limitations have usually been processing power to sample, amplify, and acquire the multiple data streams. These early investigators generally made their own electrodes in the lab, starting with fine metal wires, usually tungsten, shaping the tips, either through sharpening[12] or beveling,[13] and then insulating them. These processes were time consuming, and the yield of good electrodes was low. This low yield was often due to an incomplete insulation of the microwires.

However, successful experiments did lead to insights into neural information processing. For example, Gerstein and Perkel[7] used simultaneously recorded neurons to examine connectivity among neurons. By measuring the correlation of the time between action potentials, they began a long history of studying functional connectivity between neurons (see Chapter 8). In a second series of studies, Humphrey[14] developed a chronically implantable, multiple microelectrode system in which each electrode was individually adjustable. He then used tungsten electrodes to record simultaneously selected neurons in the motor cortex of unanesthetized monkeys while the animals performed a variety of arm movements.[15] The results suggested that information about a movement is carried not simply in the discharge patterns or spike trains of individual neurons but to a significant extent by the temporal relations between them.

This field is changing rapidly. Recent advances in silicon-based electrodes have added a new dimension to multiple single-unit recording (see below). New technologies for recording more neurons simultaneously,[16] telemetry[17] to free the animal from the experimental equipment, and the development of hybrid circuits that include amplifiers,[18-20] filters, and multiplexing[21-23] small enough to fit on the animal's headstage are currently being developed.

The general overview presented here can aid in evaluating different electrodes for specific applications. In addition, it should give some insight into troubleshooting mechanisms. However, this information is not really sufficient, and the reader should follow up with references listed at the end of this chapter. In addition we have listed company names and web sites that may be helpful. The rest of this chapter lists some advantages and pitfalls of acute vs. chronic recordings and then goes on to describe many of the electrodes that are currently available commercially.

2.1.2 Signal Conditioning

It is important for the user to understand the fundamentals of neural recording. The theory of neural recording is the same whether you are recording with a single-unit electrode or a multiple-unit electrode. First, a basic understanding of the extracellular space around neurons is necessary. In the CNS, neurons are surrounded by glial processes and a water-based solution rich in salts. This can be a very corrosive environment for many materials, and multiple-site chronic electrodes must be insulated with a material that can maintain its integrity in this environment. Since synaptic processes

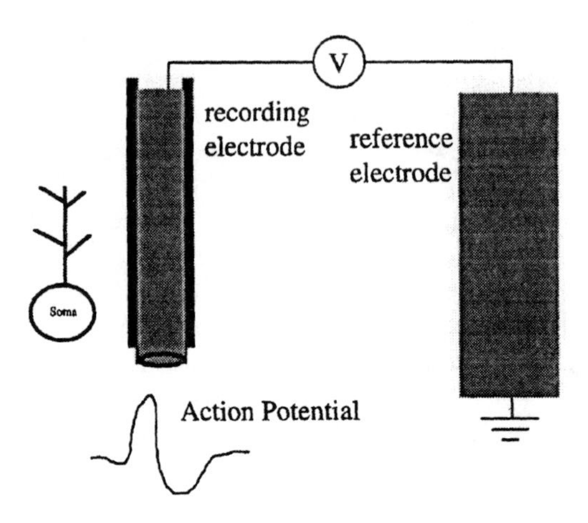

FIGURE 2.1
Schematic representation of the extracellular recording environment. The recording electrode measures the change in voltage due to an action potential. The signal is measured as a voltage difference between the recording electrode and the reference or ground electrode.

produce small currents, they cannot be recorded extracellularly. Thus, the cell's action potential is the signal recorded in multiple, single-unit recording.

The signal measured in extracellular recordings reflects the voltage change between the electrode tip and the reference electrode (Figure 2.1). The reference electrode, often called the ground wire, is generally larger that the recording electrode. The larger the surface area, the more current that can pass through the electrode tip. Therefore, a fat reference electrode is good at picking up lots of neural activity and not just from one particular neuron, making a good reference. In addition to the reference, it is often useful to have a differential electrode. This electrode is used to rid the recording of local perturbations or noise in the signal. The potential of the differential electrode can be subtracted from the source electrode and any common sources of noise can be removed from the signal. It is important that the differential electrode is not recording a neuron because this neuron will show up, inverted, on the source channel.

The reflection of the action potential in the external environment is carried by the external membrane resistance. The potential difference between the electrode and the reference is directly related to the input impedance of the electrode. The larger the input impedance, for a given current, the greater the voltage. Therefore, you can get better signal-to-noise results with high impedance electrodes. The most important issue is that the impedance of your recording equipment matches your electrode. Generally, spurious noise is generated when low impedance recording equipment is used for high impedance electrodes.

However, since these voltages are tiny, on the order of a few hundred microvolts at best, these signals must be immediately amplified (Figure 2.2). Immediate amplification is not a problem for acute recordings, and the amplifiers should be run with direct current as close as possible to the electrodes. However, in chronic recordings, where the animal is free to move, it is generally not yet feasible to have the amplifiers

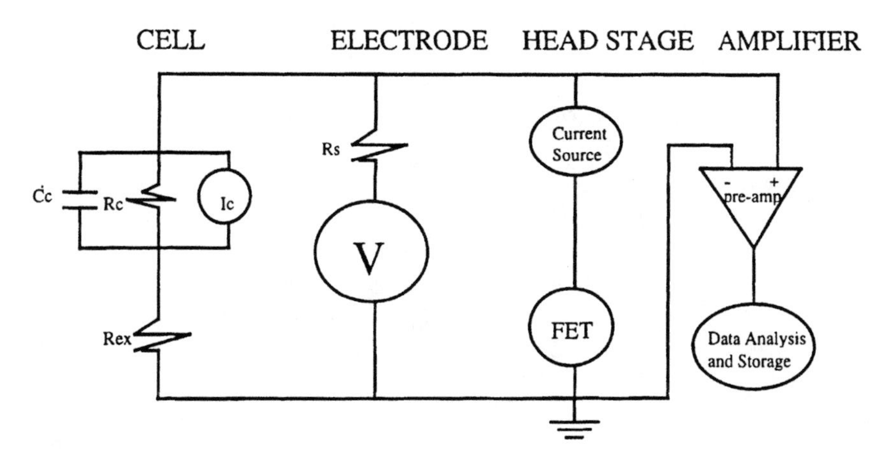

FIGURE 2.2
Schematic representation of the signal conditioning required for recording neural events from an extracellular electrode. Electrically, the cell is represented by a current source, I_C, in parallel with a resistor, R_C, and a capacitor, C_C. The resistance of the extracellular environment is represented by R_{ex}. The electrode has an impedance, R_S, and measures the voltage drop, V, between the recording electrode and a reference. The resulting current is amplified with a field effect transistor, FET, which requires a DC current source. The signal is then amplified by a precision instrumentation preamplifier. Finally, the data are analyzed and stored. The FET increases current, producing a stronger signal for the preamplifier, thereby reducing the noise.

close enough. Therefore, headstages that connect to the electrode cap on the animal's head are generally fitted with field effect transistors[24] (FETs). FETs boost the current without affecting the voltage. This allows the signal to make it to the preamplifiers or amplifiers without completely losing its strength. The FET requires DC power. Chapter 3 offers a detailed review of the electronic instrumentation required to record the neural signals after they are amplified.

2.2 Acute Recordings

Acute recordings require the animal to be anesthetized throughout the procedure and therefore can simplify some of the difficulties associated with chronic recordings. These difficulties include the possibility of controlling the animal's behavior. In addition, investigators can successively reuse electrodes after an acute experiment by rinsing them with distilled, deionized water. Currently, there are several different kinds of acute, multiple-site electrodes (see Chapter 5 for an example). In fact, it is recommended that multielectrode users become proficient with acute recording preparations before trying chronic recording preparations.

There are generally two types of recording electrodes to choose from, low and high impedance. The microwire bundle, an example of a low impedance electrode, is usually formed by 8 stainless steel wires, 12 to 50 microns in diameter, insulated with Teflon® (E.I. du Pont de Nemours and Company, Inc., Wilmington, Delaware).

The larger the diameter of the electrode tip, the lower the impedance. A bundle of tungsten wires, with insulation applied to the edge of a beveled tip is an example of a high impedance electrode. For acute experiments, high impedance electrodes are most desirable because of their better signal-to-noise, which makes discriminating each neuron's action potential much easier.

In addition to choosing the appropriate electrode, care must be taken when selecting the anesthesia, since it is well known that different anesthetics can affect the firing rate of cells. In some cases, neural firing rates can decrease to very low levels during anesthesia, making it extremely difficult to obtain viable recordings. However, there are some advantages in using anesthetized animals. For example, if the background neuronal firing rates are suppressed, neural responses due to stimulation may be easier to discriminate.

Another consideration regarding the surgical technique used to implant the electrodes, particularly the choice of whether to remove the dura before implanting the electrode, is an important one. When the dura is removed, the brain generally swells into the opening and the underlying tissue can become damaged. Generally, the less damage to the dura the better. However, there are several other considerations that have to be taken into account to design a surgical strategy. For instance, the deeper the brain structure that you are recording from, the less important it is to keep the dura intact. Moreover, it is also difficult to insert electrodes through the dura since most multiple-site recording electrodes bend at the dura surface. Users of this technology have devised several ways around this dilemma, depending on the type of electrode used (see Chapters 5 and 7). Microwires can sometimes "pop" through the dura when driven very quickly. However, this may cause damage just below the dura. Therefore, it may be better to make a small incision in the dura. Users of the University of Michigan's silicon electrodes (see Chapter 1) generally use a set of forceps covered with polyethylene tubing. The forceps gently but firmly hold the electrode in place while the tip is forced through the dura. Bionics Inc. (Salt Lake City), which has a very unique design (see below), uses a special tool to implant the electrode into the brain with a high-speed inserter. Because these electrodes are only about 1.5 to 2 mm long, without this inserter they would only press the dura down, compressing the brain without penetrating it.

2.3 Chronic Recordings

Generally, the results obtained from anesthetized, acute preparations are different from the results of recordings with awake animals. This has clearly been seen in cortical mapping studies.[25-27] The size and shape of receptive fields during chronic recordings are often different from the response properties of cells recorded during acute conditions. Clearly, there is a role for both chronic and acute recording in studying the properties of neural circuits, but chronic recording requires additional considerations before it can be successfully accomplished. For example, the behavior of the animal must be monitored during the experiment. In addition, the animal must be tethered to the recording equipment, which puts severe constraints on its movements.

Within these limitations, there are many possible paradigms. For some experiments animals can be highly trained, and the behavior of the animal is tightly controlled.[28,29] In this case, the response of the neurons across populations of cells can be compared to the state of the animal during the behavioral task. Another, perhaps more relevant paradigm, is to allow the animal more freedom to move naturally within its environment. This has recently worked well for the investigation of place cells in the hippocampus[30,31] and for a study of the barrel cortex in the rat.[32] In these cases, spatial averaging across the population of cells on a single trial basis may give more information about the animal's behavior than the more typical temporal averaging across many trials (see Chapters 7 and 10 for examples).

To date, the most successful chronic, multi-unit recordings have been obtained with low impedance microwires. Recording from 32 wires in the rat is commonplace, and now with the appropriate equipment (for example, see Plexon Inc., Texas) 128 simultaneous microwires in a monkey is quite feasible (see Chapter 7). Discriminating multiple neurons per wire is difficult and may require off-line analysis. Several manufacturers of multiple, single-unit recording equipment also sell neuron discrimination packages that run on regular PCs or Macs.

It is quite common for investigators to make their own electrodes, but this trend may be decreasing since multielectrode arrays can be purchased from commercial vendors. For instance, NBLabs in Texas (see below and Figure 2.3) builds custom electrodes for the user which are highly reliable for most experimental paradigms described in the literature.

There has been much less success in chronic preparations with high impedance, thin-film electrodes.[33,34] The reasons for this are unclear, especially since they seem to work well in acute experiments. There are two commercial sources for these electrodes: the University of Michigan Center for Neural Communication and Bionics Inc., which is supported by the University of Utah. The designs of these electrodes are very different. The Bionics electrodes have the recording site at the tip, like the microwires. However, the recording sites of the Michigan electrodes are along the shaft of the electrode and are therefore resting against tissue that may be damaged by the passage of the electrode. This may be the cause of the poor single-unit recording yield that has been consistently recorded with these electrodes (see below).

However, if these problems with chronic recordings can be solved, thin-film electrodes have the advantage of being built with integrated amplifiers and filters which could greatly improve the signal-to-noise of these electrodes. Much work has been done in parallel with the development of these electrodes, and there are several instrumentation amplifiers that are small enough to fit in bundles of 6 to 12 on the animal's headstage. However, the eventual goal of having hundreds of electrodes implanted in an animal, will require local multiplexing of the signal.

As mentioned above, the surgical technique is crucial for the users of chronic electrodes. The electrodes must be clean before implanting, and they must remain undamaged. Dried blood or brain matter on the recording site will certainly compromise the recordings. If the electrode comes in contact with blood or brain material before the final implantation is made, rinse the electrode immediately with distilled, deionized water. This should remove any material before it has a chance to dry. Many of the thin-film electrodes, described below, are very fragile, and it is easy to

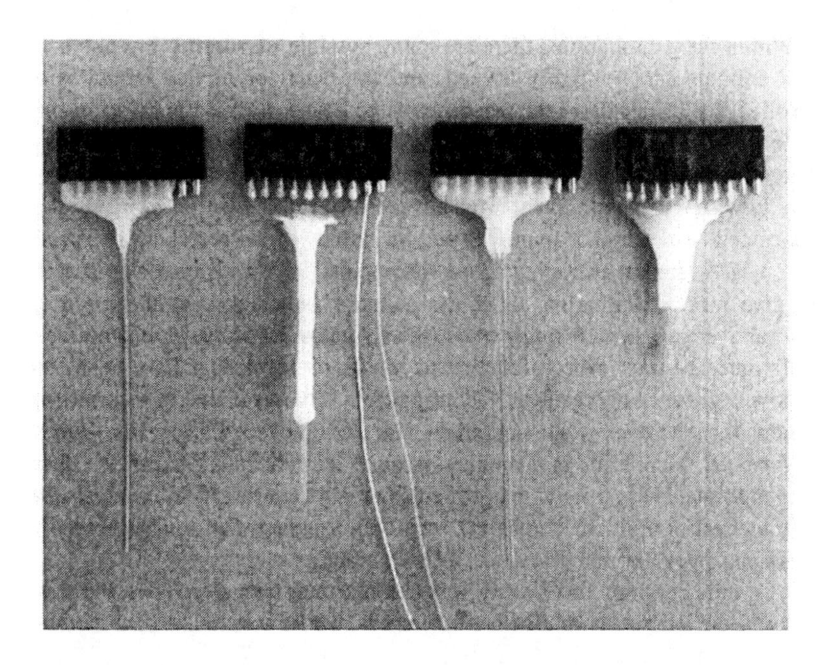

FIGURE 2.3
Multiwire electrodes produced by NBLabs, Denison, Texas. A. A sample of four different types of arrays (from left to right: an eight-wire bundle, an array of $2 \times 3 \times 3$, a double eight-wire bundle, and a 16-wire cortical array. B. A side view of an eight-wire array. C. A view from the bottom of a 16-wire cortical array. The array consists of two rows of eight wires. D. A view from the bottom of a $2 \times 3 \times 3$ array. The array consists of three rows: one row of two wires and two rows of three wires.

break them. One method for reducing breakage is to mount the electrode to a dissecting pin before insertion to give it added stability. Unfortunately, this will create much more damage to the surrounding tissue, making it more difficult to record individual neurons.

Another important rule to follow is that the more skull screws used during the surgical procedure to stabilize the electrode the better the long-term recording results. This is absolutely imperative for anchoring the electrode to the skull. It has been thought that it is best to put one screw in each skull plate. This ensures that as stresses are placed on the electrode cap, all of the skull plates will move together. Otherwise, as the plates move against each other, cerebral spinal fluid may leak out and get under the cap, which can lead to infection, and the cap may loosen from the skull surface. This leads to one final point about surgical procedure. All blood and debris must be completely removed from the skull and around the skull screws before the electrode is cemented into place. First of all, the dental cement will adhere much better to a clean skull. In addition, dried blood left on the skull surface will attract macrophages that can destroy the integrity of the cap.

One of the most fundamental issues for getting good recordings is to have a good ground. Often investigators implant screws into the skull and then tie the grounds to the screws. This is generally a good procedure. However, if tissue damage is not a problem, the ground wires can be implanted into the brain tissue far from the recording site and then wrapped around the screw for a more secure ground. In addition, a differential electrode will reduce local background noise.

Last, as in the acute case, there remains the question of whether to leave the dura intact or to remove it before inserting the electrode for a chronic preparation. Many electrodes are too flexible to break through the dura. If the electrode is used to break through the dura, the pressure exerted may cause damage just below the surface, or the tips of the electrode can become damaged. If the dura is removed surgically (as stated above under acute recording), the brain will swell through the opening, distorting your recording. This is especially true for the cortex. Therefore, completely removing the dura is probably not recommended. Instead, a small slit can be made in the dura about the size of the electrode tip.

After the electrode has been lowered into place, the surface of the brain can be covered with a 2% agar solution or gel foam which will harden immediately and protect the brain surface from the rest of the electrode cap. It is imperative to keep the agar or gel foam off the surface of the skull. This will destroy the integrity of the electrode cap because the methyl–methacrylate will only adhere to the dry skull surface. The gel foam or agar will prevent this if it gets on the surface of the skull.

2.4 Multiple Wire Electrodes Used

Multiple wire electrode arrays have been used in varied configurations since the 1960s. It is only recently that sufficient hardware and software have been developed so that large numbers (>32) of neurons can routinely be recorded simultaneously from these electrodes. The characteristics of the individual wires used remain the same. The wire must have a small diameter while still maintaining adequate stiffness and tip shape to penetrate the dura and traverse the tissue with minimal bending and mechanical disturbance to surrounding tissue. The wire must be sufficiently insulated with a material that is biologically inert in order to minimize the reactivity with the tissue. If the impedance of the electrodes is low, with minimal noise, it is likely to record multiple neurons per wire, thus greatly increasing the yield of these electrode arrays. These electrodes have generally been produced within the lab by paying particular attention to the tips.

There is some controversy over the benefits of high impedance recordings in chronic freely moving animals. Some investigators have noticed that the high impedance electrodes tend to pick up more movement artifact than the lower impedance electrodes. Some on-board electronics in the headstages, for example field effect transistors (FET), have reduced this problem but certainly not eliminated it.

2.4.1 Fabrication

There has also been considerable controversy surrounding the diameter of the wires used to make the electrode: 12 to 50 microns are the typical sizes in use today. We have found that the amount of damage is related more to the speed at which the electrode is lowered than by the size of the electrode. If the electrodes are lowered slowly enough, it is very difficult to find the electrode track without using a marking current and staining the tissue. Often, microdrives are used to lower the electrode over the course of several days into the proper position.

The benefits of the 25- to 50-micron wires include being strong enough to go through the dura of rodents and rigid enough to be lowered into the correct position. Wires smaller than 25 microns have difficulty breaking through the dura and can often be sidetracked by a fiber bundle when lowered into the brain. Therefore, they often do not end up in the intended brain structure.

Shaping the tips of these electrodes takes numerous forms. Chemical etching of the tips or beveling them are both used. The resulting shape is generally a preference of the investigator. Theory suggests that a fine, pointed tip, with insulation covering the wire as close to the tip as possible, is ideal. The smaller the surface area, the higher the input impedance and therefore the better the signal-to-noise characteristics of the electrode. However, suitable recordings have been acquired from blunt-tip electrodes.[25] The reasons for the divergent theories about tip shape arise because it is not clear what happens to the electrode after it is implanted. Clearly, a foreign object introduced into the brain will cause a tissue reaction no matter how biologically inert the substrate. The brain's reaction will have an effect on the recording parameters of the electrode, making the relationship between recording surface area and electrode impedance less relevant.

The two most popular types of wires are tungsten,[35] stainless steel,[25] and platinum,[36] though carbon has been used with success.[37] In addition, there are several types of insulating material, and pre-insulated wires are now readily available. Insulating material ranges from polyimide to Teflon. Teflon works very well, as do several commercial versions of polyimide, or polydicholoroparaxylene.[38-43] The integrity of the insulation is very important. If the insulation is breached and the wires come in contact with each other, the recordings will be shorted out. It is also possible that a breach in the insulation can occur somewhere along the length of the electrode that is embedded in a different region than the tip. This exposed metal surface becomes a conducting medium, and there will be cross-talk noise within your signal from this spurious site. The effects of the brain microenvironment on the integrity of the insulation is also an important consideration. The electrode is essentially soaking in a 0.9% salt solution. Very few insulating materials are impermeable to the extracellular fluid, and this is an important problem for long-term recording experiments.

2.4.2 Applications

Microwire bundles and arrays come in many shapes and sizes (Figure 2.4). These unique architectural designs can be used to gain an advantage over the structure of the brain. Depending on the regions recorded from and the type of data the investigator wants to collect, these electrodes can be shaped and positioned to record from multiple dimensions. They can be shaped to follow the curvature of a structure

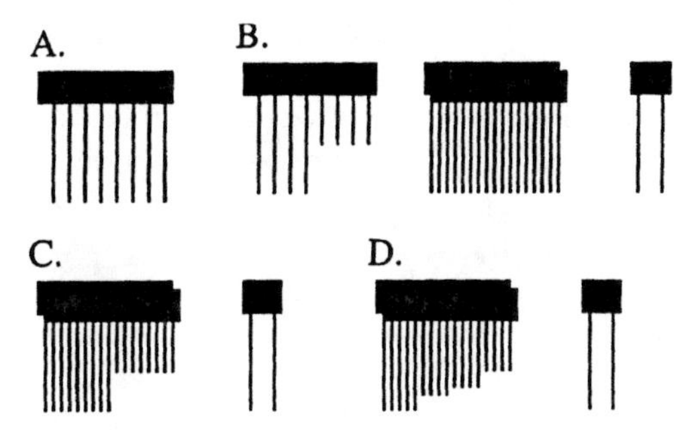

FIGURE 2.4
Multidimensional wire arrays designed to exploit the geometry of the brain. A. A one-dimensional array. B. Two examples of two-dimensional arrays. C. A three-dimensional array. D. A multidimensional array.

within the brain and lowered during surgery at various angles to ensure proper placement in the brain area of interest or avoid trouble spots such as the ventricles.

In addition to straight, individual microwires bundled together, tetrodes[44] or stereotrodes[45] have been used to separate neurons whose action potentials have very similar amplitudes or neurons that burst with action potentials of varying amplitude. Tetrodes are four wires, and stereotrodes are two wires, twisted together. The tips of the electrodes are placed very close together so that they see the same information but from a slightly different position. The need for tetrodes comes from the practical limitations of multiple single-unit recording. Most neural discrimination is based on the amplitude of the action potential (AP). The width of the AP, the slope of the falling phase, and the after-hyperpolarization in combination with the amplitude can often aid discrimination but are rarely useful by themselves.

There are two cases where tetrodes can be useful. In the first case, when a bursting cell has a variable amplitude, which is common with hippocampal pyramidal cells for example, it is not possible to discriminate more than one neuron, based on amplitude, at a single electrode. Differing amplitudes could be from the same cell. However, if the spikes in question can be viewed on at least two recording sites simultaneously, which is possible if two insulated wires are twisted together, then the amplitudes will be slightly different on each electrode. If the varying heights of the spikes are from the same cell, then the ratio of the heights of the spikes at the two sites will be the same. However, if the ratio of heights is different, then the spikes come from neurons at two different locations in space.

In the second case, with small cells that are tightly packed, layer four cortical cells for example, sometimes the shape of the APs at the same electrode, with the baseline noise contributing, is not enough to discriminate them. By using a stereotrode, or perhaps a tetrode, the ratio of the amplitudes from different cells recorded at different recording sites should be significantly different such that in the amplitude space, two distinct centers will be found. Obviously, because of noise in the signal and noise in the baseline, there may be some overlap. It is at this point that cluster cutting becomes crucial. However, this technique is beyond the scope of this chapter. See Chapter 4 for a detailed review of methods for automatic discrimination of single units.

A distinct advantage of this method then is that more neurons can be discriminated per electrode. Given that there is certainly some damage to tissue when any sized electrode is implanted, the more neurons that can be recorded per recording site the better. In addition, it is becoming obvious that small local circuits, for example those in cortical columns, may be processing information streams from several sources. This method allows the investigator to improve the yield of recorded neurons in a very small region. The disadvantage of this method is that it requires a large amount of data storage. Standard microwire recordings discriminate on-line and therefore only store the time of each action potential. Tetrodes and stereotrodes require off-line discrimination, and therefore the analog signal for each electrode must be stored. Lack of storage space can severely limit the number of neurons that can be recorded. Moreover, no study has demonstrated that tetrodes can be used for long-term chronic recordings (weeks or months).

Commercially Available Microwires

NBLabs
1918 Avenue A
Denison, TX 75021-6751
(902)465-2694
www.nblabslarry.com

FHC Inc.
9 Main Street
Bowdoinham, ME 04008
www.fh-co.com

2.5 Thin-Film Microelectrodes

Thin-film circuits are made by photofabrication, which involves the use of light-sensitive polymers and photoresists, to produce printing templates. The University of Michigan Center for Neural Communication has been developing a silicon-based microelectrode for several years.[46] These processes require specialized equipment and cannot generally be made by individual investigators.[47] Although the basic processes used are typically used in the semiconductor industry, collaboration between physiologists and engineering laboratories producing these electrodes is essential to overcome some of the unique problems associated with the biological medium. Initial engineering concerns behind the design of these electrodes did not sufficiently address most of the needs of the single-unit physiologists, which are only now beginning to be resolved. Progress has only recently been made in implanting the electrodes, designing appropriate electrode caps and increased reliability so that they are readily available to investigators.

The greatest advantage of thin-film microelectrodes is the potential for sampling many sites in a small tissue area. The electrodes can be as small as 15 microns thick by 50 microns wide,[48] and they can have approximately 10 recording sites. Several of these electrodes implanted in the brain can increase the yield of recorded neurons by an order of magnitude, with minimal tissue damage.

An additional advantage is the ability to know the distances between recording sites more accurately. These can be used for stereotrode analysis, for highly structured areas of the brain such as the layers of the cortex, or for current source density analysis in freely moving animals. It is impractical to attempt the simultaneous recording of electrical activity at a large number of spatially specific tissue locations with conventional metal wire or pulled microelectrodes. Using thin-film technology, the characteristics of the array can be standardized for any given application. Therefore, more precise information about the spatial relationships between the cells being recorded is more readily available than with conventional electrodes. The scale of these spatial dimensions can range from microns to centimeters across the brain. The photofabrication process is extremely versatile, and there is little restriction to the designs that can be produced.

Last, the far-reaching developmental potential of thin-film electrodes is the promise that these simple electrodes can be integrated with existing industrial thin-film circuits to give on-site signal conditioning. Such hybridizations can significantly reduce noise, cross-talk, attenuation, and distortion of the signal. This can increase the yield per recording site. Eventually, these electrodes can be interfaced with telemetric circuits to facilitate studies concerned with relating neural activity to more complex behaviors.

However, there are some drawbacks to the currently available electrodes. For example, silicon electrodes tend to be very brittle. The University of Michigan has reduced this problem by making them more flexible, though the flexibility makes it very difficult to penetrate the dura. In addition, most of the electrodes have the recording sites along the length of the substrate so that after insertion, some of the sites are in contact with tissue that has been damaged during insertion. This can render the recording site useless and greatly reduce the yield of recorded neurons. Bionics Inc. has eliminated this problem by only recording at the tips of the electrodes.

2.5.1 Fabrication of Thin-Film Microelectrodes

The process to produce thin-film electrodes is a standard photolithography process. The process begins by creating a photomask which is an image of the electrode design. These masks can then be used repeatedly and are relatively inexpensive. Therefore, creating unique designs for a particular experimental paradigm is generally not a concern. The process of creating the electrodes from the masks is generally the same, regardless of the design printed on the mask (Figure 2.5).

A substrate, usually silicon or ceramic, is polished and cleaned and coated with a photoresist material. The choice of substrate is critical. Some applications require the resulting electrode to be quite flexible. Silicon and glass tend to be very brittle when the substrate is made thin enough for microelectrodes. Ceramic is stronger than silicon and makes an excellent rigid electrode. The University of Michigan has developed a technique using boron ion-doped silicon substrates[49] that produce a very flexible silicon and thin (about 15 microns) electrode.

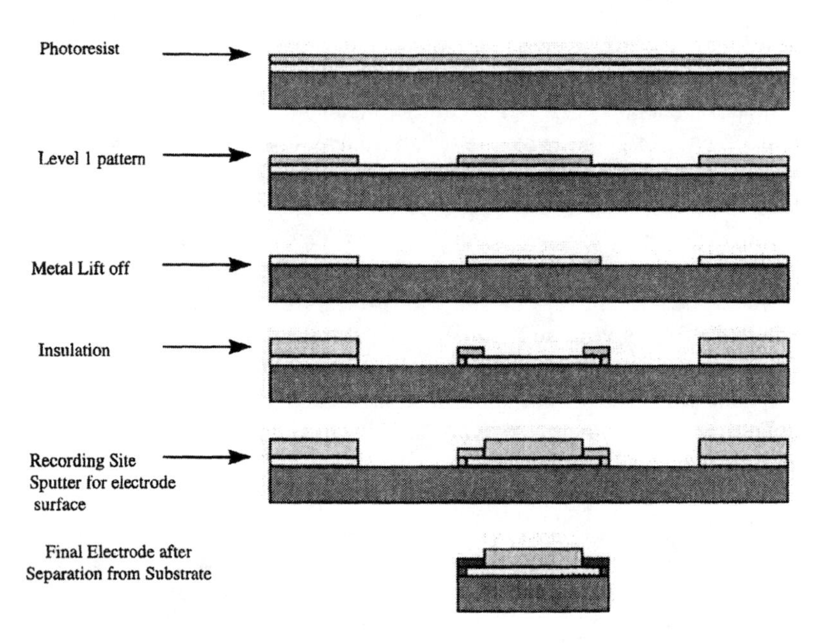

FIGURE 2.5
A schematic representation of the steps involved in the development of a thin-film electrode. A substrate, usually silicon or ceramic, has an adhesion material, usually titanium, coated to it. This allows the metal, applied in the next step, to adhere to the substrate. During the metallization step, a metal — gold, platinum or carbon — is deposited on the substrate. Next a photoresist is applied and patterned using the photomask. The excess metal is then removed. The metal is then insulated with a second level pattern. Last, the recording sites are exposed and can be sputter coated with a final metal layer to create the recording surface.

The photoresist is a light-sensitive resin that can coat the surface of the substrate to guide the next metallization layer. After drying, the coated substrate is covered by the photomask and exposed to UV radiation. This process converts the exposed resist to a hard resin, outlining the electrode circuit. The unexposed resist is easily removed.

The next step is the metallization process. The metal that will constitute the electrode circuit is uniformly applied to the substrate. Generally, the metal does not stick well to the substrate, and a very thin layer of an adhesion material, such as titanium, is used. The metallization is generally achieved by evaporation, sputtering, or some combination of the two. In sputtering, ions are electrically generated in a low pressure gas, such as argon. The ions are accelerated toward the metal target, causing atoms to be ejected and subsequently driven to the substrate surface. In evaporation, the metal is suspended in a gaseous state and deposited on the surface of the substrate by lowering the temperature. The metal will be deposited uniformly across the photoresist and the desired electrode circuit. Subsequent dissolution of the photoresist removes unwanted metal, leaving the circuit intact (lift-off method).

Alternatively, a substrate can be metallized before application of the resist. In this case, unwanted metal is removed by etching away areas not protected by the photoresist. Chemical etching is the simplest way to remove the metal. The rate and accuracy of the etching is controlled by doping the substrate or by removing the reaction product (i.e., the metal ions) from the surface of the substrate. The etchant can be sprayed, stirred, or bubbled with nitrogen to increase turbulence at the interface, which further increases control of the etchant process. There are several difficulties with this process, however. The etching process is not uniform because in addition to etching vertically through the metal layer, there is a lateral etch occurring where the desired metal circuit meets the photoresist. This can generally be compensated for in the photomask. Microetching can be accomplished by physically bombarding the substrate with etchant. This minimizes lateral etch. RF sputtering, plasma etching, and ion beam milling are all examples of this process.

There are many choices for the metal conductor — nickel, stainless steel, tungsten, gold, or platinum. Gold can cause difficulties in the manufacturing process because it is very soft. When the lift-off or etching process is conducted, it may be difficult to leave behind a quality metal line. Platinum has been shown to cause the least histological damage in long-term implants.[50] However, because of the nature of the thin-film process, one conductor can be used for the circuit, while a final layer of the material most desirable as an electrode–biological interface can be sputtered only on the recording sites.

Next the metal circuits must be appropriately insulated, leaving only the recording sites and bonding terminals exposed. For chronic, long-term recordings this has been one of the most difficult challenges. Because of the close proximity of the metal lines that carry the signal from the recording sites to the bonding terminals, any breach in the insulation will greatly increase noise, cross-talk, and distortion in the signal. Some materials that have been known to be protective against fluid penetration in other applications are not suitable for microelectrodes. The general measure of the insulating property of a material is its pin-hole density. The thinner the layer of material, the greater the pin-hole density in many cases. While insulating materials such as Parylene® C and Silastic® have been used to insulate and encapsulate microelectronic implants and cardiac pacemakers, the films are relatively thick compared to thin-film application of insulators for microelectrodes.[51] To date, silicon nitride[52] is the most commonly used.

The insulation procedure is generally similar to the metallization procedure. A second photomask is produced that leaves only the recording sites and the bonding terminals exposed. Photoresists are applied over the entire circuit and developed using this photomask so that the terminals and bonding pads are protected. The entire substrate is then layered with an insulator, and finally the resist over the recording sites and bonding terminals is removed, removing the insulator as well.

The individual electrodes are then released from the substrate by sawing, laser cutting, or through an etching process. Laser cutting is the most accurate; however, if the original substrate is appropriately preprocessed to control the etching process, back etching of the substrate can produce very thin electrodes. The University of Michigan Center uses this method. The electrode shape is patterned into the substrate using deep boron diffusion to heavily dope the intended electrode area before metallization. The temperature and time of this diffusion determine the final thickness of the probe. The final process is to subject the substrate to an ethylenediamine-pyrocatechol-water etch that separates the electrodes and stops at the boron boundary.

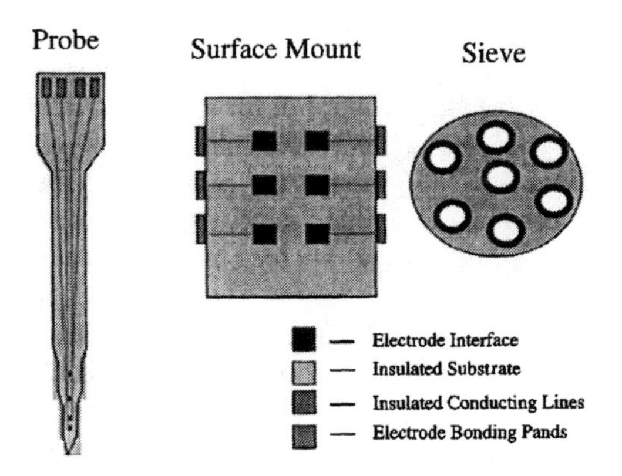

FIGURE 2.6
Different configurations of thin-film electrodes. The probe type electrodes can be inserted into brain tissue to record from deep structures. The surface mount electrodes can have nerve tissue grow onto the plates, or they can record from the surface of slice preparations. The sieve type electrode can be used to record from peripheral nerve tissue. The nerve is cut and then grows back through the electrode.

The bonding terminals are then attached to second-stage recording equipment. The integrity of these bonds affects the recording capability of the entire electrode. The small size of the electrodes make them difficult to manipulate and most of the yield is lost during this process. A common bonding procedure is ultrasonic wire bonding that can be accomplished with gold or platinum bonding terminals. For acute recordings, the electrodes are generally connected to a larger PC board that may or may not contain other integrated circuits and a set of connectors to third-stage recording equipment.

For chronic recording, a more flexible cable is required. Several different kinds of flexible cables are available. In addition, Hetke et al.[53] have developed a silicon ribbon cable that is extremely flexible and can be from 2 to 20 microns in thickness. The advantage is that the same process that is used to make the electrodes is used to make the cable. Therefore, by applying appropriate masks, the cable and electrode can be integrated during fabrication, eliminating the need to bond the electrode to the cable.

2.5.2 Applications

The flexibility of the thin-film fabrication process, produces enormous variety in the types of applications in which these electrodes can be used (Figure 2.6). Llinas et al.[54] first proposed a gold on silicon dioxide wafer with more than one hundred 25-micron-sized holes that could be sandwiched between the cut ends of a nerve bundle in such a way that regenerating axons would grow through the holes and ultimately regain their original innervation sites. A gold annuli around the holes produces an *in situ* recording of nerve impulses. Many fruitful preparations have been developed from this idea including enclosures of regenerating axons.[55] Surface electrodes have also been developed and used to record brain surface transients,[56] to record from cell cultures,[57] and for recording insect brains.[58]

Probe type electrodes have made advances for multiple, single-unit recording in central nervous tissue. Because of the flexibility of the manufacturing process, the terminals of arrays of electrodes can be patterned in a wide variety of ways. Circuits with a straight line of terminals (forked shaped) can be produced in the plane of a conducting path or at right angles to them. Two-dimensional patterns can be produced where the terminals of the arrays are of different lengths. By printing recording circuits on both sides of the substrate, the density of sites can be increased. In fact, several multidimensional probes have been devised.

Commercially Available Thin-Film Electrodes

University of Michigan Center For Neural Communication Technology
1301 Beal Avenue
Ann Arbor, MI 48109-2122
(313)764-8040
www.engin.umich.edu/facility/cnct

Bionic Technologies, Inc.
1763 East 900 South
Salt Lake City, Utah 84108-1333
(801)583-5832
www.bionictech.com

The University of Michigan Center Integrated Circuits and Sensors provides electrodes to investigators free of charge. They have both acute and chronic electrode styles. There are also many different designs available, and the users can design their own electrodes if their application requires it. The recording surfaces are made of iridium and are positioned along the length of the electrode. Although they have been used extensively for field potential recordings, so far isolation of large populations of single neurons has proved to be difficult with these probes.

Bionic Technologies provides a unique design. The arrays consist of up to 100 sharpened needles that project from a silicon substrate. The needles are electrically isolated by a nonconducting glass, and the tips are coated with platinum to make them electrically active recording sites. The tips are usually 50 to 100 microns in length and have electrode impedances from 0.1 to 0.2 megaohms. The needles are either 1.0 or 1.5 mm long. This limits their use in *in vivo* brain preparations to cortical regions. However, they are also ideal for spinal cord, peripheral tissue, and brain slices. For acute recordings all 100 sites can be connected to a thin-film card for connection to second-stage electronics. The high density of recording needles makes it difficult to implant the arrays. They provide a pneumatically actuated impulse inserter that implants the array at high velocity with little tissue damage. For chronic recordings, they provide a twelve-lead connector to an electrode cap. The number of active sites is currently limited by the electronics of the cap. The twelve leads provide for 11 active recording sites and one reference wire.

References

1. Brindley, G. S., Electrode arrays for making long-lasting electrical connection to spinal roots, *J. Physiol.* (Lond.) 222, 135, 1972.
2. Ninomiya, I., Yonezawa, Y., Wilson, M. F., Implantable electrode for recording nerve signals in awake animals, *J. Appl. Physiol.*, 41, 111, 1976.
3. Mannard, A., Stein, R. B., Charles, D., Regeneration electrode units: implants for recording from single peripheral nerve fibers in freely moving animals, *Science,* 183, 547, 1974.
4. Loeb, G. E., Marks, W. B., Beatty, P. G., Analysis and microelectrode design of tubular electrode arrays intended for chronic, multiple single-unit recording from captured nerve fibres, *Med. Biol. Engng. Comput.*, 15, 195, 1977.
5. Thomas, C. A., Springer, P. A., Loeb, G. E., Berwald-netter, Y., Okun, L. M., A miniature microelectrode array to monitor the bioelectric activity of cultured cells, *Exp. Cell Res.*, 74, 61, 1972.
6. Pine, J., Gilbert, J., Regehr, W. Microdevices for stimulating and recording from cultured neurons, in *Artificial Organs*, J. D. Andrade et al. (Eds.), VCH Publishers, Berlin, 573, 1987.
7. Gerstein, G. L., Perkel, D. H., Simultaneously recorded trains of action potentials: analysis and functional interpretation, *Science,* 164(881), 828, 1969.
8. Wise, K. D., Angell, J. B., A low-capacitance multielectrode probe for use in extracellular neurophysiology, *IEEE Trans. Biomed. Eng.*, BME-17(3), 238, 1970.
9. Kennedy, P. R., Bakay, R. A. E., Sharpe, S. M., Behavioral correlates of action potentials recorded chronically inside the cone electrodes, *Neuroreport,* 2, 605, 1992.
10. Perkel, D. H., Gerstein, G. L., Moore, G. P., Neuronal spike trains and stochastic point processes. II. Simultaneous spike trains, *Biophy. J.*, 7(4), 419, 1967.
11. Caspersson, T., Zech, L., Modest, E. J., Predicting measures of motor performance from multiple cortical spike trains, *Science,* 170, 758, 1970.
12. Kaltenback, J. A., Gerstein G. L., A rapid method for production of sharp tips on perinsulated microwires, *J. Neurosci.*, 16, 283, 1986.
13. Baldwin, D. J., Dry beveling of micropipette electrodes, *J. Neurosci. Meth.*, 2, 153, 1980.
14. Humphrey, D. R., A chronically implantable multiple micro-electrode system with independent control of electrode position, *Electroencephalography and Clinical Neurophysiology,* 29, 616, 1970.
15. Humphrey, D. R., Schmidt, E. M., Thompson, W. D., Predicting measures of motor performance from multiple cortical spike trains, *Science,* 170(959), 759, 1970.
16. Campbell, P. K., Jones, K. E., Normann, R. A., A 100 electrode intracortical array: structural variability, Proc. 27th Annu. Rocky Mountain Bioeng. Symp. and the 27th Int. ISA Biomed. Sci. Inst. Symp., Denver, CO, Apr. 6-7:161, 1990.
17. Fryer, T. B., Sandler, H., A review of implant telemetry systems, *Biotelemetry,* 1, 351, 1974.
18. Wise, K. D., Angell, J. B., Starr, A., An integrated-circuit approach to extracellular microelectrodes, *IEEE Trans. Biomed. Eng.*, 17(3), 328, 1970.
19. Wise, K. D., Integrated sensors: key to future VLSI systems, *6th Sensors Sympos. Digest,* 6, 109, 1986.

20. Najafi, K., Wise, K. D., An implantable multielectrode array with on-chip signal processing, *IEEE J. Solid-State Circuits,* SC-21(6), 1035, 1986.
21. BeMent, S. L., Wise, K. D., Anderson, D. J., Najafi, K., Drake, K. L., Solid-state electrodes for multichannel multiplexed intracortical neuronal recordings, *IEEE Trans. Biomed. Eng.,* 33, 230, 1986.
22. Ji, J., Najafi, K., Wise, K. D., A low-noise demultiplexing system for active multichannel microelectrode arrays, *IEEE Trans.,* BME-38, 75, 1991.
23. Ji, J., Wise, K. D., An implantable CMOX circuit interface for multiplexed microelectrode recording arrays, *IEEE J. of Sol. State Circuits,* 27, 433, 1992.
24. Cobbold, R. S. C., *Theory and Applications of Field-Effect Transistor,* John Wiley, New York, 1970.
25. Nicolelis, M. A. L., Baccala, L. A., Lin, R. C. S., Chapin, J. K., Sensorimotor encoding by synchronous neural ensemble activity at multiple levels of the somatosensory system, *Science,* 268, 1353, 1995.
26. Eggermont, J. J., Functional aspects of synchrony and correlation in the auditory nervous system, *Concepts in Neuroscience,* 4, 105, 1993.
27. Deadwyler, S. A., Hampson, R. E., Ensemble activity and behavior: What's the code?, *Science,* 270, 1316, 1995.
28. Evarts, E. V., Precentral and postcentral cortical activity in association with visually triggered movement, *J. Neurophysiol.,* 37(2), 373, 1974.
29. Mountcastle, V. B., Lynch, J. C., Georgopoulus, A., Sakata, H., Acuna, C., Posterior parietal association cortex of the monkey: command functions for operations within extrapersonal space, *J. Neurophysiol.,* 38(4), 871, 1975.
30. Wilson, M. A., McNaughton, B. L., Dynamics of the hippocampal ensemble code for space, *Science,* 261, 1055, 1996.
31. Wilson, M. A., McNaughton, B. L., Reactivation of hippocampal ensemble memories during sleep, *Science,* 265, 6761, 1996.
32. Nicolelis, M. A. L., Lin, R. C. S., Woodward, D. J., Chapin, J. K., Dynamic and distributed properties of many-neuron ensembles in the ventral posterior medial thalamus of awake rats, *Proc. Natl. Acad. Sci. U.S.A.,* 90, 2212, 1993.
33. Ylinen, A., Bragin, A., Nadasdy, D., Jando, G., Szabo, I., Sik, A., Buzaki, G., Sharp wave-associated high-frequency oscillation (200 Hz) in the intact hippocampus: network and Intracellular Mechanisms, *J. of Neurosci.,* 15(1), 30, 1995.
34. Bragin, A., Jando, G., Nadasdy, D., Hetke, K., Wise, K., Buzaki, G., Gamma oscillation (40-100 Hz) in the hippocampus of the behaving rat, *J. of Neurosci.,* 15(1), 47, 1995.
35. Jaeger, D., Gilman, S., Aldridge, J. W., A multiwire microelectrode for single unit recording in deep brain structures, *J. Neurosci. Meth.,* 32, 143, 1990.
36. Sonn, M., Feist, W. M., A prototype flexible microelectrode array for implant-prosthesis applications, *Med. Biol. Eng. Comput.,* 12(6), 778, 1974.
37. Fox, K., Armstrong-James, M., Millar, J., The electrical characteristics of carbon fiber microelectrodes, *J. Neurosci. Meth.,* 3, 37, 1980.
38. Hahn, A. W., Yasuda, H. K., James, W. J., Nichols, M. F., Sadir, R. K., Sharma, A. K., Pringle, O. A., York, D. H., Charlson, E. J., Glow discharge polymers as coatings for implanted devices, *Biomed. Scis. Instrumentation,* 17, 109, 1981.

39. Sadir, R. K., James, W. J., Yasuda, H. K., Sharma, A. K., Nichols, M. F., Hahn, A. W., The adhesion of glow-discharge polymers, Silastic and Parylene to implantable platinum electrodes: results of tensile pull tests after exposure to isotonic sodium chloride, *Biomaterials,* 2, 239, 1981.
40. Yasuda, J., Gazicki, M., Biomedical applications of plasma polymerization and plasma treatment of polymer surfaces, *Biomaterials,* 3, 68, 1982.
41. Yasuda, H. K., Sharma, A. K., Hale, E. B., Hames, W. J., Atomic interfacial mixing to create water insensitive adhesion, *J. Adhesion,* 13, 269, 1982.
42. Levy, B. P., Campbell, S. L., Rose, T. L., Definition of the active area of a microelectrode tip by plasma etch removal of Parylene, *IEEE Trans. Biomed. Eng.,* 33, 995, 1986.
43. Nichols, M. F., Hahn, A. W., Electrical insulation of implantable devices by composite polymer coatings, *ISA Trans.,* 26, 15, 1987.
44. Gray, C. M., Maldonado, P., Wilson, M., McNaughton, B., Tetrodes markedly improve the reliability and yield of multiple single-unit isolation from multi-unit recordings in cat striate cortex, *J. Neurosci. Meth.,* 63, 43, 1995.
45. McNaughton, B. L., O'Keefe, J., Barnes, C. A., The stereotrode: A new technique for simultaneous isolation of several single units in the central nervous system from multiple unit records, *J. Neurosci. Meth.,* 8, 391, 1993.
46. BeMent, S. L., Wise, K. D., Anderson, D. J., Najafi, K., Drake, K. L., Solid-state electrodes for multichannel multiplexed intracortical neuronal recording, *IEEE Trans. Biomed. Eng.,* 33(2), 230, 1986.
47. Pickard, R. S., A review of printed circuit microelectrodes and their production, *J. Neurosci. Meth.,* 1, 301, 1979.
48. Najafi, K., Ji, J., Wise, K. D., Scaling limitations of silicon multichannel microprobes, *IEEE Trans. Biomed. Eng.,* 37, 1, 1990.
49. Najafi, K., Wise, K. D., A high yield IC-compatible multichannel recording array, *IEEE Trans. Electron. Dev.,* 32, 1206, 1985.
50. White, R. L., Gross, T. J., An evaluation of the resistance to electrolysis of metals for use in biostimulation microprobes, *IEEE Trans. Biomed. Eng.,* BME-21, 487, 1974.
51. Pickard, R. S., A review of printed circuit microelectrodes and their production, *J. Neurosci. Meth.,* 1, 301, 1979.
52. James, K. J., Normann, R. A., Low-stress silicon nitride for insulating multielectrode arrays, Proc. 16th Ann. Int. Conf. *IEEE Eng. Med. Biol. Soc.,* Baltimore, Nov., 306, 836, 1994.
53. Hetke, J. F., Lund, J. L., Najafi, K., Wise, K. D., Anderson, D. J., Silicon ribbon cables for chronically implantable microelectrode arrays, *IEEE Trans. Biomed. Eng.,* 41(4), 314, 1994.
54. Llinas, R., Nicholson, C., Johnson, K., Implantable monolithic wafer recording electrodes for neurophysiology, in *Brain Unit Activity During Behavior,* M. I. Philips, (Ed.), C.C. Thomas, Springfield, IL, 238, 1973.
55. Loeb, G. E., Marks, W. B., Beatty, P. G., Analysis and microelectronic design of tubular electrode arrays intended for chronic multiple single-unit recording from captured nerve fibres, *Biol. Eng. Comput.,* 15, 195, 1977.

56. Campbell, P. K., Jones, K. E., Huber, R. J., Horch, K. W., Normann, R. A., A silicon-based, three-dimensional neural interface: manufacturing process for an intracortical electrode array, *IEEE Trans. Biomed. Eng.,* 38(8), 758, 1991.
57. Robert, F., Correges, P., Dupart, S., Stoppini, L., Combined electrophysiology and microdialysis on hippocampal slice cultures using the Physiocard System, *Current Separations,* 16(1), 3, 1997.
58. Pickard, R. S., Welberry, T. R., Printed circuit microelectrodes and their application to honeybee brain, *J. Exp. Biol.,* 64, 39, 1976.

Chapter 3

Trends in Multichannel Neural Ensemble Recording Instrumentation

Koichi Sameshima and Luiz Antonio Baccalá

Contents

3.1 Introduction

Techniques for instrumentation have become unthinkable without electronic computers, and modern multichannel neural ensemble recording (MNER) is no exception. It is

0-8493-3351-2/99/$0.00+$.50

not difficult to see why. Computer-based instrumentation technology, whose value was recognized rather early[1-3] in neuroscience, evolved from the need to accurately monitor and control complex industrial processes, viz. chemical plants, nuclear reactors, etc., which is a problem similar to the recording of the extracellular activities of tens to hundreds of single neurons, as both involve the measurement of large numbers of variables in real time.

In terms of systems for multichannel single-unit recording, computers have allowed the following neurobiological questions to be addressed:

- Observing and characterizing the responses of large populations of neurons;
- Relating the neural responses of various brain structures;
- Monitoring neural ensemble activity in awake animals and correlating this activity with particular behaviors;
- Measuring the neural ensemble response to a variety of sensory stimuli (peri-stimulus time histogram or evoked potentials).

To perform these tasks, an ideal multichannel recording system must:

1. Store and display accurate representations of the analog signal that represents each neuron action potential (**spatial sampling**);
2. Discriminate valid action potentials from large populations of isolated neurons from many days to several months (**chronicity**);
3. Acquire and maintain simultaneous recordings of as many single-units (N) as possible for as long as possible (T);
4. Keep accurate track of the timing relationships between many action potentials, behavioral events, and stimuli (**high temporal resolution**);
5. Coordinate electrophysiological recording with the generation of a variety of stimuli.

These requirements have important implications for MNER instrumentation design. Mass memory storage capacity, for instance, is a function of the product $N \times T$ in a given experiment. Also the method of signal representation, with or without data compression, bears decisive influence on the overall capacity of an acquisition system. In fact, one may classify computer-based MNER systems according to the form of their data storage:

a. **Digital Storage Systems** (DSS) which store faithful discrete time representations of the analog neural signals. These systems are basically general purpose multichannel signal acquisition systems which are very similar to the digital recording systems used for voice, EKG, or EEG recordings.
b. **Event Timing Systems** (ETS) which store only the time of occurrence of each action potential in each one of its multiple channels. These systems take advantage of the stereotypical nature of the action potentials in the CNS and differ from the DSS by having a built-in, albeit simple, data compression mechanism.

In what follows, we briefly discuss the main components and issues of MNER systems.

3.2 Extracellular Single-Unit Signal Characteristics

Extracellular recordings are the only practical choice in experiments that intend to establish correlations between neural ensemble responses and behaviors involving awake animals.[4-8] Extracellular single-unit potentials measured by chronically implanted electrodes result from ionic current exchanges across the membrane of nearby neurons. These currents modify charge concentrations and alter the electrical potential sensed by microelectrodes immersed in extracellular neural tissue. Thus, extracellular recordings involve the activity of many adjacent single neurons which can usually be resolved into independent action potentials. This is possible because voltage variations due to a neuron depend on the inverse of its distance to the microelectrode,[9,10] implying that action potentials due to a cell located near the microelectrode (or hopefully to only very few cells) predominate in comparison to neurons farther away.

While the amplitude of intracellularly recorded action potentials ranges between 60 and 100 mV, extracellular action potentials can be two orders of magnitude smaller from hundreds of μV to tens of mV,[9] making their observation especially prone to degradation due to noise and artifacts.

Action potentials usually last from 0.2 ms to no more than a couple of milliseconds. This allows us to treat them as events of a point process whereby considerable data reduction is achievable by only storing action potential time markers (or time stamps).

The nature of extracellular action potentials defines the methods employed in their detection (see Chapter 4). Efficient detection severely depends on how much noise is present, whether from electronic (thermal random conduction), from biological sources (distant spiking neurons, field potentials, and movement artifacts), or from electromagnetic interference. Other factors influencing detectability include action potential amplitude, which is a function of cell shape, the conductive properties of the extracellular medium, microelectrode configuration and localization with respect to cell axis, and the type and quality of the microelectrodes.[9,11,12]

3.2.1 Digital Representation — Analog to Digital Conversion

Internally, computers deal *only* with finite numbers. Thus, neural signals must be mapped onto discrete quantities both in amplitude and in time. For amplitudes, the mapping process consists of *quantization*, i.e., the measured voltage values become represented via discrete levels. To represent the temporal evolution one records instantaneous quantized amplitude values at regular periodic time intervals called the sampling period T_s. The inverse of the sampling period, $1/T_s$, is known as the *sampling rate*.

Amplitude quantization introduces an error referred as quantization noise. This error is a function of how fine the amplitude grid is. In other words, for a signal with a given dynamic range (difference between its maximum and minimum attainable values), the finer the grid the smaller the error. Usually the number of levels is chosen so that quantization noise is much smaller than that of the noise due to other sources. The accuracy of a sampled waveform representation is a function of the sampling rate and is governed by *Nyquist's criterion,* which provides a condition for exact reconstruction of the analog signal from its sampled version by suitable interpolation. Usually, the criterion is that a rate at least twice that of the maximum relevant frequency (f_{max}) present in the Fourier representation of the waveform should be used.[13] Typical extracellular spike values for f_{max} lie in the range between 5 kHz and 10 kHz, depending on the type of neuron monitored, leading to sampling rates of at least 20 kHz to meet the Nyquist criterion.*

3.2.1.1 Spike detection

Spike detection is the process of recognizing and recording the timing of discharge of a particular neuron. Since extracellular microelectrodes may record more than a single cell, the spike detector has to distinguish the cell that actually fired on the basis of the measured waveform. Long before computers were used, the most popular method of spike detection was through analog delay window discriminators (easy to implement with analog circuitry).[14,15] These devices work because the usual spike shapes are characterized by crossing a pre-established threshold with a sufficiently large derivative and by returning to baseline through characteristic level ranges after typical delays. The use of these detectors requires careful oscilloscope monitoring of action potentials to adjust suitable window parameters, a burdensome task for the experimenter, especially when many channels are involved. Setting up an experiment by this procedure consumes a significant amount of time and ultimately contributes to reduce the number of neurons one can simultaneously record.

Analog detectors are being phased out in favor of digital ones for the latter's increased flexibility. Among the benefits is the use of more reliable pattern recognition algorithms and of unsupervised learning techniques that could, at least in principle, ease the burden of setting up an experiment. (For a complete review see Chapter 4.)

Whether analog or digital, the chief limitation to spike detection is the amount of noise/interference present in the signal (signal-to-noise power ratios of 3 to 4 are not uncommon). This is so because the signal-to-noise-ratio (SNR) parameter determines the probability of incorrect detection. In fact, much of the interest and effort placed on digital detection alternatives focus on the development of methods to improve the probability of correct spike detection.[16]

Among digital approaches, many are simple and mimic analog windows, while others involve sophisticated template matching,[17] including discriminant analysis and ANN (artificial neural networks) learning algorithms.[18] The matter of which approach is best is not yet settled but some issues such as computational complexity and the need for "training" are especially important and often render a particular approach impractical in a given application. For instance, in "on-line" experiments,

---

* Usually, sampling rates larger than twice f_{max} are employed for improved accuracy. Also to ensure a well-defined f_{max} associated with a signal, it is customary to impose bandwidth limitation through low pass analog filtering prior to sampling. This is called the anti-aliasing filter, which not only sets a well-defined value for f_{max} but also rejects noise above that frequency.

methods that require extensive pattern learning by the classifier are impracticable. Also, complex algorithms requiring "off-line" processing may tax data storage heavily, thereby severely limiting experiment duration (*T*) and the number of simultaneously recordable microelectrodes (*N*).

A distinct advantage of digital spike detection algorithms is in situations where microelectrodes pick up spikes of more than one cell with more or less the same strength and/or frequency characteristics (i.e., they may not be reliably separated by filtering). In this case more sophisticated alternatives are required, as the signal is a superposition of several simultaneous spikes due to different cells sensed by a single microelectrode. Another situation involves multiple discharge bursts that alter the spike waveform of a given cell (due to residual ionic charge in the medium)[19] and which may lead to incorrect decisions on spike presence when more usual detectors are used.

3.3 Data Acquisition Instrumentation

The overall schematic for an MNER system containing four signal processing layers comprising the a) electrode layer, b) the analog layer, c) the digital preprocessing layer, and d) the computer layer is summarized in Figure 3.1.

MNER systems require an array of sensing microelectrodes (electrode layer; see Chapters 1 and 2 for reviews) used to capture the raw analog signal of action potentials from many neurons. The analog signal amplitude must be amplified and processed to reduce noise and interference. Once suitably conditioned, the signal is ready for conversion to digital form, with the result relayed to a computer for storage, display, and further analysis.

In many ways, this layered structure reflects only a "logical" division of the tasks involved, as some of the layers actually end up embedded into one another in most practical systems. For instance, only more recent data acquisition approaches have incorporated a digital preprocessing layer physically separate from the computer layer itself.[6,20]

3.3.1 Electrode Layer

Many types of microelectrodes have been built and tested for use in chronic preparation.[21,22] (See Chapters 1 and 2 for reviews.) The most straightforward and probably easiest way to manufacture microelectrodes is through a small caliber insulated metal wire arranged in bundles or matrices. Metal microelectrodes provide sufficient stiffness to allow brain tissue penetration, making it possible to reach both superficial and deep neural structures. Several designs of microdrives have been proposed that allow manipulation of individual or groups of microelectrodes.[23-26] These microdrives allow independent adjustment of depth so that the electrodes can be positioned at desired target structures. Extensive reviews may be found elsewhere. (See references 9, 11, and 12, and Chapters 1 and 2 of this book.)

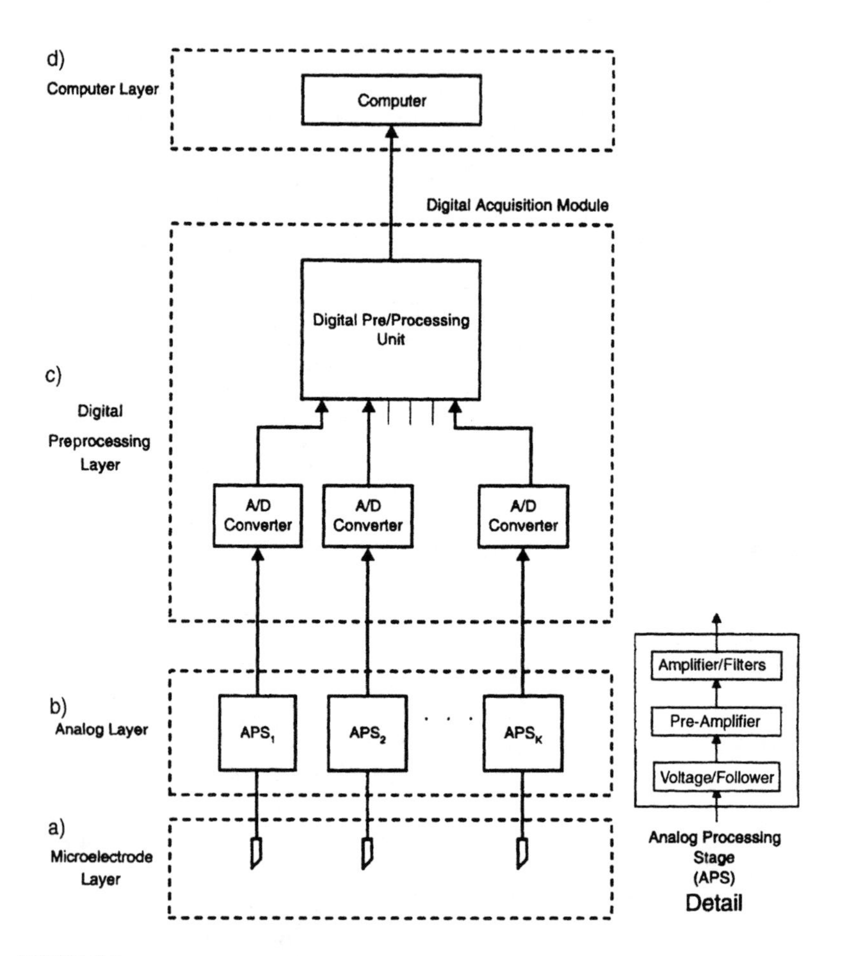

FIGURE 3.1
A modern multisingle unit data acquisition system can be divided into a number of functional levels: (a) the microelectrode layer, (b) the analog signal processing layer (Shown in further detail on the right), (c) the digital signal preprocessing layer whose internal structure varies according to whether DSS or ETS systems are employed and finally (d) the computer layer.

Microelectrode properties depend not only on the electrical conductivity of its sensing tip but also on the type of its shank insulation whose capacitance sets the microelectrode's high frequency cut-off properties.[9,10] From an experimenter's standpoint, the electrical features of microelectrodes can be summarized by an equivalent circuit model (Figure 3.2). For instance, glass microelectrodes are the electric equivalents of low-pass filters, whereas their metal counterparts can be modeled as band-pass filters. Both microelectrodes have frequency dynamic ranges that are adequate for recording action potential whose main signal power lies between 0.2 and 5 kHz.

Among a microelectrode's electrical properties, the most important characteristic is its output impedance because it determines the amount of thermal noise that will be recorded along with the signal. As a rule of thumb, glass microelectrodes produce 20 to 40 μV of noise, a value that can be reduced by as much as 60% by

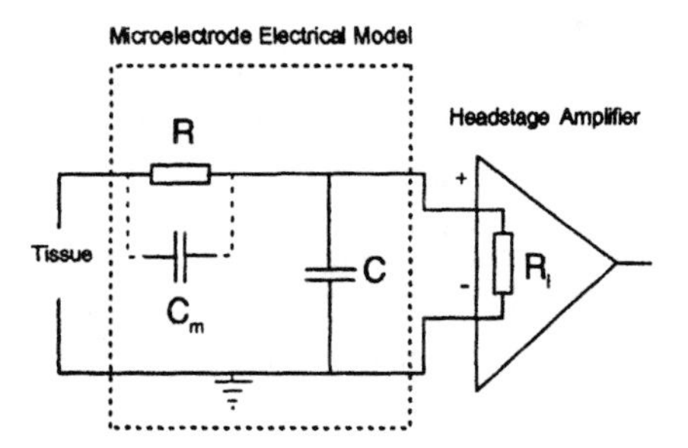

FIGURE 3.2
The equivalent circuit model for glass (metal when capacitor C_m is included) microelectrodes describes a low (band) pass filter with high frequency cut-off at 5 kHz (plus low pass cut-off at 200 Hz).

using metal microelectrodes,[9] which are preferable when probing cells that produce small spikes or whenever it is important to boost the signal-to-noise ratio to improve spike detection. Another important microelectrode parameter is the dimension of the exposed tip. Tip dimensions between 25 and 50 microns are recommended to help prevent recording the overlapping simultaneous activity of a large number of neurons, and also to minimize the probe's shunting effect on the extracellular medium, which results, in smaller recorded spikes whose detection is more difficult.

3.3.2 Analog Layer

This layer performs three essential functions (see details on the right inset of Figure 3.1): (1) impedance matching by the headstage amplifier; (2) signal amplification to a dynamic range compatible with later A/D conversion; and (3) filtering to reject noise, movement artifact, and unintended biological signals that act as interference (such as EKG, residual field potentials, or mioelectric signals). The filtering stage also guarantees observance of Nyquist's Criterion.

The input headstage amplifier is used in its "differential" mode — i.e., all that is recorded is the difference between the microelectrode signal and a ground wire, usually looped around steel screws implanted in the animal's skull. Thus, disturbances common to nearby measuring microelectrodes and the ground wire are effectively nullified. A parameter used to evaluate amplifier performance in this configuration is its common mode rejection ratio, i.e., how much of the voltage common to both amplifier inputs is rejected with respect to the voltage difference between the inputs; a value between 90 and 100 decibels is considered adequate.[27] Usually, the input of headstage amplifiers employs a field-effect transistor (FET), in voltage follower configuration, which contributes very low electronic noise when used with little or no gain.[28] FETs have the additional advantage of possessing an input impedance approximately matched to microelectrode impedance,* thus ensuring maximum signal power transfer with minimal degradation due to noise.

* This value is mainly determined by the microelectrode geometry and lies between 0.5 and 10 MΩ at 1kHz.

To minimize external noise induction the headstage amplifier should be placed as close as possible to the microelectrode.[11,27] Considerable advancement has been made in this area thanks to integrated circuit technology where several amplifiers packed on single silicon wafers can be stockpiled on the animal's head, thereby giving access to many more microelectrodes simultaneously.[27] Later amplifier stages provide the bulk amplification between 100 to 100,000 times, and usually incorporate filtering with adjustable low and high pass cut-off frequencies of 1 to 500 Hz and 3 to 10 kHz.

In some older systems, spike detection also takes place in the Analog Layer via delayed window level discriminators[14,15] whose timing is digitally recorded. The present trend leaves detection to the digital pre-processing layer.*[20,29]

3.3.3 Digital Preprocessing Layer

This layer begins with the analog-to-digital (A/D) conversion of the output of signal of the analog layer. Signal samples are stored at a rate of at least twice that of f_{max}. In practice, rates a few times higher are employed, usually between 20 to 60 kHz. Typical voltage resolutions lie between 2.4 and 1.5 mV levels depending on whether 12- or 16-bit A/D with $\pm$ 5 V input-range converter chips are used.

The organization of this layer is what determines whether a system is DSS or ETS. In DSS systems, besides A/D conversion, only very elementary signal processing operations take place in this layer. An example is linear filtering whose digital implementation is often more convenient when done digitally. Because this layer lacks significant computational power in DSS systems, spike detection is deferred to the computer layer,[2,30,31] where it is performed off-line if many multiple single-units are being monitored simultaneously. By contrast, in ETS systems, a significant portion of the computational power is concentrated in this layer so that on-line spike detection becomes feasible even if many neurons are being observed simultaneously. This alleviates the main computer allowing an increase in the number of simultaneously recorded microelectrodes.

In ETS systems, the computational power of this layer is achieved by the use of independent digital processors, usually DSPs (digital signal processors) as in the MNAP (Many Neuron Acquisition Processor, Plexon, Dallas TX).[20] This architecture has the added advantage of allowing distributed control of stimuli in awake animal contexts.

Some rough figures help illustrate the difference between DSS and ETS systems. Suppose 1 Gbyte of storage is available and that signal sampling is performed at 64 kHz using a 16-bit A/D converter on signals that contain an average of 10 spikes

* For correct spike detection, amplifier class A operation is required, i.e., it must occur in the amplifier's linear regions so as to avoid spike shape distortion that could degrade spike detection.

The 50/60 Hz power line noise is not a problem in spike recording experiments (as opposed to EEG measurements) due to analog band-pass filtering with low frequency cut-off around 300 Hz. Power line harmonics, which is usually introduced by the power supply converters (AC to DC), can be a problem that requires special methods.[11]

Large amplification can result in artifacts due to induction by external electromagnetic fields. To prevent this, careful shielding is often necessary through the placement of metal sheets (e.g., aluminum foil) to provide good grounding and through the use of Faraday caging.[11]

per second recorded via 32-bit-long time stamps. In that case, the signal due to one single microelectrode would fill the mass storage after 2 hours for DSS compared to 60,000 hours using ETS systems!*

The inherent parallel structure of ETS systems helps reduce the total signal processing time when compared to DSS systems, whose processing is usually done off-line and sequentially at the computer layer. The number of microelectrodes each DSP can handle is an important design issue for ETS systems and is related to typical firing rates, DSP speed (clock), spike detection algorithm complexity, and sampling rate. Current technology allows one DSP to control 8 electrodes and up to 4 units per electrode.[20]

3.3.4 Computer Layer

This layer integrates, stores, displays, and summarizes the acquired information.

3.3.4.1 Interface to the digital preprocessing layer

Computer layer architecture (hardware and software) varies according to whether an ETS or a DSS system is desired. Most DSS systems are comprised of a computer interface board (or boards depending on computer speed) where the signal is digitized. The computer processor oversees directly all the tasks, including filtering, spike detection, signal display, and storage. Therefore, the computer processor is an important factor in the recording setup since it may limit the number of neurons than can be recorded simultaneously, even if spike detection is done off-line. By contrast, on ETS systems, the computer processor can concentrate on more sophisticated tasks, such as display and signal correlation procedures.

Unlike DSS, where full control is under the computer processor, distributed processing of many units in ETS systems introduces new challenges for the interface between the digital preprocessing layer and the computer. The solution to the interfacing problem is not trivial. One possible solution involves DMA (direct memory access).[29] In this case, compressed neural data are transferred under control of the preprocessing unit, which gains autonomous access to the computer memory without the main computer's intervention. (This type of interface may also be used in DSS systems to transfer sampled signals from a buffer.) In another approach, memory addresses are shared between the main computer processor and the digital signal preprocessors. This option is adopted in some standards for instrumentation control (VMXbus and MXIbus for PCs from National Instruments) and is employed by the MNAP system.[20]

3.3.4.2 Man–machine interface

This layer controls data flow in the experiment. As such, the software must give the experimenter easy access to changing channel calibrations (filter gains and bandwidths) and spike discrimination parameters. Since calibration must be repeated for

* In some systems, the savings in storage are a little less dramatic because spike waveforms are also stored in addition to their time stamps.

each channel, a friendly software-user interface should be part of the design. This interface should furnish adequate visual feedback to the experimenter and help him or her implement parameter changes by minimizing the number of steps in each procedure through automatization and by facilitating interruption and resumption of data acquisition.

In experiments involving behaving animals, one often has to change experimental protocols quickly, either by producing different stimuli or by monitoring different sets of neurons or channels. This task can be simplified by using previously designed and edited configuration files which the software must help prepare.* Automatic sequential signal file naming is another useful feature.

3.4 Typical Instrumentation-Related Tasks in Setting Up an Experiment

For an overall view of what is involved in multisingle neuron data acquisition, consider a neuroscientist who wishes to monitor the cortical response of an awake behaving rat subject to sensory stimulation. Figure 3.3 shows a schematic diagram of the experimental setup.

Assuming that the brain areas of interest were already implanted, the experimenter must perform a number of tasks ranging from animal preparation, probe headstage connection, choice of the number of viable observation channels, which involves the calibration of analog filters (gain and bandwidth choice), selection of adequate spike discrimination parameters, verification of the adequacy of noise shielding, and finally, quality control of the acquired signal for each and every channel.

As behavioral studies are also intended in this case, in addition to the data acquisition aspects, the experimenter must also be concerned with setting up behavioral equipment which must be properly interfaced to the main computer. Hence, it is essential that time stamps for stimulus signals and spike data also be adequately synchronized and stored (Figure 3.3).

Other useful instruments, such as oscilloscopes** and loudspeakers,*** can help the experimenter in providing additional feedback cues on spike occurrence.

* A typical situation where saved configurations are of interest is when studying chronically implanted animals in a series of experiments spanning several days or weeks.

** A standard two-channel digital oscilloscope helps in checking the overall quality of recorded signal and also in debugging malfunctioning channels in the data acquisition system. Its freely running traces at convenient sweep speeds permit fast visualization of several signal characteristics hint at the signal-to-noise ratio attained. Properly triggered, the oscilloscope's digital storage option can produce a still image of the recorded waveform where several sequential spikes can be superimposed to indicate the presence of multiple single-units on a given channel.

*** An audio monitor provides a continuous and convenient feedback of neuronal spiking activity. Cycling through different microelectrodes chosen by a multipositional mechanical switch, or, more conveniently, via the data acquisition system provides the experimenter a nice idea of single-unit activity, freeing his visual attention for other experimental setup aspects. Audio feedback is invaluable while adjusting the depth of microelectrode surgical implantation.

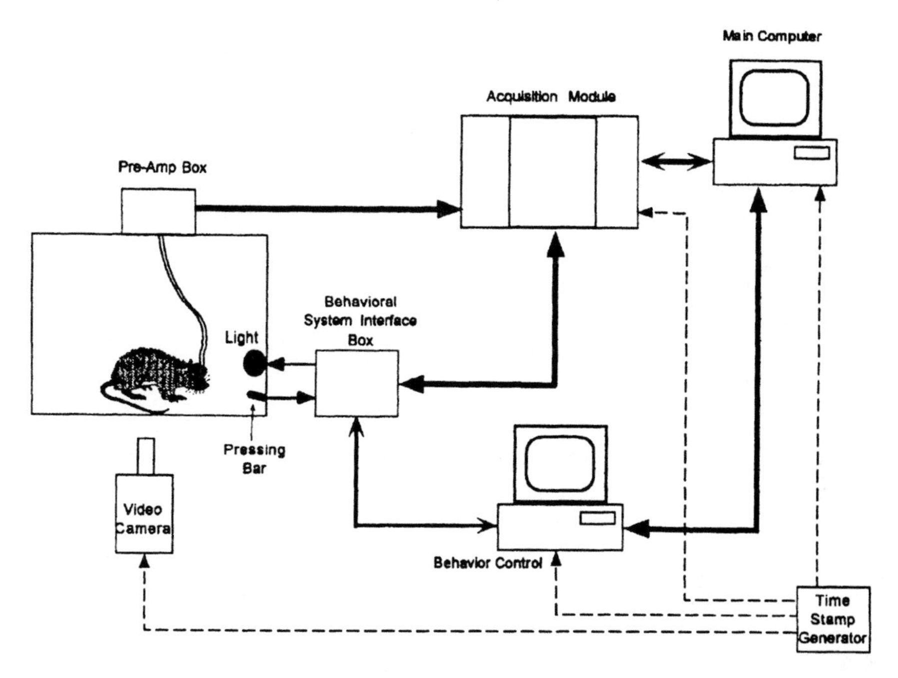

FIGURE 3.3
Schematic instrumentation diagram of a typical behavioral experiment setup. The acquisition module contains the analog and the digital preprocessing layers. In this example the behavioral stimulus control is overseen by a separate computer. Stimulus control may be alternatively handled by the main computer, depending on computing power compromises. Because of its importance, synchronization is displayed separately via dashed lines connecting all instruments. In practice the time stamp generator may be contained in one of the modules — usually in the acquisition module.

3.5 Characteristics of Some Commercial and Academic Data Acquisition Systems

Table 3.1 summarizes the main characteristics of some of the available MNER systems, classifying them with respect to the type of their architecture and number of channels. The relative rarity of ETS systems reflects both the novelty of the concept and the added technological difficulties involved.

3.6 Conclusion and Future Perspectives

Most neurobiological systems of interest have billions of neurons, which naturally raises the question of how many neurons must be simultaneously monitored to understand how the brain functions. Obviously, the larger the number of neurons, the more faithful the experimental representation. At present the maximum attained number of simultaneously monitored neurons is 128 by an ETS system. Traditionally,

TABLE 3.1
Classification and Characteristics of Some Commercial and Academic Multisingle Unit Data Acquisition Systems, Listed in Alphabetical Order. (Academic systems are referenced through the name of their first author.)

System	Maximum number of channels	Spike detection method	Stimulus generation	Classification
DATAWAVE[a]	8	Off-line detection by various methods.	Yes	DSS
Gädicke[29]	16	On-line single electrode DSP discrimination window with template matching	No	ETS
Wilson[7]	48	Tetrode based spike detection	No(?)	DSS
MNAP[b]	128	On-line single electrode DSP discrimination window + principal component template matching	No	ETS
SPIKE 2[c]	8	Off-line template matching	Yes	DSS
TEMPO[d]	9	External spike discriminator	Yes	DSS
Peterson[32]	4	Off-line window and template discrimination	Yes	DSS

[a] DataWave Technologies — 380 Main Street, Suite 209, Longmont, CO 80501. Tel. (303) 776-8214, Fax (303) 776-8531, e-mail sales@dwavetech.com.

[b] Plexon, Inc. — 6500 Greenvile Ave Ste 730, Dallas TX 75206. Tel. (214) 369-4957, Fax (214) 369-1775, e-mail info@plexoninc.com.

[c] Cambridge Electronic Design Ltd. — Science Park, Milton Road, Cambridge CB4 4FE. ENGLAND. Tel. +44(0)1223 420186, Fax: +44 (0)1223 420488, e-mail info@ced.co.uk.

[d] Reflective Computing — 917 Alanson Dr., St. Louis, MO 63132. Tel. (314) 993-6132, Fax (314) 993-3316, e-mail sheldon@crl.com.

the electronic instrumentation posed the most stringent limitation on the number of simultaneously recordable neurons. As ETS system technology evolves, more and more probes will be simultaneously recordable, and one should expect this limitation to shift to surgical rather than instrumentation issues.

In both DSS or ETS, setting up experiments is difficult and time consuming because of the need to adjust spike detector parameters manually. This is ever more crucial as the number of monitored neurons grows. One should also expect considerable future effort in the research to automate adjustment procedures for spike detection (see Chapter 4).

In summary, this chapter has described and compared two possible computer-based instrument layout philosophies for monitoring the activity of multisingle units. Each technique has its domain of interest. The older, simpler DSS approach, consisting of fully digital signal storage, is useful if not essential in some experimental settings of interest such as spike detection methodology research. This approach is now giving way to the new data-compression-based ETS approach that exploits the point process nature of action potentials.

Acknowledgments

K. S. was supported by FAPESP 96/12118-9, PRONEX 41.960925.00, CNPq 301059/94-2 and FFM, and L. A. B. by FAPESP 97/01690-6 and CNPq 301273/96-0.

References

1. Adey, W. R., Computer analysis in neurophysiology, *Computers in Biomedical Research*, 1, 223, 1965.
2. Mishelevich, D. J., On line real-time digital computer separation of extracellular neuroelectric signals, *IEEE Trans. Biomed. Eng.*, 17, 147, 1970.
3. Schoelfeld, R. L., The role of a digital computer as a biological instrument, *Annals of the N. Y. Acad Sci.*, 115, 915, 1964.
4. Deadwyler, S. A., Bunn, T. and Hampson, R. E., Hippocampal ensemble activity during spatial delayed-nonmatch-to-sample performance in rats, *J. Neurosci.*, 16, 354, 1996.
5. Georgopoulos, A. P., Lurito, J. T., Petrides, M., Schwartz, A. B. and Massey, J. T., Mental rotation of the neuronal population vector, *Science*, 243, 234, 1989.
6. Nicolelis, M. A. L., Ghazanfar, A. A., Faggin, B. M., Votaw, S. and Oliveira, L. M. O., Reconstructing the engram: simultaneous, multisite, many single neuron recordings, *Neuron*, 18, 520, 1997.
7. Wilson, M. A. and McNaughton, B. L., Dynamics of the hippocampal ensemble code for space, *Science*, 261, 1055, 1993.
8. Wilson, M. A. and McNaughton, B. L., Reactivation of hippocampal ensemble memories during sleep, *Science*, 265, 676, 1994.
9. Humphrey, D. R. and Schmidt, E. M., Extracellular single-unit recording methods, in *Neurophysiological Techniques: Applications to Neural Systems*, Boulton, A. A., Baker, G. B. and Vanderwolf, C. H., Eds., Humana Press, Clifton, 1990, 1.
10. Leung, L.-W. S., Field potentials in the central nervous system — recording, analysis, and modeling, in *Neurophysiological Techniques: Applications to Neural Systems*, Boulton, A. A., Baker, G. B. and Vanderwolf, C. H., Eds., Humana Press, Clifton, 1990, 277.
11. Millar, J., Extracellular single and multiple unit recording with microelectrodes, in *Monitoring Neuronal Activity: a Practical Approach*, Stamford, J. A., Ed., Oxford University Press, Oxford, 1992, 1.
12. Robinson, D. A., The electrical properties of metal microelectrodes, *Proc. of the IEEE*, 56, 1065, 1968.
13. Brigham, E. O., *Fast Fourier Transform and Its Applications,* Prentice-Hall, Englewood-Cliffs, 1988, 448.
14. Bak, M. J. and Schmidt, E. M., An improved time-amplitude window discriminator (for neuronal spike trains), *IEEE Trans. Biomed. Eng.*, 24, 486, 1977.
15. Dyer, R. G. and Beechey, P., A window discriminator with visual display, suitable for unit recording, *Electroencephalogr. Clin. Neurophysiol.*, 31, 621, 1971.
16. Dinning, G. J. and Sanderson, A. C., Real-time classification of multiunit neural signals using reduced feature sets, *IEEE Trans. Biomed. Eng.*, 28, 804, 1981.

17. Goodall, E. V. and Horch, K. W., Separation of action potentials in multiunit intrafascicular recordings, *IEEE Trans. Biomed. Eng.*, 39, 289, 1992.
18. Mirfakhraei, K. and Horck, K., Classification of action potentials in multi-unit intrafascicular recordings using neural network pattern-recognition techniques, *IEEE Trans. Biomed. Eng.*, 41, 89, 1994.
19. Gray, C. M., Maldonado, P. E., Wilson, M. and McNaughton, B., Tetrodes markedly improve the reliability and yield of multiple single-unit isolation from multi-unit recordings in cat striate cortex, *J. Neurosci. Methods*, 63, 43, 1995.
20. Plexon, *MNAP — Technical Specification Sheet*, Dallas, 1996.
21. Hoogerwerf, A. C. and Wise, K. D., A three-dimensional microelectrode array for chronic neural recording, *IEEE Trans. Biomed. Eng.*, 41, 1136, 1994.
22. Nordhausen, C. T., Maynard, E. M. and Normann, R. A., Single unit recording capabilities of a 100 microelectrode array, *Brain Res.*, 726, 129, 1996.
23. Aldridge, J. W., Gilman, S. and Jones, D., A microdrive positioning adapter for chronic single unit recording, *Physiol. Behav.*, 44, 821, 1988.
24. Goldberg, E., Minerbo, G. and Smock, T., An inexpensive microdrive for chronic single-unit recording, *Brain Res. Bull.*, 32, 321, 1993.
25. Jaeger, D., Gilman, S. and Aldridge, J. W., A multiwire microelectrode for single unit recording in deep brain structures, *J. Neurosci. Methods*, 32, 143, 1990.
26. Korshunov, V. A., Miniature microdrive for extracellular recording of neuronal activity in freely moving animals, *J. Neurosci. Methods*, 57, 77, 1995.
27. Szabo, I. and Marczynski, T. J., A low-noise preamplifier for multisite recording of brain multi-unit activity in freely moving animals, *J. Neurosci. Methods*, 47, 33, 1993.
28. Evans, A., *Designing with Field-Effect Transistors*, McGraw-Hill, New York, 1981.
29. Gädicke, R. and Albus, K., Real-time separation of multineuron recordings with a DSP32C signal processor, *J. Neurosci. Methods*, 57, 187, 1995.
30. Sclabassi, R. J. and Harper, R. M., Laboratory computers in neurophysiology, *Proc. of the IEEE*, 61, 1602, 1973.
31. McCann, G. D., Interactive computer strategies for living nervous system research, *IEEE Trans. Biomed. Eng.*, 20, 1, 1973.
32. Peterson, B. E. and Merzenich, M. M., EXP: a Macintosh program for automating data acquisition and analysis applied to neurophysiology, *J. Neurosci. Methods*, 57, 121, 1995.

Chapter

Automatic Discrimination of Single Units

Bruce C. Wheeler

Contents

0-8493-3351-2/99/$0.00+$.50

4.1 Introduction

Action potentials (AP) are often considered the basic units of neural signaling. They can be recorded extracellularly as potential waveforms, called "spikes." These are remarkably similar in overall shape, permitting them to be distinguished from noise, yet possessing modulations of amplitude and form, the consequences of cell type, size, and electrode position, which permit them to be sorted into multiple classes. Often it is assumed that all the spikes from a class are the result of the activity of a single neuron which, since it usually cannot be physically localized, is known by the anonymous name of a "unit." By recording the behavior of multiple neurons simultaneously, the discrimination of single units or neurons, also known as spike sorting, offers experimenters the promise of a multiplication of their experimental efforts and the gaining of insight into the role of spatio-temporal correlation of neuronal firing.

In this decade automated spike sorters will see a new prominence because of the advent of multichannel electrodes, which make imperative automated processing of spike data, and the abundance of inexpensive computational power. Three examples illustrate the need: the planar microelectrode arrays of Gross at the University of North Texas have sensed 64 channels of spontaneous spike activity from cultures of mouse spinal cord neurons[1]; Nicolelis and Chapin recorded from 48 neurons simultaneously with 16-microwire arrays[2]; the probe arrays developed by Wise at the University of Michigan show real promise for chronic recording in the cortex.[3]

This chapter serves to review techniques for the detection and classification of spikes, whose principles are well established in the fields of signal processing and statistical pattern recognition, but often reinvented in the neuroscience community. A spike sorting example is given with the intent of illustrating the need for attention to the statistical performance on real, not simulated, data. The discussion argues that the more pressing issues in spike sorting include what constitutes a good and usable system design, how we are to work with the relative statistical certainty that results from the sorting process, and how to store, manage, analyze, and visualize data from upwards of 100 channels.

4.2 Classical Signal Detection and Pattern Recognition

A classical approach to spike sorting includes three phases: (1) a learning phase in which the parameters governing the decisions are made; (2) the detection of the spikes; and (3) the classification of the spikes according to a set of features extracted from each spike. In the discussion below classification is discussed first because it serves also to introduce concepts common to detection and learning phase algorithms. Background can be found in many textbooks.[4,5]

In the context of neural spike sorting, these algorithms are most easily understood with reference to three commonly used approaches to spike sorting: the use

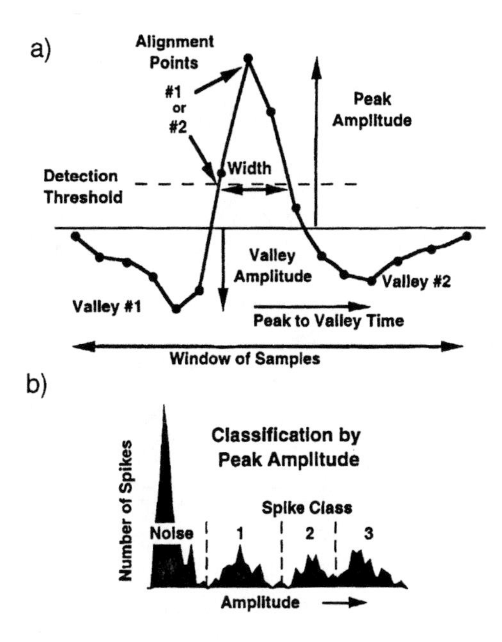

FIGURE 4.1
Waveform parameters and epoch detection. (a) A typical action potential or spike, and (b) a peak amplitude histogram of recorded spikes.

of various parameters extracted from the waveform (Figure 4.1); template matching, in which the errors between a new waveform and known waveforms are computed (Figure 4.2); and the method of principal components (Figure 4.3).

4.2.1 Classification

Classification assumes that a spike has been detected and aligned, after which a set or vector of features $\underline{f}$ (such as peak amplitude and width, Figure 4.1a) are extracted from each new action potential waveform and used to find the best match among each of the classes of action potentials. In Figure 4.1b, peak amplitude determines which range (labeled 1, 2, 3, or noise) best matches the new spike. In one commercial approach the user graphically defines a box around each cluster of features (see Figure 4.4a) putatively corresponding to a single neuron. This approach greatly simplifies real-time implementation to a rapidly executed series of comparisons. In principle the acceptance "box" could be of arbitrary shape (see Figure 4.4d).

"Best match" can be defined as most likely to have occurred. For equally likely classes and white, Gaussian feature distributions of equal variance, it is optimal to find the minimum Euclidean distance

$$d_j^E = \sqrt{\sum_k \left(f_k - m_{jk}\right)^2} = \sqrt{\left(\underline{f} - \underline{m}_j\right)^T \left(\underline{f} - \underline{m}_j\right)} = \left\|\underline{f} - \underline{m}_j\right\|_E \qquad (1)$$

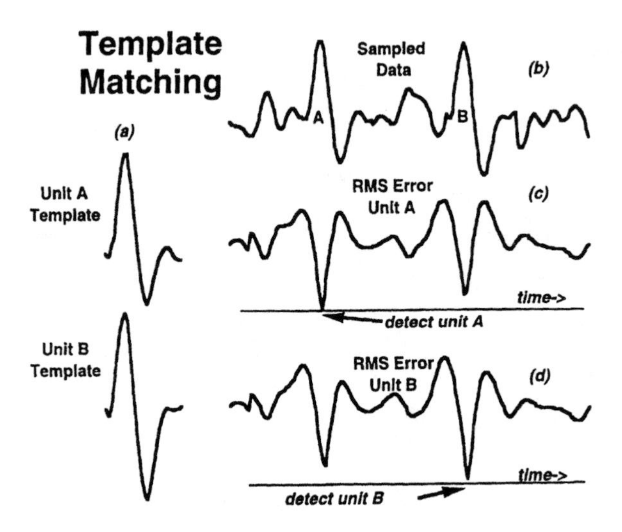

FIGURE 4.2
Continuous time template matching. Two AP waveforms (a) are continuously compared to incoming data (b) with computation of the rms error (Equation 1) (c) which can be used to detect APs and classify according to best match. If the epochs have been determined independently, then it is necessary to compute the template match error only at the event times, which appear here as the minima in the rms errors (d).

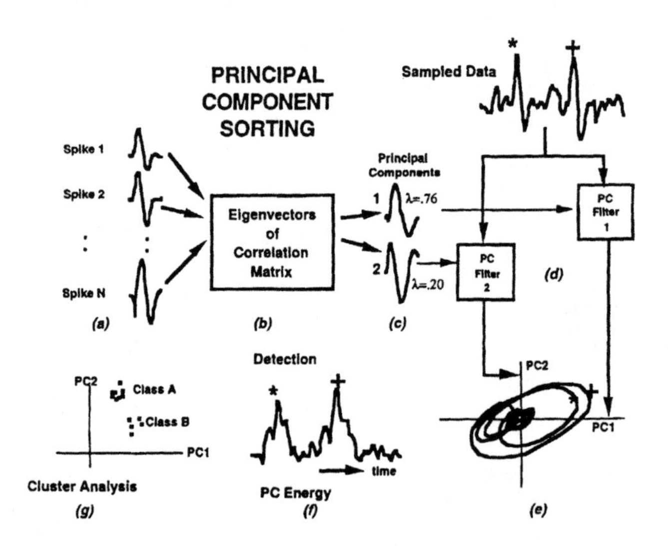

FIGURE 4.3
Continuous time principal component sorting. Refer to Equation 7 in the text for references to variables. (a) In the learning phase a set of spikes are accumulated. (b,c) From these are computed the correlation matrix and its eigenvectors, also known as principal component vectors, p_1 and p_2. (d) These are used in a real-time digital filter. (e) The projections (PC1, or g_1, and PC2, or g_2) show maximal excursion from the origin at the time of the epochs (*, +). (f) Plotting the projected energy $e^{PC}(t)$ shows how the epochs can be detected as a peaks (*, +) above threshold. (g) The outputs of the filters (PC1 and PC2) may be used to classify the AP. Alternatively, if the detection is done independently, the filter values need be calculated only at the epochs.

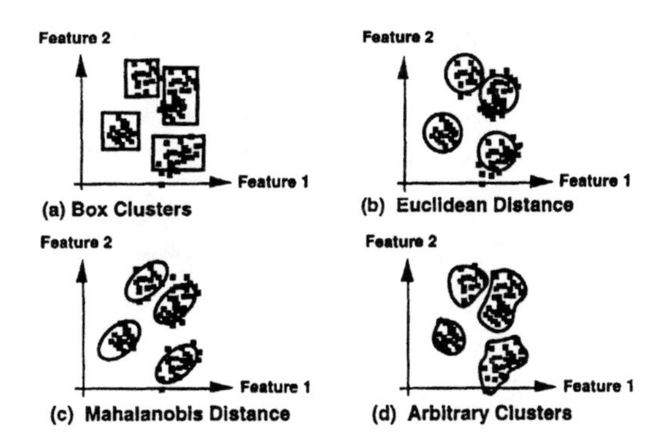

FIGURE 4.4
Examples of cluster definitions for spikes in two-dimensional feature space.

between the new feature vector ($\underline{f} = [f_1 \ldots f_n]^T$) and any of the known mean feature vectors ($\underline{m}_j = [m_{j1} \ldots m_{jn}]^T$), provided d_j^E is less than a user defined acceptance radius. Decision regions using the Euclidean norm are illustrated in Figure 4.4b. This approach makes sense intuitively and is valuable even if the noise in the system is not easily characterized. The case where the classes are not equally likely is discussed below.

If the system noise is Gaussian but not white, one can either use the inverse of the covariance matrix $\underline{C}$ of the features to compute the Mahalanobis distance:

$$d_j^M = \sqrt{\sum_k \left(\underline{f} - \underline{m}_j\right)^T \underline{C}^{-1} \left(\underline{f} - \underline{m}_j\right)} \tag{2}$$

which results in the decision regions shown in Figure 4.4c. One can also whiten the distribution by computing new features $\underline{f}^*$ and means $\underline{m}_j^*$:

$$\underline{f}^* = \underline{E}\underline{\Lambda}^{-1/2}\underline{f};\ \underline{m}_j^* = \underline{E}\underline{\Lambda}^{-1/2}\underline{m}_j^* \tag{3}$$

or

$$f_i^* = \left(\underline{e}_i^T \underline{f}\right)\lambda^{-1/2};\ m_{ij}^* = \left(\underline{e}_i^T \underline{m}_j\right)\lambda^{-1/2} \tag{4}$$

and then compute the Euclidean distances between $\underline{f}^*$ and the $\underline{m}_j^*$. Here $\underline{E}$ is the matrix of eigenvectors of $\underline{C}$ and $\underline{\Lambda}^{-1/2}$ is a diagonal matrix whose entries are $\lambda_i^{-1/2}$ where the λ_i are the eigenvalues of $\underline{C}$. Either procedure provides a measure which directly reflects the likelihood of the features being measured, given that each class j had occurred.

The derived features $\underline{f}^*$ are of interest in a related context if the vector $\underline{f}$, and therefore the vector $\underline{f}^*$, is of dimension N, but one wants to use only a small set of features to be obtained by linear transformation. If the eigenvalues, which are necessarily non-negative, are ordered such that $\lambda_1 \geq \lambda_w \geq \ldots \geq \lambda_N$, then the optimal features are $f_1^*, f_2^*, \ldots. f_N^*$. For action potential data, it is likely that only two features are needed for classification — nearly as good as with all of the original samples.[6]

One other popular distance measure, which does not use multiplication and which may execute faster on some microprocessors, is the city block distance (CBD)

$$d_j^{CBD} = \sum_k \left| f_k - m_{jk} \right| \tag{5}$$

4.2.2 Epoch Detection

To detect spikes, some feature of the waveform must be computed, or derived in analog fashion, at every sample interval. Further, the waveform must be aligned temporally (determination of the epoch) so that in subsequent processing features are extracted in identical manner. Two simple, easily computed, and effective alignment points are the threshold crossing and the peak above threshold (see Figure 4.1a). One empirical investigation found threshold detection to be superior to the more sophisticated detection techniques below.[7]

Traditional signal processing theory recommends the use, in white noise, of a matched filter which correlates the known action potential signal $\underline{w}$ with the current set of samples $\underline{v}(t)$. Detection and epoch determination occur at the local maximum above threshold of

$$y(t) = \underline{w}^T \underline{v}(t) \tag{6}$$

If the noise is not white, then the filter is a whitened form of $\underline{w}$ (see Equation 3).

If one is trying to detect any of the members of the class of action potentials, a multidimensional matched filter (Figure 4.3d) can be defined using principal component vectors (Figure 4.3c) commonly used for data compression. The first M of these eigenvectors of the correlation matrix $\underline{R}$, ordered according to the size of their associated eigenvalues, are the optimal orthonormal basis for approximating the action potential waveform. It is often reported[6,8,9] that one or two of the principal component vectors ($\underline{p}_1$ and $\underline{p}_2$) are sufficient to represent 80% to 90% of the action potential signal energy. In the detection problem, it can also be shown[10] that the signal-to-noise ratio can be optimized if one uses the first M components in which the signal energy exceeds the noise energy. This recommends the strategy of computing an energy $e^{PC}(t)$ and detecting and aligning the waveform at its local maximum above a threshold (Figure 4.3f). For the practical case where two components suffice,

$$e^{PC}(t) = g_1^2 + g_2^2 = \left[\underline{p}_1^T \underline{v}(t)\right]^2 + \left[\underline{p}_2^T \underline{v}(t)\right]^2 \tag{7}$$

There are a number of alternative approaches to signal detection. One is to threshold the derivative of the signal (or the output of a high frequency emphasis filter). Also, one can compute the total energy in the set of samples $\underline{v}(t)$, or

$$e^{SQ}(t) = \underline{v}(t)^T \underline{v}(t) \tag{8}$$

and detect its peak above threshold. The maximum of the ratio of $e^{PC}(t)$ to the "mismatch energy" $e^{SQ}(t) - e^{PC}(t)$[10] is also effective. In conjunction with template matching, one could compute $d_j(t)$ (Equation 1) at each sample time and for each class of waveform. Detection, alignment, and classification occur at the local minima of the $d_j(t)$ (Figure 4.2c,d).

4.2.3 Learning Phase and Adaptive Classification

More difficult, and considerably less often reported in the literature, is the problem of determining the number of classes, i.e., the number of discriminable neurons, and the parameters, such as mean feature values and thresholds, used to make the classification and detection decisions. The experimenter may determine the parameters directly, or establish a set of examples of correctly identified spikes from which the parameters may be computed, or let a computer algorithm make the decisions. There exist *ad hoc* techniques as well as recommendations based on statistical pattern recognition theory.

Several proven *ad hoc* techniques are easily comprehensible because of graphical presentation and feedback to the experimenter. Examples are detection in the manner of an oscilloscope trigger and classification by amplitude window levels, which are conveniently superimposed on traces of the analog or sampled data. It is very useful to display clusters of points corresponding to the detected waveforms in a two-dimensional feature space (Figure 4.4). Among the many possible features could be height vs. width, or principal component values (g_1 and g_2 in Equation 7). Often readily apparent to the eye are clusters of points, each hopefully corresponding to one neuron or class. With graphical feedback, the experimenter could define boundaries for each cluster, such as rectangles, circles, ellipses, or arbitrary shapes. Multiple 2-D displays permit more than two features of the clusters to be viewed.

K-means clustering is a straightforward iterative method of automatically partitioning a data set into clusters.[11] One specifies the number K of clusters which are expected, seeds the algorithm with K cluster means, and then assigns each point to the closest cluster. Each cluster mean is then updated as the mean of the points assigned to the cluster. The process is repeated until there is no change in the assignments. One can then combine or split clusters depending on measures of the quality of either the individual cluster or of the overall partitioning.

The most straightforward measure of the separation of any pair i and j of clusters is the distance d_{ij} of their means $\underline{m}_i$ and $\underline{m}_j$, assuming whitened features:

$$d_{ij} = \left\| \underline{m}_i - \underline{m}_j \right\|_E \tag{9}$$

One can choose features so as to minimize the sum of the d_{ij}.[12] An alternative measure is to sum distances between pairs of points in opposite clusters, or, as was done by Fee et al.,[13] an energy-like exponential function of the distances:

$$d_{ij} = \sum_{k_i} \sum_{k_j} \exp\left\{ -\frac{\left\| \underline{f}_{k_i} - \underline{f}_{k_j} \right\|_E^2}{d_o} \right\} \tag{10}$$

where k_i and k_j index over the points in cluster i and j, respectively. This measure (Equation 10) is very useful in determining whether or not two clusters should be combined.

A more classical approach[14] is to compute measures based on the scatter matrices $\underline{S}_i$, the within class scatter matrix $\underline{S}_w$, and the between class scatter matrix $\underline{S}_B$, all derived from the feature vectors $\underline{f}_{ki}$ and the mean feature vector $\underline{m}_i$ for each class of N_i spikes, and the overall mean $\underline{m}$:

$$\underline{S}_W = \sum_i \underline{S}_i = \sum_i \sum_{k_i} \left(\underline{f}_{k_i} - \underline{m}_i \right) \left(\underline{f}_{k_i} - \underline{m}_i \right)^T \tag{11}$$

$$\underline{S}_B = \sum_i N_i \left(\underline{m}_i - \underline{m} \right) \left(\underline{m}_i - \underline{m} \right)^T \tag{12}$$

Here the index i ranges over the set of classes, while j ranges over the set of points within the ith class. Commonly used measures of class separability are:[14]

$$J_1 = \sum_i \sum_{k_i} \left\| \underline{f}_{k_i} - \underline{m}_i \right\|_E^2 = \mathrm{trace}\left[\underline{S}_W \right] \tag{13}$$

$$J_2 = \sum_i N_i \left\| \underline{m}_i - \underline{m} \right\|_E^2 = \mathrm{trace}\left[\underline{S}_B \right] \tag{14}$$

$$J_3 = {}^{J_2}\!/_{J_1} = {}^{\mathrm{trace}\left[\underline{S}_B \right]}\!/_{\mathrm{trace}\left[\underline{S}_W \right]} \tag{15}$$

$\| \cdot \|_E$ is the Euclidean distance, and the trace of a matrix equals the sum of its eigenvalues. J_1 reflects average distance of points from their class mean and is minimized for compact classes. J_2 reflects the average distance between classes and is maximized for well-separated class means. (If all classes have the same number N of points, J_2 equals N times the sum of the d_{ij} in Equation 9.) Thus, the ratio J_3 is maximized for well-separated, compact classes. When adjusted for the number of classes and spikes, J_3 is the pseudo-F statistic reported in many statistical clustering packages.[11] If the features have been whitened using $\underline{S}_w$ as the estimate of the correlation matrix, then one need compute only $\underline{S}_B$ and J_2.

Measures of the overall cluster separation quality are used as objective functions to be optimized in a variety of clustering algorithms, each involving a tradeoff of speed and accuracy.[14] The K-means algorithm has been used in automated spike

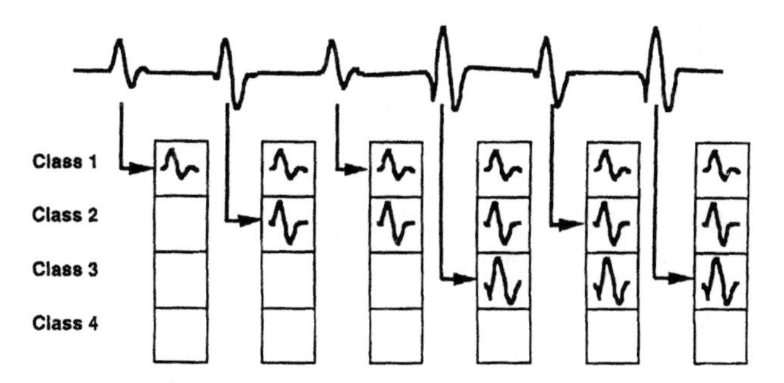

FIGURE 4.5
Sequential learning of classes. Each new spike is compared to all previously defined classes. If a match is possible, the features of the class are updated; if there is no match, a new class is created. A means of pruning the list (not shown) is needed.

clustering.[15-16] In one commercial implementation the experimenter specifies the likely range of the number of classes, and the program returns the sorting quality measures with a graphical display of superimposed waveforms to help the user determine the adequacy of the partition.

This approach has been extended in an attempt to completely automate the clustering process.[17] Clustering quality measures were computed as a function of the number of clusters assumed to exist. For each measure, theory predicts an optimal number of clusters, permitting one to determine the number of distinct neural waveforms, as well as their identity. Practically, the approach produced good results which, when combined with user intervention, are the basis for practical spike sorting systems for large numbers of channels.

Unsupervised sorting can be approached dynamically and adaptively as outlined in Figure 4.5. Each newly detected spike is matched, if possible, to one of the existing classes, and the class mean waveform is updated. If no match is found, a new class is created. The algorithm naturally tracks gradually changing spike shapes and amplitudes. An application to muscle unit action potentials with real-time implementation has been reported.[18]

4.3 Features for Spike Sorting

There is surprisingly good, but not complete agreement on a limited set of features which are useful for spike detection and classification. The entire collection of pattern recognition techniques can be applied to each set of features, including the use of *ad hoc* cluster boundaries vs. the calculation of the various distance measures, or the use K-means clustering vs. adaptive sorting techniques.

4.3.1 Waveform Parameters

Figure 4.1a illustrates a few of the *ad hoc* waveform parameters which offer the advantage of simplicity: they are easily understood and very rapidly computed. Far and away most effective is amplitude, either measured as peak-above-zero or the peak-to-peak distance. When sorting spikes traveling along nerve fibers with two separated electrodes, the delay time (or conduction velocity) is next most effective. Spike width may be measured in a number of ways, and usually makes only a modest contribution to spike separability, but does help to filter out noise spikes similar to neural spikes in amplitude. The ratio of spike amplitudes on two or four closely positioned electrodes (stereotrodes or tetrodes) is relatively robust when amplitudes change due to electrode movement or relative refractory period (see below).

4.3.2 Template Matching

Here (Figure 4.2) template matching is taken to mean that the waveform sample values constitute the features, and that the Euclidean distance (Equation 1) is computed. No linearly derived feature set can outperform template matching, provided that the noise is white and uncorrelated with the spike class. The success of template matching is dependent on the length of the "window" of samples (Figure 4.1). In general it is advantageous to define the window to be as narrow as possible to avoid overlap with visually easily distinguishable interfering spikes.[7] A related technique is that of reduced feature set discrimination, in which samples are retained at a small number of sample times optimally chosen for discrimination.[15,16] This greatly reduces computational overhead and can be quite effective.[19]

4.3.3 Principal Components

The method of principal components is also very effective. Insight into its performance may be obtained by recognition of the fact that the first component equals the normalized average of the AP waveforms. The computation of $g_1 = \underline{p}_1^T \underline{v}(t)$ (Equation 7) is the quantitative answer to the question: "How much does waveform $\underline{v}(t)$ look like prototypical action potential $\underline{p}_1$?" In many recordings it provides an excellent estimate of the amplitude of the action potential. The second component measures differences in waveform shape, and often appears similar to the temporal derivative of the first component. Vectors $\underline{p}_1$ and $\underline{p}_2$, the eigenvectors of the correlation matrix $\underline{R}$, are often very similar to $\underline{e}_1$ and $\underline{e}_2$, the eigenvectors of the covariance matrix $\underline{C}$ (see Equation 3). This reflects the fact that the principal variation among action potentials is the same as the variation which distinguishes them from noise — a change in amplitude of a generic AP waveform. The features g_1 and g_2 are thus nearly equivalent to f_1^* and f_2^*, the optimal pair of features for classifying the detected spikes. Real-time implementation is readily accomplished with digital signal processing chips.[20]

4.3.4 Fourier Coefficients

Although Fourier techniques are very powerful in many biological signal processing applications, they are inefficient for sorting because the AP waveforms, which are of very limited temporal duration, have their energy spread widely across the frequency domain. The differences among waveforms are not concentrated in a few features of Fourier space, making it more difficult or inefficient to execute most classification algorithms. However, if all the Fourier components are used, one neither gains nor loses information because the procedure is a reversible linear transformation, permitting efficient implementation of epoch alignment (temporal shifting) via phase rotations in the Fourier domain.[18]

The wavelet transformation approach is similarly inefficient in helping to analyze an event which is easily localized in time by amplitude thresholding techniques. Wavelet filters could be used to separate a spike channel from an EEG or EMG channel, however, and might be useful in cases where spike widths vary by factors of two, similar to the typical wavelet dilation constant. An approach using Haar transforms has been reported.[21]

4.3.5 Tetrodes, Stereotrodes, and Multiple Peripheral Electrodes

There has been recent, justifiable excitement over the use of tetrodes and stereotrodes in spike sorting.[22-24] The tetrode is a set of four closely spaced electrodes which are lowered into the brain such that they record simultaneously from a set of neurons. Whereas it may occur that two neurons have spikes of the same amplitude on a single electrode, it is highly unlikely that they will also have the same amplitude on the remaining electrodes, since the amplitudes are functions of the electrode-to-neuron distances. Additionally, the ratios of the amplitudes of a single neuron's AP, as recorded on the different electrodes, are much more likely to be constant from spike to spike than are the amplitudes themselves. Both factors lead to improved spike sorting. The principal disadvantage to the technique is increased complexity of electrode assembly and a greater amount of supporting instrumentation. However, it has been reported that the tetrode yields on the order of four times as many discriminable units, thereby compensating for the additional instrumentation, with added reliability.[24]

The theoretical approach to two or four channel spike sorting is little different from single channel recording. All the techniques described above can be applied to a single data vector $\underline{w}$ which is the concatenation of all the individual waveform data vectors captured when a spike occurs.[13,25] The covariance of the stereotrode data can be exploited for superior discrimination,[13] or one can extract the peak amplitudes to form a reduced feature vector $\underline{f}$ which is powerful for discrimination.[24] The technique strongly suggests using ratios of amplitudes as additional features.

4.4 Other Spike Sorting Techniques and Issues

4.4.1 Optimal and Matched Filters

There have been approaches which explicitly use multichannel data to effect optimal detectors of the individual spikes. The approach has been used with peripheral nerve recordings with two to ten electrodes. The basic approach is to derive an optimal linear filter $\underline{v}$ (see Equation 6) with the property that the output y is maximum when the desired unit is present in the recording and the expected value of the output is minimized when either noise or other units are present. The optimization can be formulated as a multichannel signal detection problem and is usually solved in the Fourier domain.[26,27] It has been found to work well given one electrode per neuron,[27] and to fail with fewer electrodes.[25]

4.4.2 Neural Net Classifiers

Neural nets can be substituted for the classical classifiers mentioned early in this chapter, operating on any set of features, such as peak amplitudes or data samples. Neural networks enjoy tremendous popularity as pattern recognizers, partially because their ease of use makes knowledge of statistical pattern recognition nonessential. The primary advantage is that, if the underlying statistics of the features are known, either explicitly or implicitly by training examples, then a neural net can be trained easily to perform discrimination tests which are inherently difficult to program with classical techniques, such as those suggested by Figure 4.4d. The primary disadvantage is that neural nets do not have natural, recognized statistical figures of merit with which to estimate levels of confidence in the discrimination process.

Unfortunately, it is rare that an experimenter has a means of independently verifying the identity of training data. For example, one popular means of constructing a training set is to collect one sample waveform from each class and then create large numbers of training samples by adding random noise. In this case the neural net will learn to distinguish among classes with Gaussian distributed noise, a task more easily done with a least-distance classifier such as is illustrated in Figure 4.4a,b. Another approach is to collect many spikes and to cluster them into hyperspheres or ellipses or some other compact volume in the feature space, in which case the neural network will do no better than to replicate the classical statistical judgment which was used to create the clusters. Often it will do worse because insufficient training examples exist; also, unless trained to recognize noise spikes, it will recognize them as neural spikes.[7]

An exception occurs when one can identify the spikes independently, such as might occur when recording from peripheral nerves and exciting individual receptors at the beginning of an experiment.[28] In this case, one can generate independent training data, which may have a peculiar distribution which a neural network will learn better than any other technique.

4.4.3 Overlapping Spikes

The resolution of overlapping spikes has been a target of many researchers for decades. It is most often based on a least squares fit of the sum of pairs of templates at varying temporal offset. This is straightforward in theory and simulation, but difficult and very often inaccurate in practice. For an early reference, see Keehn.[29]

A narrow window of samples may succeed in eliminating large numbers of overlapping spikes, although the remainder are likely to be very difficult to discriminate. Conversely, if the window of samples is wide, then the incidence of overlapping spikes is larger. Overlap resolution algorithms tend to be most successful in disambiguating the spikes which are easily identified visually as only partially overlapping.

4.4.4 Refractory Period

There is nothing inherent in the techniques described above that prevents a spike discriminator from reporting that a single unit fired during its own refractory period. This can be prevented by a variety of techniques. Clusters may be split if there are significant numbers of spikes which violate refractory period assumptions, or clusters which would otherwise be combined are kept separate if their temporal firing patterns make clear that two units are involved.[13] There are more reports in the motor unit action potential literature in which the classification of a unit is conditioned on recent past firing activity.[30]

4.5 A Spike Sorting Example

Some of the promise and difficulties in spike sorting are illustrated in some results obtained in recording APs from the giant interneurons (GIs) of the cockroach ventral nerve cord.[31] This preparation has a signal-to-noise (SNR) much higher than representative, but permits simultaneous intracellular recordings for absolute identification of the neurons, which have high burst rates typical of sensory neurons. Amplitude variation is due to inherent variability of spike amplitude (presumably partly due to the relative refractory period), overlap with much smaller spikes from undetectable neurons, partial overlap with other GIs, occasional electromyogram (EMG) signals, and a very small amount of instrumentation noise. Figure 4.6 schematizes the overlapping probability distributions of the amplitudes of the GIs which lead to sorting difficulties. In each experiment, recordings were made from one of the two connectives, each of which has three large ventral GIs, three not quite as large dorsal GIs, and a large number of smaller neurons including one identifiable but smaller neuron GI 4. In a series of experiments none of the sorting and clustering techniques described above was able to achieve greater than 75% correct single unit discrimination (with less than 25% false classification) on more than two of twelve data sets, even when obviously overlapping spikes were removed. However, in nine cases this accuracy was exceeded when the spikes were sorted using the amplitude histogram as large, medium, and small, corresponding, anatomically, to ventral GIs, dorsal GIs, and other neurons.

This example is chosen to illustrate the idea that the most sophisticated sorting criteria are sometimes but not always best. Here manually controlled amplitude discrimination outperformed all of the other techniques described in this chapter. The useful classes are composed of action potentials from three neurons, not the

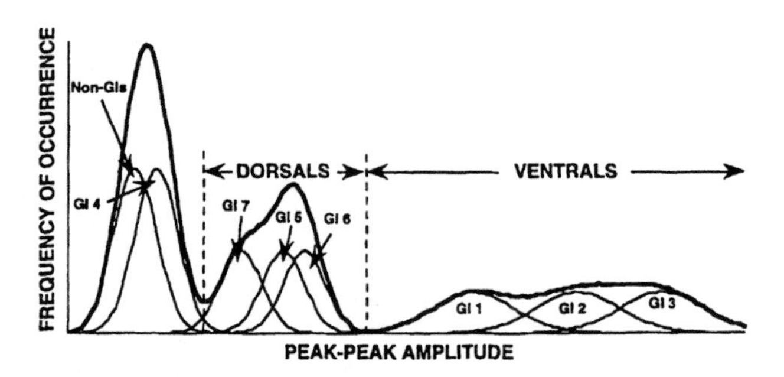

FIGURE 4.6
Idealized composite amplitude histogram of cockroach giant interneuron spikes. The ventral nerve cord of the cockroach has three large ventral giant interneurons (GIs 1, 2, and 3), three large dorsals (GIs 5, 6, and 7), and one smaller identified ventral (GI 4). Although it is very difficult to distinguish individual neurons, it is easy to distinguish anatomically relevant classes due to the AP amplitude distribution.

single neuron assumed in derivation of many of the classification criteria above. Fee et al.[13] similarly recognized that spike sorting techniques often result in multiunit classes which are nonetheless useful in neuroscience investigations. Thus it is of great practical importance to remain flexible in the use of the results of spike sorting.

A second lesson from this example is that spike sorting is likely to result in only a small number of classes with non-negligible error rates. In accordance with other reports,[6,8,19,32] the typical number of discriminable classes per electrode was three, varying from two to six, and the typical success rate was 85%, varying from 60% to 100%. This performance level is dramatically lower than that claimed by most simulation and theoretical analyses.[6,12]

The third lesson is that, despite the fact that it was not possible to accurately sort the spikes into classes corresponding to individual neurons, the resulting data were useful because the classes correspond to anatomically and physiologically significant groupings of the neurons. In other preparations, functional distinctions, perhaps as shown by correlation with varying stimuli, may be attributable to different classes, even though the spike feature distributions indicate multiple neurons within each class. In the case of prosthetic control using neural activity one does not in principle need to discriminate individual units; it suffices to discriminate activity patterns with sufficient degrees of freedom to control a desired function.

4.6 Discussion

If spike sorting is to be done on a large scale, it will have to be done in real time because there is no affordable analog or digital instrument capable of storing signals from a large number of channels for a long time. Data acquisition system architecture will have to change from the popular approach wherein a personal computer or workstation executes spike sorting algorithms for all channels, to one in which a large number of processors are dedicated to one or a few channels. Spike sorting is a computationally intensive operation well-suited to digital signal processing (DSP) chips. Earlier estimates showed that any of the spike classification algorithms above can be executed on eight channels simultaneously and in real-time[33]; when assigned one per channel, they are capable of real-time adaptation[18] and perhaps clustering. With rapidly improving technology it is clear that the technology bottleneck is in the creation of systems and software to service the tiny multichannel neuroscience market.

It should be emphasized that visualization of the spike sorting and of the statistics of the spike trains is critical to both the performance of the system and the confidence of the user. Separate processors (or displays for off-line use) might be dedicated to display superimposed spikes of each class, clusters in feature space, sorting quality measures, and physiological information, such as firing rates, peri-stimulus time histograms, cross-correlograms, and spectral composition, during the course of the experiment.

Acceptance of the results of spike sorting depends on standard procedures and common description of the results. Such a description should include sampling speed, spike window size, detection and alignment criteria, features used, clustering technique (manual, K-means, etc.), and quality measures J_1, J_2, and J_3; for each spike class, the SNR, range of firing rates, class scatter $\underline{S}_i$, and pairwise separations d_{ij}. Although probabilistic significance levels cannot be ascribed to these processes, they provide a common language which gives independent laboratories a basis on which to judge the quality of the sorting achieved.

Above it is argued that the approaches appropriate to spike sorting are well understood. The most fruitful efforts will be in the development of more cost-effective computer instrumentation, especially when accompanied by excellent process visualization. Further, it is necessary to accept the statistical nature of the sorting process, including the idea that errors are necessary but not fatal to the scientific hypothesis to be tested. Further development is needed in the visualization and understanding of data from many channels, perhaps using the approach of Gerstein (see Chapter 8), where the connectivity patterns of perhaps dozens of neurons are inferred from spike trains.[34] Finally, and most important, it is necessary to show that significant physiological inferences can be made using sorted multiple neuron data.

References

1. Gopal, K. V. and Gross, G. W., Auditory cortical neurons *in vitro*: cell culture and multichannel extracellular recording, *Acta Otolaryngol.*, 116, 690, 1996.
2. Nicolelis, M. A. L., Baccala, L.A., Lin, R. C. S., Chapin, J. K., Sensorimotor encoding by synchronous neural ensemble activity at multiple levels of the somatosensory system, *Science*, 268, 1353, 1995.
3. Hoogerwerf, A. C. and Wise, K. D., A three-dimensional microelectrode array for chronic neural recording, *IEEE Trans. Biomed. Eng.*, BME-41, 1136, 1994.
4. Duda, R. O. and Hart, P. E., *Pattern Classification and Scene Analysis*, Wiley, New York, 1973.
5. Therrien, C. W., *Decision, Estimation and Classification*, Wiley, New York, 1989.

6. Wheeler, B. C. and Heetderks, W. J., A comparison of techniques for classification of multiple neural signals, *IEEE Trans. Biomed. Eng.*, BME-29, 752, 1982.
7. Fang, L., Multiple unit action potential sorting: investigation of features, algorithms, and parameters for a fully automated system, M.S. Thesis, ECE Dept., Univ. Illinois, Urbana-Champaign, 1994.
8. Abeles, M. and Goldstein, M. H., Jr., Multispike train analysis, *Proc. IEEE*, 65, 1977.
9. Schmidt, E. M., Computer separation of multi-unit neuroelectric data: A review, *J. Neurosci. Methods*, 12, 95, 1984.
10. Abeles, M., A journey into the brain, in *Signal Analysis and Pattern Recognition in Biomedical Engineering*, G. F. Inbar, Editor, Wiley, New York, 1975, 41.
11. Hartigan, J. A., *Clustering Algorithms*, Wiley, New York, 1975.
12. Heetderks, W. J., Criteria for evaluating multiunit spike separation techniques, *Biol. Cybernetics*, 29, 1978.
13. Fee, M., Mitra, P. and Kleinfeld, D., Automatic sorting of multiple unit neuronal signals in the presence of anisotropic and non-Gaussian variability, *J. Neurosci. Meth.*, 69, 175, 1996.
14. Spaeth, H., *Cluster Analysis Algorithms*, Ellis Horwood, New York, 1980.
15. Dinning, G. J. and Sanderson, A.C., Real-time classification of multiunit neural signals using reduced feature sets, *IEEE Trans. Biomed Eng.*, BME-28, 804, 1981.
16. Salganicoff, M., Sarna, M., Sax, L. and Gerstein, G. L., Unsupervised waveform classification for multineuron recordings: a real-time, software based system. I. Algorithms and implementation, *J. Neurosci. Meth.*, 25, 181, 1988.
17. Zardoshti-Kermani, M. and Wheeler, B. C., Finding the number of neurons in extracellular recordings: A comparison of the effectiveness of popular clustering indices, in *Proc. Comp. Neurosci. Symp.*, Indianapolis, IN, 1992, 229.
18. McGill, K. C., Cummins, K. C., and Dorfman, L. J., Automatic decomposition of the clinical electromyogram, *IEEE Trans. Biomed. Eng.*, BME-32, 470, 1985.
19. Sarna, M. F., Gochin, P., Kaltenbach, J., Salganicoff, M. and Gerstein, G. L., Unsupervised waveform classification for multineuron recordings: a real-time, software-based system. II. Performance comparison to other sorters., *J. Neurosci. Meth.*, 25, 189, 1988.
20. Smith, S. R. and Wheeler, B. C., A real-time processor system for acquisition of multichannel data, *IEEE Trans. Biomed. Eng.*, BME-35, 875, 1988.
21. Yang, X. and Shamma, S. A., A totally automated system for the detection and classification of neural spikes, *IEEE Trans. Biomed. Eng.*, 35, 806, 1988.
22. O'Keefe, J. O. and Reece, M. L., Phase relationship between hippocampal place units and the EEG theta rhythm, *Hippocampus*, 3, 317, 1993.
23. Wilson, M. A. and McNaughton, B. L., Dynamics of the hippocampal ensemble code for space, *Science*, 261, 1055, 1993.
24. Gray, C. M., Maldonado, P., Wilson, M., McNaughton, B., Tetrodes markedly improve the reliability and yield of multiple single unit isolation from multiunit recordings in cat striate cortex, *J. Neurosci. Meth.*, 63, 43, 1995.
25. Wheeler, B. C., The evaluation of neural multi-unit separation techniques, Ph.D. Thesis, EE Dept., Cornell University, Ithaca, New York, 1981.
26. Oguztorelli, M. N. and Stein, R. B., Optimal filtering of nerve signals, *Biol. Cyber.*, 27, 41, 1977.

27. Roberts, W. M. and Hartline, D. K., Separation of multi-unit nerve impulse trains by a multi-channel linear filter algorithm, *Brain Res.*, 94, 141, 1975.
28. McNaughton, T. G. and Horch, K. W., Action potential classification with dual channel intrafascicular electrodes, *IEEE Trans. Biomed. Eng.*, BME-41, 609, 1994.
29. Keehn, D. G., An iterative spike separation technique, *IEEE Trans. Biomed. Eng.*, BME-28, 19, 1966.
30. Stashuk, D. and de Bruin, H., Automatic decomposition of selective needle-detected myoelectric signals, *IEEE Trans Biomed. Eng.*, BME-35, 1, 1988.
31. Smith, S. R., Extraction, processing, and analysis of multineuron data, Ph.D. Thesis, ECE Dept., University of Illinois at Urbana-Champaign, 1991.
32. Vibert, J.-F., Albert, J.-N., and Costa, J., Intelligent software for spike separation in multiunit recordings, *Med. & Biol. Eng. & Comput.*, 25, 366, 1987.
33. Willming, D. A. and Wheeler, B. C., Real-time multichannel neural spike recognition with DSPs, *IEEE EMBS Magazine*, 9, 37, 1990.
34. Gerstein, G. L. and Aertsen, A.M.H.J., Representation of cooperative firing activity among simultaneously recorded neurons, *J. Neurophysiol.*, 54, 1513, 1985.

Chapter 5

Multielectrode Recording from the Cerebellum

John P. Welsh and Cornelius Schwarz

Contents

5.1 Introduction

Among the many central nervous system nuclei and cortices, the cerebellar cortex may be the most amenable, but at the same time one of the most challenging, regions for meaningful multielectrode recording. This seemingly paradoxical opening statement is due to the fact that, although the unique anatomical organization of the cerebellar cortex completely lends itself to array recording, the location and fragility of its cellular components make it highly liable to disruption by placing microelectrodes into its circuitry. Thus, the purpose of this chapter is to highlight the features

0-8493-3351-2/99/$0.00+$.50

of the cerebellar cortex that lend themselves to multielectrode recording, to describe experimental methods that we have used for doing so, and to describe analytical methods and results obtained from multielectrode recording of cerebellar cortex during movement.

Perhaps the most fundamental question that must be considered prior to designing any multielectrode experiment is: What is to be achieved by simultaneous recording with multiple microelectrodes that cannot be achieved with successive recordings with a single electrode? For the cerebellum, the tissue is clear. It has become increasingly appreciated that motor function within the cerebellum is not necessarily encoded in the firing frequency of single neurons, but is likely to be encoded in the spatial organization of temporally coupled firing among populations of neurons. This appreciation was fed by the finding of compartmentation, rhythmical firing, and electrotonic coupling of cells giving rise to cerebellar inputs. Thus, in a metaphorical sense, the cerebellum can be viewed as a highly distributed and overcomplete system whose function is an emergent property of the activity within populations of neurons and not the single elements themselves.[21] To test the validity of this hypothesis, it is essential to use a method that can reveal spatial activity of multiple neurons with millisecond precision, and this is the basis for employing multiple microelectrode recording in the cerebellum.

5.2 Organization of the Cerebellar Cortex

The cerebellar cortex is notable for its simplicity, its extreme regularity, and its conservation throughout phylogeny. There is only a single type of output neuron in the cerebellar cortex — the Purkinje cell — and it is this neuron toward which most attention in multielectrode neurophysiology has been directed. The Purkinje cell is an intricate structure that is notable for its elaborate dendritic tree and its very large soma, which is approximately 30 μm in diameter (Figure 5.1A). In gross morphology, the Purkinje cell dendritic arbor is akin to a flat plate extending in one direction from the soma, being approximately 350 μm tall and 350 μm wide, but only 9 to 35-μm-thick.[1,14] It is quite a remarkable feature of the cerebellar cortex that the Purkinje cell somata are organized in a sheet that is precisely one somatic diameter thick (Figure 5.1C) and that their dendritic trees extend right to the pial surface. The dendritic trees of the Purkinje cells are aligned precisely perpendicular to the long axis of the cerebellar folium within which they are located (Figure 5.1B). The Purkinje cell somata are regularly spaced such that there are about 1000 Purkinje cells per square mm in the rat[11] and approximately 500 per square mm in cats, monkeys and humans.[1,8,9] The axon of the Purkinje cell emerges from the basal pole of the soma and projects to the deep cerebellar nuclei, which form the final output station of the cerebellum and project to motor and premotor nuclei in the brain stem and thalamus. Thus, the monolayer organization of the Purkinje cells, their surface location and regular spacing, the precision of the orientation of their dendritic trees, and the fact that they are the sole output neuron of the cerebellar cortex make them a compelling cell type for multielectrode neurophysiology.

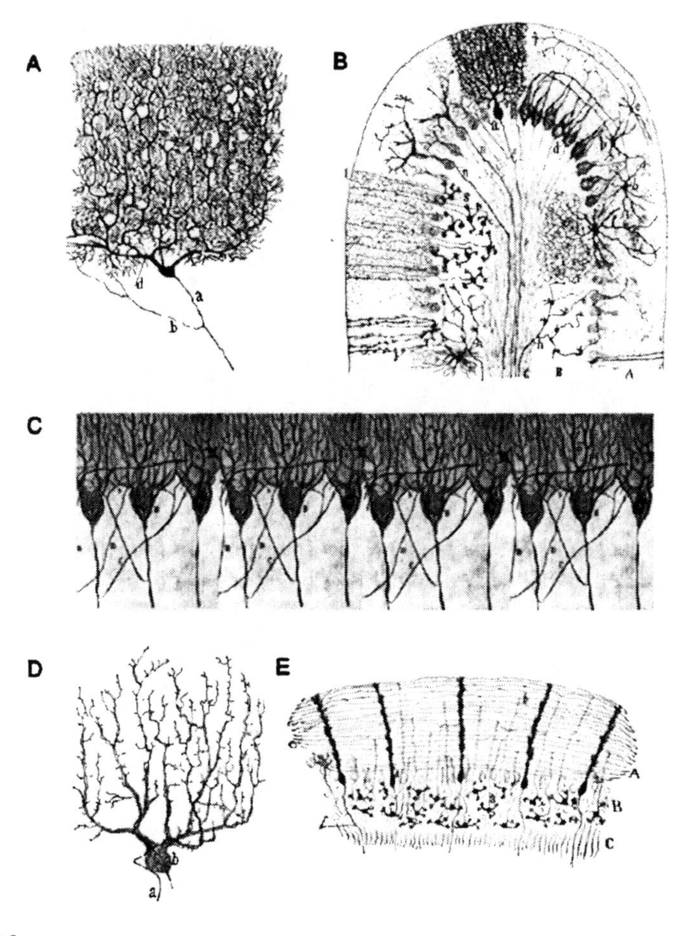

FIGURE 5.1
The fundamental elements of the cerebellar cortex as drawn by Ramon y Cajal (1933). A. The Purkinje cell. B. The cerebellar cortex: *a* shows the Purkinje cell whose dendritic arbor extends to the pial surface; *b* shows the basket cell which sits above the Purkinje somata and whose axons course longitudinally to innervate Purkinje cell somata; *e* shows a stellate cell close to the pial surface whose axon innervates Purkinje cell dendritic arbors; *f* shows a Golgi cell, whose dendrites extend toward the pial surface and whose highly ramified axonal plexus innervates the underlying granule cells; *g* shows granule cells and their ascending axonal branches that extend toward the pial surface; *h* shows a mossy fiber which branches in the granular layer and forms rossettes; *n* shows climbing fibers coursing over Purkinje somata to cover the Purkinje cell dendritic arbor; *o* shows a Purkinje cell axon which leaves the cerebellar cortex and bifurcates to innervate neighboring Purkinje cells. C. A composite of drawings showing the monolayer arrangement of Purkinje cell somata and their regular spacing. Purkinje cell dendrites extend to the pial surface and their axon leaves the basal pole of the soma. The drawing also indicates basket cells and their innervation of the Purkinje cell somata in unique basket configurations. D. Innervation of a Purkinje cell by a climbing fiber. Note the nearly complete coverage of the Purkinje cell dendritic arbor by the climbing fiber. E. Innervation of Purkinje cells by parallel fibers. Granule cells underlying the Purkinje cell somata extend axons (a) toward the pial surface which bifurcate (b) in the molecular layer to form the parallel fibers. The parallel fibers course parallel to the long axis of the folium and through the dendritic arbor of the Purkinje cells.

The Purkinje cell receives two types of excitatory afferents that are morphologically quite distinct from each other. The first afferent is the climbing fiber, so named because it "climbs" over the Purkinje cell soma and up the dendritic tree where it forms approximately 200 synaptic contacts with the smooth dendritic branches of the Purkinje cell (Figure 5.1D). The climbing fiber represents the axonal terminal arborization of a neuron in the contralateral inferior olive. In the adult, each Purkinje cell receives one, and only one, climbing fiber. The second excitatory afferent of the Purkinje cell are axons of the underlying granule cells of the cerebellar cortex, which have an ascending branch that forms en passant synapses with the lower dendritic spines of the Purkinje cell before it bifurcates into a T within the molecular layer at the level of the Purkinje cell dendrites to provide the so-called parallel fibers (Figure 5.1E). Parallel fibers course parallel to the long axis of the cerebellar folia and orthogonal to the plane of the dendritic arborization of the Purkinje cell. Each parallel fiber extends for up to 3 mm in each direction after the bifurcation and may cross the dendritic trees of up to several hundreds of Purkinje cells. Approximately 400,000 parallel fibers traverse the dendritic tree of a single Purkinje cell, and as many as half may form synapses on the dendritic spines of the Purkinje cell. Nearly every dendritic spine on the Purkinje cell contains a parallel fiber synapse.

As mentioned above, the granule cells lie densely packed in the so-called granular layer of the cerebellar cortex, immediately below the Purkinje cell somata. Granule cells are the receiving neurons of the mossy fibers, so named because of their moss-like terminals, whose somata of origin reside in a wide variety of brain stem and spinal nuclei. The mossy fibers form unique structures called rosettes which are either en passant or terminal enlargements that extend finger-like appendages in the granular layer (Figure 5.1B). The rosettes are surrounded with granule cell dendrites as well as Golgi cell axon terminals to form glomeruli. Granule cells are the most numerous in the central nervous system and there are about 2 to 3 million granule cells per cubic mm of granular layer in mammals.[11]

There are three classes of inhibitory interneurons of the cerebellar cortex. The first is the Golgi cell, whose somata lie in the granular layer (Figure 5.1Bf). The Golgi cell has both descending and ascending dendrites. The former reside in the granular layer and receive mossy fibers and recurrent axons of Purkinje cells, and the latter extend toward the pial surface in order to receive parallel fibers. The axons of Golgi cells give rise to a highly ramified plexus which forms synapses with the dendrites of granule cells in the glomeruli. A second inhibitory neuron, the basket cell, lies immediately above the sheet of Purkinje cell somata and is so named because its axon forms a distinctive pericellular "basket" around the Purkinje soma (Figure 5.1Bb). Basket cell dendrites form a hemiellipsoid oriented parallel to the dendritic arbors of the Purkinje cells that extends close to the pial surface and receives the parallel fibers. Their axons extend perpendicular to the long axis of the cerebellar folium, a distance that allows them to innervate eight or nine Purkinje cell somata in the rostro-caudal plane. Additionally, the axons of basket cells form a unique pinceau structure around the initial segment of the Purkinje cell providing powerful inhibition of Purkinje output. Ascending collaterals of the basket cell axon contact dendrites of Purkinje, other basket, and stellate cells. Stellate cells lie in the outer aspect of the molecular layer and form synapses on the dendritic shafts of Purkinje cells (Figure 1Be). They receive parallel fibers and may receive a collateral of a climbing fiber.

5.3 Electrophysiological Signatures of Purkinje Cell Afferents

The two major afferents of the cerebellar cortex — the climbing fibers and the mossy fibers — have distinct electrophysiological consequences on the Purkinje cells which can be recognized in extracellular records taken from the Purkinje cell dendritic tree. In 1966, Eccles, Llinás, and Sasaki[3-7] provided a comprehensive and definitive analysis of the electrophysiological organization of the cerebellar cortex which forms the foundation of modern cerebellar neurophysiology. Each climbing fiber has a direct influence via its monosynaptic connection with a Purkinje cell, while the mossy fibers influence Purkinje cell activity disynaptically via the granule cells. Activation of a climbing fiber evokes a large excitatory postsynaptic potential (EPSP) in the Purkinje cell, which is followed by a sustained depolarization upon which two to six partial spike responses are superimposed. In a series of studies both *in vivo* and *in vitro*, it was established that these spikes are dendritic in origin, are mediated by voltage-dependent calcium channels, and are electrotonically conducted to the initial segment, where the multiple spike bursts are conducted down the axon to the terminals in the deep cerebellar nuclei. Recorded extracellulary, the climbing fiber response can be recognized as a fast initial spike followed by a slow wave lasting 10 to 20 ms upon which small spikes are superimposed. Band-pass filtering the extracellular potentials from 0.3 to 10 kHz removes the slow wave and leaves only the initial fast spike followed by a multiphasic, high frequency spike burst at about 500 Hz. It is this spike complex, the so-called complex spike, that indicates a climbing fiber activation of a Purkinje cell.

In contrast to the climbing fiber which elicits a massively large EPSP and a sustained spike burst in Purkinje cell dendrites, parallel fibers elicit much smaller EPSPs which summate and induce the Purkinje cell to fire spikes with a simpler configuration. Recorded extracellularly, parallel fiber volleys trigger a distinctly triphasic potential known as the simple spike. Unlike the complex spike, which indicates the firing of a single climbing fiber, the occurrence of a simple spike indicates the convergence of an unknown number of parallel fiber activations of the Purkinje cell and/or the excitation of the Purkinje cell by an ascending branch of a granule cell axon sufficient to bring the initial segment over its threshold to fire an action potential.

Figure 5.2 shows extracellular records of complex and simple spikes obtained from a single Purkinje cell at different depths below the pial surface. As can be seen from this figure, insertion of the microelectrode 40 μm below the pial surface already allowed complex and simple spikes to be detected. The complex spike was immediately identifiable as a multispike burst lasting about 10 to 15 ms, while simple spikes could be identified as low amplitude triphasic potentials. Gradually lowering the electrode to 160 μm below the pial surface increased the signal-to-noise ratio

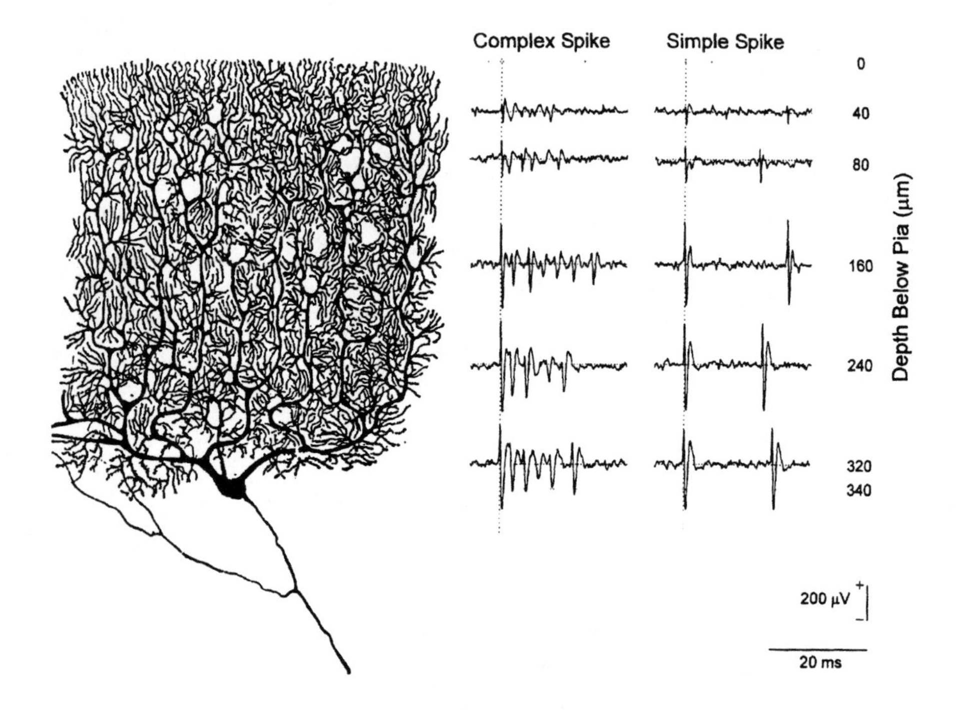

FIGURE 5.2
Extracellular records (tungsten microelectrode, 3 MΩ, bandpass filtered at 0.3 to 10 kHz) of complex and simple spikes obtained from a single Purkinje cell at different depths below the pial surface. Insertion of the microelectrode 40 μm below the pial surface allowed the detection of complex and simple spikes. Gradually lowering the electrode along the dendritic tree to 160 μm below the pial surface increased the signal and provided an easier detection of simple and complex spikes. In the molecular layer, simple and complex spikes began with a fast positive potential, most probably reflecting the somatic sodium spike. The potentials indicate that the Purkinje cell dendrites act as a current source for fast somatic sodium currents with a maximum at a depth of 240 μm. Further lowering of the electrode reduced the positive potential compatible with the approach of a current sink in the soma. The potential following the fast potential in complex spikes also have a fast positivity but are dominated by broad negative potentials. These potentials are likely to represent interference of dendritic calcium spikes with somatic sodium spikes. At 340 μm below the pial surface, the electrode contacted the soma or proximal dendrite of the Purkinje cell and destroyed it. This demonstrates that sharpened metal microelectrodes can be smoothly moved close to a dendrite of a Purkinje cell and demonstrates the spatial range along the vertical axis within which useful Purkinje cell recordings can be obtained.

and provided an easier detection of the complex and simple spikes. Further lowering of the electrode toward the soma did not appreciably increase spike height, although slight changes in spike width are often observed.

It is important to note that the proper and unambiguous identification of cerebellar cortical unit as a Purkinje cell requires identifying a complex spike, which is uniquely the membrane response of a Purkinje cell to climbing fiber activation. Without the clearly identifiable presence of a complex spike, it remains ambiguous as to whether a recorded unit in the cerebellar cortex is a Purkinje cell as opposed to a nearby stellate or basket cell in the molecular layer of the cerebellar cortex.

In summary, extracellular recording of Purkinje cell activity provides information regarding activity in the two major afferent pathways of the cerebellum. Climbing fibers, which are the axonal terminal arborizations of neurons in the inferior olive, innervate Purkinje cells directly and trigger a multiphasic spike burst known as the complex spike. Each Purkinje cell receives one, and only one, climbing fiber. Thus, the occurrence of a complex spike indicates that a neuron in the contralateral inferior olive fired an action potential and each Purkinje cell is a faithful transducer of an olivary neuron. Mossy fibers, on the other hand, originate from a variety of brain stem and spinal nuclei and only influence the Purkinje cells indirectly via their synaptic relationship with the granule cells. The granule cells provide the parallel fibers, each of which innervates hundreds of Purkinje cells and provides an ascending axonal branch that contacts more proximal Purkinje cell dendrites. While each Purkinje cell receives only one climbing fiber, it may receive as many as 200,000 parallel fibers. While the complex spike provides an unambiguous indication of activity in an inferior olive neuron, the simple spike provides a less clear message since it may be triggered by many possible combinations of parallel fiber inputs or possibly by the ascending branch of a granule cell axon. Finally, because the Purkinje cells are the only output neuron of the cerebellar cortex, their activity provides a measure of cortical output. Thus, the Purkinje cell is a unique neuron for functional neurophysiology *in vivo*, because it provides distinct information regarding two major afferent systems and simultaneously provides a measure of cortical output.

5.4 Electrode Profiles and Insertion Methods for Cerebellar Multielectrode Neurophysiology

It must always be kept in mind that the process of single neuron physiology begins with the displacement and destruction of neural tissue resulting from the introduction of a microelectrode into the brain. Successful multielectrode recording from the cerebellar cortex critically depends upon choosing an electrode profile and an insertion method that minimizes destruction to the greatest extent possible. That the Purkinje cell dendritic arbor may be as thin as 9 μm already places a definite limit on the diameter and width of a usable microelectrode for recording Purkinje cell activity. The most important consideration is that the profile of the electrode must be as slim as possible in order to prevent destruction of the dendritic arbor of the Purkinje cell which is the critical electroactive structure that mediates Purkinje cell responsiveness.

A second consideration is that an array of microelectrodes produces a displacement of brain tissue simply by virtue of introducing a volume into the brain. This can compress tissue in the interelectrode spaces which can alter local circuitry or even the cable properties of the neurons whose activity is to be measured. Because one goal of multiple microelectrode neurophysiology is to measure meaningful spatio-temporal activity patterns within a local region of the brain, it is important that the electrode array does not significantly distort the brain tissue within which it is placed. This can be accomplished by employing microelectrodes in arrays that occupy the smallest possible fraction of brain volume.

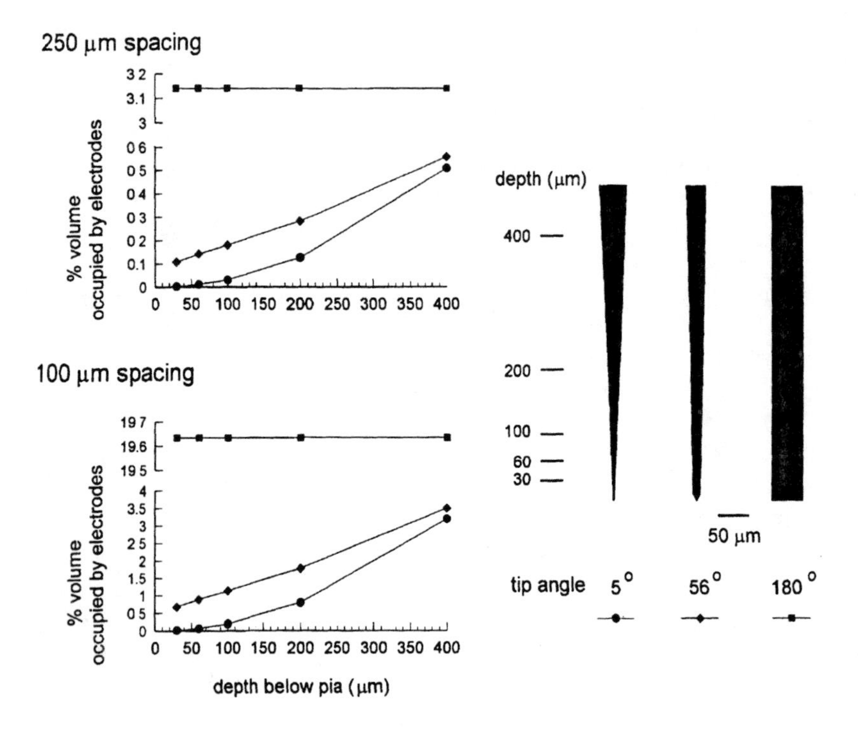

FIGURE 5.3
Percentage volume of brain tissue within the space of a two-dimensional array of electrodes occupied by electrodes as a function of insertion depth and electrode spacing for three profiles of microelectrode. Gradually tapered microelectrodes having a tip angle of 5 degrees (circles) occupied less than 0.1% and 0.5% of array-space volume at insertion depths up to 200 μm for 250 μm and 100 μm interelectrode distances, respectively. Tapered electrodes with a tip angle of 56 degrees occupy a greater volume of array space immediately upon insertion into the brain while nontapered 50-μm-diameter electrodes occupy a very large percentage of array-space volume.

Figure 5.3 shows profiles for three typical electrode geometries used for extracellular neurophysiology fashioned from cylindrical substrates. The first electrode is a gradually tapered profile shaped from 100-μm-diameter substrate and having a tip angle of 5 degrees. Such a profile is typical of a pulled glass micropipette or a finely etched microwire. The second electrode is also tapered but has a more complex geometry. Its tip is abruptly tapered from a diameter of 10 μm, has a tip angle of 56 degrees, and has a base diameter of 80 μm 1.5 mm away from the tip. Such a profile is similar to some silicon-based microelectrodes. The third profile is an untapered 50 μm microwire. From these geometries, the percentage of brain volume within the array space occupied by a two-dimensional array of such electrodes can be calculated. The left panels of Figure 5.3 plot the percentage volume of brain tissue occupied by electrodes at various tip depths for each of the electrode geometries configured in a two-dimensional array having either 250 μm or 100 μm interelectrode spacing. As can be seen from these plots, an array fabricated from the most gradually tapered electrodes occupies the smallest volume of brain tissue from 30 to 400 μm insertion depths. The impact of such an array in terms of tissue

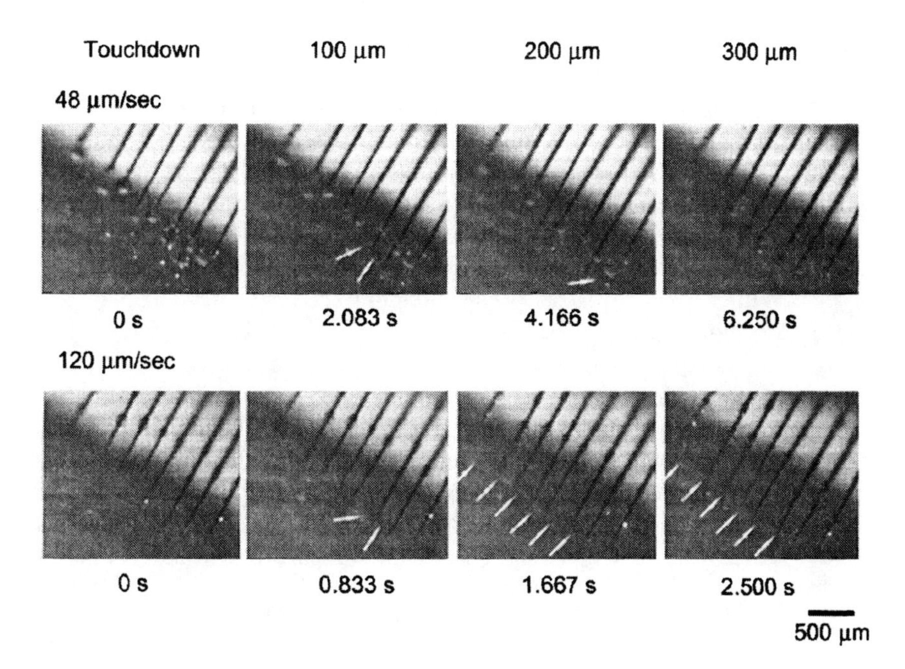

FIGURE 5.4
High-speed video imaging (125 frames/sec) of the surface deformation of the cerebellar cortex during an insertion of a linear array of eight microelectrodes with an interelectrode spacing of 250 μm. The electrodes had the 5 degree tip angle profile shown in Figure 5.3 and were inserted at 48 μm/sec (top panels) and 120 μm/sec (bottom panels). Lowering the array at 48 μm/sec produced noticeable dimpling of the cortical surface at insertion depths of 100 μm and 200 μm (white arrows). At 300 μm of insertion depth, no obvious dimpling is present and the cortical surface is without distortion. Lowering the array at 120 μm/sec produced similar dimpling at 100 μm insertion depth without subsequent recovery of surface morphology. At 120 μm/sec, dimpling is exacerbated when the electrodes are 200 μm below the pial surface and a distinct indentation of the surface is produced at 300 μm insertion depth. The beads on the electrodes are saline.

compression is negligible up to 200 μm of insertion even with 100 μm interelectrode spacing because it occupies less than 0.7% of the brain volume within the array space. The abruptly tapered electrode occupies a significantly greater volume of array space immediately upon introduction into the brain such that its insertion to 30 μm occupies as much brain volume as does insertion of the fine tip array 200 μm into the brain. Because of the differing taper profiles of the two electrodes, the fine tip array maintains a distinct advantage up until 400 μm depth. Finally, the nontapered 50 μm wire occupies more than an order of magnitude more brain volume and would be expected to significantly distort the local circuitry within the region of recording.

Another important issue for successful multielectrode recording of the cerebellum is that the electrodes must be introduced through the pia in a manner that minimizes dimpling and compression of the cerebellum. Figure 5.4 shows video images taken from the surface of a cerebellar folium during the insertion of an array of microelectrodes 300 μm below the pial surface. In this experiment, the electrodes were eight finely etched microwires having a 5 degree tip angle and fixed in a linear array with an interelectrode distance of approximately 250 μm. The electrode array was lowered via a hydraulic drive at either 48 μm/sec or 120 μm/sec into a vermal lobule whose dura had been previously removed. As indicated by the white arrows in the top row of Figure 5.4, 48 μm/sec hydraulic insertion produced noticeable dimpling of tissue around some of the electrodes. These dimples had a radius of approximately 40 μm and did not appear to interact with one another. Importantly, the dimples were present only transiently, as the cortex recovered its original surface contour even during continued insertion of the electrodes. When the electrode tips reached 300 μm below the pial surface, no dimpling could be observed. It appeared, then, that the cortex behaves elastically and that its intrinsic restoring force allowed it to return to its equilibrium position even during continued insertion of the electrodes at this slow rate.

In contrast, inserting the linear array at a faster rate of 120 μm/sec prevented the cortex from returning to its equilibrium position during the insertion process. This resulted in broader dimples that clearly overlapped and interacted with each other to produce a linear indentation of the cortical surface. This difference was visible starting at a depth of 200 μm (Figure 5.4, lower panel) where severe dimpling was clearly seen around every electrode in the array. At a depth of 300 μm the separated dimples have merged to form a large elongated hollow in the cerebellar cortex around the whole array (Figure 5.4, lower panel, white arrows).

We have quantitatively modeled the effect of dimpling on the placement of electrode arrays into the cerebellar cortex. Here we defined a dimple as a deformation of the cortical surface that is maximal at the tip of the electrode and which recovers as a logarithmic function of distance away from the site of maximal deformation. The 40-μm-radius dimples produced by the sharpened electrodes at the slow insertion rate of 48 μm/sec (Figure 5.4, top panel) allow electrode spacings as small as 100 μm to have a negligible impact on surface morphology, and hence the underlying structure of the Purkinje cell dendrites. Faster insertion rates, or the use of blunt-tipped electrodes (Figure 5.5), will increase dimple radius and dimple depth. When interelectrode distances are less than the dimple diameter, there will be a summation of dimpling along their lateral margins which will, in turn, reduce the restoring force of the brain. Thus, our experiments indicate that fast insertion rates or a high dimple radius to interelectrode distance ratio will produce surface compression and distortion of the underlying tissue.

In a final study on surface effects of electrode implantation, we tested whether very high velocity insertion of an electrode array would solve the problems of surface dimpling, distortion, and compression. We used high-speed video imaging at 500 frames/sec to visualize the effect of implanting the linear array of eight microelectrodes ballistically at 14 mm/sec, 292 times faster than the optimal slow rate determined by the previous experiment (48 μm/sec). To the eye, the ballistic insertion appeared to produce no dimpling and to have negligible impact on surface morphology, while allowing the electrodes to efficiently enter the brain. High-speed imaging, however, indicated that the very high rate of electrode insertion severely compressed the cortical surface prior to the implantation of the electrodes. Figure 5.6 shows six video images taken in the first 109 msec of the lowering of the array. These images revealed that the brain was compressed 770 μm in the first 88 msec of the implantation

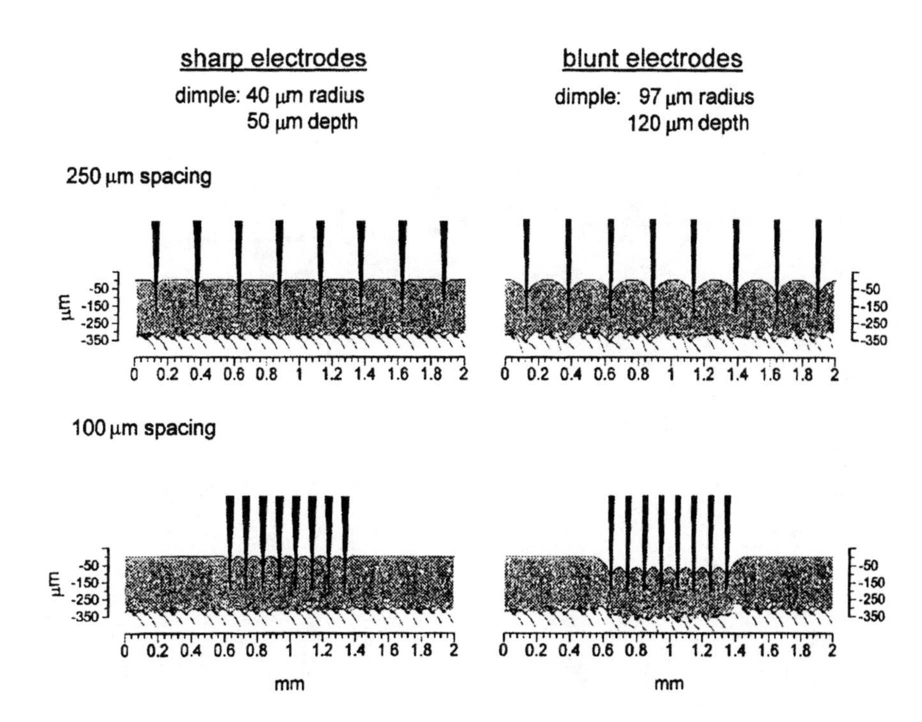

FIGURE 5.5
The effect of dimple radius and depth on cerebellar cortical morphology as a function of interelectrode spacing. Fine-tipped electrodes (5 degree) inserted slowly produce 40-μm-radius dimples that do not overlap when a linear array of microelectrodes having either 250 μm or 100 μm spacing is lowered into the cerebellum. In contrast, blunt electrodes produce wider and deeper dimples that severely distort the surface as the interelectrode spacing is reduced. Compression of the cortical surface results when the interelectrode spacing is reduced to less than twice the dimple radius, resulting in distortion of the underlying circuitry. Purkinje cell somata with their overlying dendrites are shown 340 μm below the surface. All aspects of the figure are drawn to scale.

before the electrodes even penetrated the surface. This was apparent as a prominent fissure on the cortex that remained after the electrodes penetrated the surface. It can reasonably be presumed that such severe and rapid surface compression will produce a compression wave within the brain and distort the underlying neural structures (Figure 5.7). The experiment indicates that high-velocity, ballistic insertion of electrode arrays into the cerebellar cortex is not a solution to, and in fact exacerbates, the problem of surface dimpling and brain compression.

5.5 Fabrication Methods and Procedures for Cerebellar Multielectrode Neurophysiology

Two methods have been employed for successful multiple electrode neurophysiology of the cerebellar cortex: arrays of glass pipettes and arrays of etched tungsten microelectrodes. Here we describe the methods and advantages and disadvantages of each.

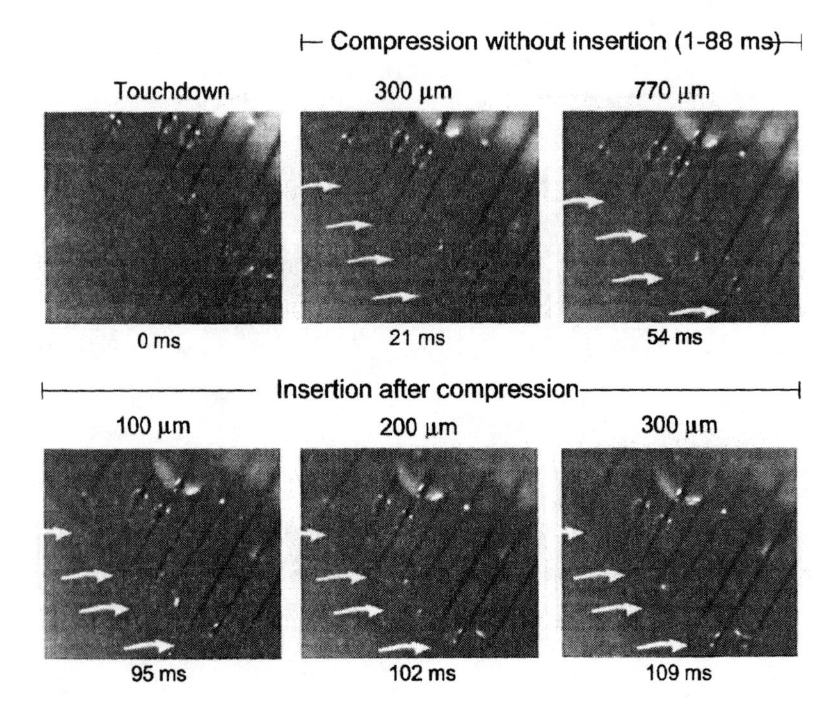

FIGURE 5.6
High-speed video imaging (500 frames/sec) of the surface deformation of the cerebellar cortex during the ballistic insertion of a linear array of eight sharpened microelectrodes at 14 mm/sec. The electrodes compress and deform the cortical surface for the first 88 msec after they contact the brain. Only subsequent to the cortical distortion do the electrode tips penetrate the surface and enter the brain. The same electrode array was used as in Figure 5.4.

Any procedure for cerebellar multielectrode neurophysiology must overcome two constraints. First, Purkinje cells are extremely sensitive to subdural bleeding. Our experience is that any blood on the cerebellar cortical surface will immediately prevent recording normal Purkinje cell activity. Second, cerebellar cortical folia are curved and are mechanically unstable; they are susceptible to rhythmic movements related to the respiration and heart beat. Opening the dura, which is essential for optimal Purkinje cell recording, exacerbates these problems. Thus, a procedure for stabilizing the cortex facilitates success. Furthermore, because the Purkinje cell layer follows the contour of the surface, gently flattening the cortex and bringing the array space into a true two-dimensional configuration allows precise spacing of multiple recorded Purkinje cells.

5.5.1 Arrays of Individually Implanted Glass Pipettes

A procedure for successively implanting very fine glass pipettes into the cerebellar cortex was developed and described in detail by Sasaki, Bower, and Llinás.[15] This method has been successfully employed by a number of investigators to reveal the

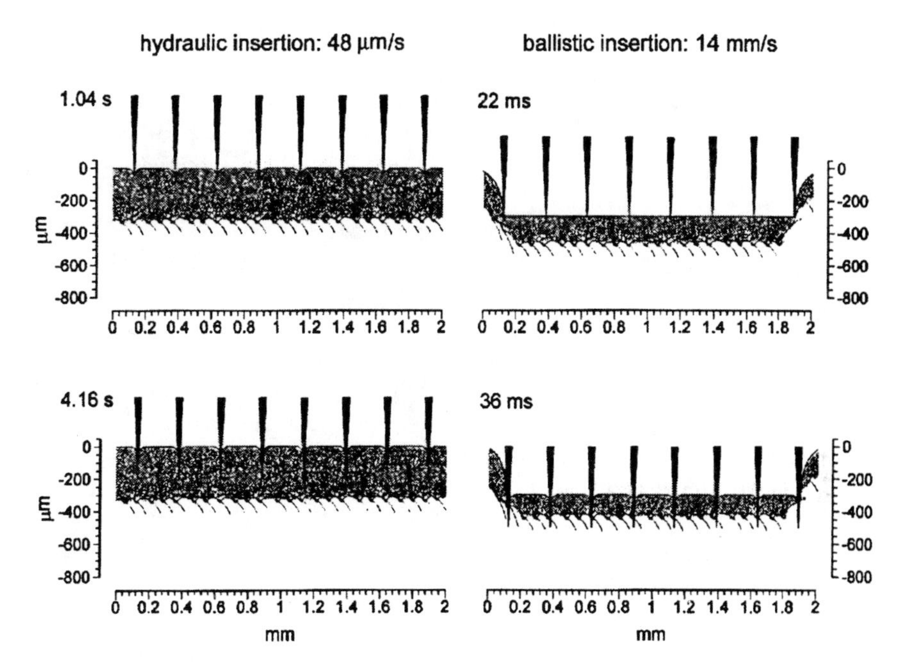

FIGURE 5.7
A schematic comparing the effect of slow hydraulic insertion with ballistic insertion of an array of microelectrodes into the cerebellar cortex. Controlled hydraulic insertion (left panels) produces minimal dimpling and tissue distortion allowing the tips of the microelectrodes to be precisely placed within the dendritic trees of the Purkinje cells. Ballistic insertion compresses the cortical surface before the electrodes enter the brain. This might be expected to compress the Purkinje cell dendrites, causing the electrodes to bypass the Purkinje cells altogether.

spatial organization of olivocerebellar activity and its neurochemical and neuroanatomical regulation.[12,13,17,18] Such studies were the first generation of experiments to employ multielectrode recording of cerebellar activity and have had an important effect on current thinking about the functional organization of the cerebellar systems. Here we briefly review the method and point readers to Sasaki, Bower, and Llinás[15] for further details.

The method centers around a unique microelectrode carrier system that is fabricated from a titanium electron microscopy grid having a mesh of 250 µm. The grid is stabilized by a loop of stainless steel wire before it is coated in a silicon elastomer (Sylgard®, Dow Corning) and shaved with a microtome to have a thickness of approximately 200 µm. The primary role of the grid assembly is to hold glass pipettes that can be inserted through the silicon once the device is surgically implanted on a folium of the cerebellar cortex. The device provides the additional advantage of providing a biocompatible substrate that can be used to gently flatten the cerebellar cortex in order to stabilize it and reduce its curvature.

Glass micropipettes can be pulled from capillary tubing (A-M Systems, 6030), filled with 2 *M* NaCl, and cut to a length of approximately 3 mm before a 25 µm, Teflon®-coated, platinum–iridium wire is inserted into the cut end of the microelectrodes. The cut end of the electrode should be approximately 100 µm in diameter, and the impedance of the pipettes should be 2 to 4 MΩ. A major advantage of this method is that it allows single electrodes to be inserted individually, through the silicon-encased grid, in order to isolate a single Purkinje cell with each electrode. Also, the electrodes can be pulled with very fine tips so that dimpling is reduced. A motorized microdrive (Burleigh) facilitates and is probably essential for precise electrode implantation without destroying the array. Although the method is extremely labor intensive, the experienced experimenter can isolate six Purkinje cells per hour. A limitation of the method is that the array is extremely fragile and can be dislodged easily if the preparation is unstable or by the slightest experimenter error. Furthermore, the array cannot be cemented into position and is probably not useful for chronic recording from awake and freely moving animals making forceful movements of their limbs and neck. Pushed to its limit, the method has allowed for successful multineuron recording of awake but heavily restrained animals in an acute experiment. The method was successfully employed for head-restrained rats making tongue and jaw movements as well as head restrained rats making synergistic movements of the arm and tongue (see Figure 5.11). However, it should be noted that the yield of such experiments involving limb movements is low due to the instability problems mentioned above.

5.5.2 Etched Tungsten Microwire Assembly

Because of the constraints of the glass pipette method, namely fragility and the inability to use it in chronic implantation, we have developed methods for multi-electrode recording of the cerebellum using metal microelectrodes.[16] These methods combine the grid carrier system, originally developed by Sasaki et al.,[15] which can be used to stabilize and gently flatten the cerebellar cortex, with finely etched, high impedence tungsten microelectrodes that are generally employed for single unit recording *in vivo*. The use of slim-profile microelectrodes with sharp tips is necessary to address the compression problem, to minimize destruction of the Purkinje cell dendrites, and to ensure that single Purkinje cells are recorded (see above).

Figure 5.8 shows a 32-microelectrode carrier assembly that we have employed for cerebellar recording. The electrodes are fabricated from standard 100 μm tungsten wire and are electrochemically etched to have a tip profile of 5 degrees. They are insulated with Epoxylite® (Epoxylite Corp., Irvine, CA). Only electrodes having an impedance within 2 to 6 MΩ are used. The electrodes are held in place by two silicon elastomer-coated electron microscopic grids as employed by Sasaki et al.[15] which are mounted on the open ends of a glass tube (diameter 4 mm, length 10 mm). Only the front grid is shown in Figure 5.8. The electrodes are back-loaded into the grid assembly in rows of eight; their tips are aligned under a microscope; and their back ends are glued and covered with polyimide tubing. The back ends of the microwires are soldered to a 9-pin microplug, allowing room for a reference channel.

Prior to surgery, the electrodes are retracted within the assembly so that less than 50 μm of the tip extrudes. Then, the entire grid-electrode assembly is gently pressed against the cortex and cemented in place with the electrodes in the retracted position. Next, each linear array of 8 electrodes can be inserted under hydraulic

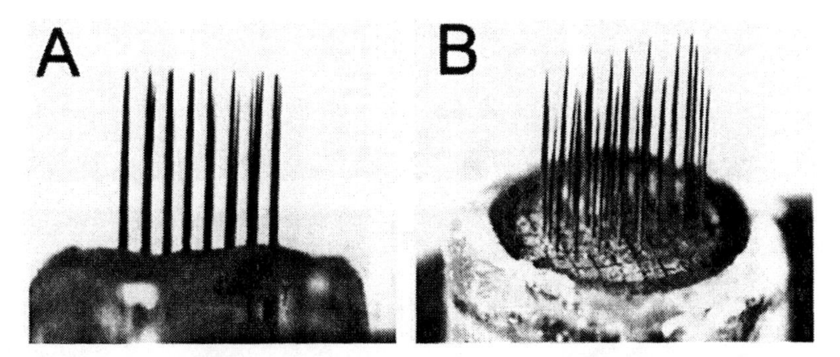

FIGURE 5.8
An etched tungsten microelectrode assembly used for multiple electrode recording from cerebellar cortex. Thirty-two microelectrodes are configured in a 4-by-8 array and held in place by a series of silicon-encased grids. The interelectrode distance is 250 μm. The system is designed to allow the electrodes to be advanced or retracted within the assembly interactively during implantation in linear sets of eight electrodes. The assembly is implanted with electrodes in the retracted position, after which they can be slowly and gently inserted into the cerebellar cortex with a minimum of surface dimpling and compression. For demonstration purposes, the electrodes are extruded 2 mm from the grid.

control independent of the other electrodes. Provided the tips are properly aligned and the cerebellar cortex is properly stabilized by the grid assembly, the electrode tips in each linear array can be gradually and precisely lowered into the optimal position of the Purkinje cell dendritic arbors to record simple and complex spike activity. After positioning the electrodes, the back end of the assembly, including the wires and microplugs, can be cemented in place.

5.6 Recording Methods and Algorithms for Spike Separation

We have used the MultiNeuron Acquisition Processor (MNAP) hardware developed by Plexon Incorporated (Dallas, TX) for multineuron recording from the cerebellum. This hardware allows for 96 channels of on-line filtering, amplification, and spike separation. The digital signal processing capabilities of the MNAP device allow four waveforms to be discriminated from each of the 96 channels. For recordings of cerebellar Purkinje cells, in practice, we are only interested in discriminating two waveforms: the complex spike and the simple spike.

The real-time separation of simple from complex spikes is a unique problem in spike separation. This is due to two factors. First, the complex spike waveform is quite variable from cell to cell and is unlike the simple uni- or biphasic potentials typically recorded extracellularly from the vast majority of neurons. Second, and more important, because the complex spike is a multispike burst whose initial component may be nearly identical to a simple spike, the discrimination of the two waveforms must be based on the presence or absence of the later components of the complex spike waveform. These components occur within 2 to 15 msec after the initial spike of the complex spike and thus, the time window for spike separation

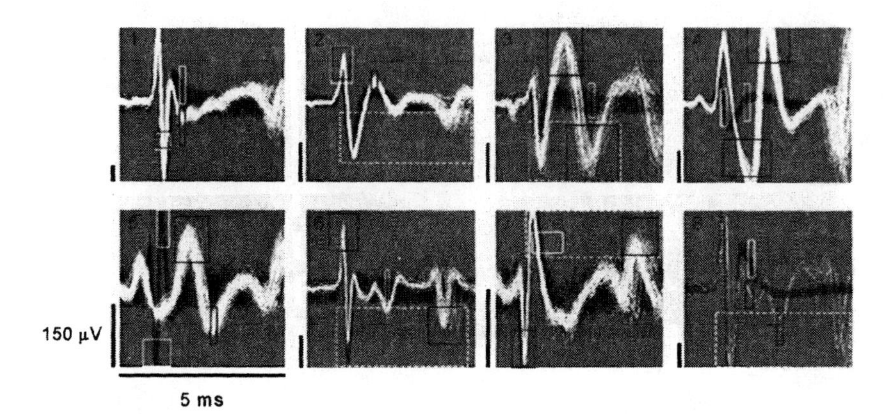

FIGURE 5.9
Examples from eight Purkinje cells of the separation of simple (black) and complex (white) spikes. The potentials must cross a threshold (not shown) before they must enter the white boxes to be classified as a simple spike or the black boxes to be classified as a complex spike. In certain cases (2, 3, 6, 7, 8) it is not possible to select only the simple spike with a standard set of boxes. This may be because background activity on the trailing edge of a simple spike interferes with detection of the later components of the complex spike when the leading edges of the two waveforms are identical. In these cases, a hierarchical analysis is employed in which a dashed box not only selects for simple spikes but also deselects for complex spikes (see text). In all cases, the potentials were recorded with high impedance, etched tungsten microwires.

should be as long as possible. However, the length of the time window should not be greater than the minimal interspike interval of either spike type. This is a point of concern, because simple spikes can occur as fast as 180 to 200 Hz so that the time window for detecting the later components of the complex spike must be limited to 5 msec. Nevertheless, because both waveforms are produced by the same cell, the spike sorting algorithm does not have to account for the detection of synchronous waveforms, a typical problem in waveform analysis.

We have developed a spike separation algorithm in collaboration with Plexon Incorporated that provides a faithful separation of simple and complex spikes (Figure 5.9). This algorithm is based on a hierarchical analysis which first requires a waveform to cross an amplitude threshold. After crossing the threshold, the waveform must pass through two boxes which are drawn by the experimenter into the time–voltage plot in order to be classified as a spike. In many cases, when the initial components of the simple and complex spike are very similar, or when the signal-to-noise ratio is low, an additional, hierarchically organized set of criteria must be used. Here, an order is assigned to the two sets of boxes used to discriminate the waveforms such that a waveform that satisfies the criteria for the first set of boxes is discounted for analysis by the second set. Figure 5.9 shows five examples of the hierarchical analysis (cases 2, 3, 6, 7, and 8). In these examples, one of the two boxes in the first set had a unique characteristic that allowed it to specifically exclude complex spikes. This box, represented by the dashed lines in Figure 5.9, required the waveform to enter one and only one time. The essential aspect of this box is that it could be spread along the time axis in order to deselect the multiphasic complex spikes which would necessarily enter two or more times. This allows unambiguous classification of a waveform as a simple spike and, if not, thereafter as a candidate for a complex spike with two boxes used in the standard way.

5.7 Functional Analysis of Olivocerebellar Synchrony

One of the important issues in olivocerebellar neurophysiology has been the spatial patterning whereby the inferior olive activates the cerebellar cortex. It is well established that inferior olivary neurons are electrotonically coupled to one another via dendrito–dendritic gap junctions and that this is a substrate for synchronous firing within that nucleus. In an important study, Llinás and Sasaki[13] were the first to use multiple microelectrode recordings to show that the inferior olive synchronously (within 1 ms) activates Purkinje cells in distinct spatial patterns, providing the first evidence that electrotonic coupling within the inferior olive regulates cerebellar output. Analyses of synchrony entailed calculating zero-time cross-correlation coefficients among the spike activity of all possible pairs of Purkinje cells. With this procedure, a matrix of cross-correlation coefficients is generated for every Purkinje cell recorded in the microelectrode array, and, provided the relative locations of the neurons are known, these matrices can be placed in anatomical space to provide plots of synchronous cerebellar activation by the inferior olive within a particular cerebellar folium. This method has been invaluable in understanding the functional significance of the inferior olive. Nevertheless, a significant drawback of the method is that it only provides a measure of population activity relative to single neurons, and is not a population measure *per se.* That is, the analysis is always related to a referent or "master" cell for the generation of a matrix and as many cross-correlation matrices are generated as there are cells in the array. It becomes somewhat arbitrary which set of cross-correlation matrices represents the population, especially when very large numbers of cells are recorded.

A method was developed[20] to derive population measures of synchrony from sets of cross-correlation matrices, thereby removing the final necessity of a "master" cell from the population analysis (Figure 5.10). The goal of this procedure was to generate "images" of the cerebellar cortex showing its preferred pattern of synchronous activation by the inferior olive at particular moments in time.

The method retains as its essential foundation the calculation of all possible combinations of zero-time cross-correlation coefficients, which provides a measure of the degree of synchronous firing (within 1 msec) between cell pairs. A matrix of zero-time cross-correlation coefficients is generated for every cell in the array (serving as a "master") for an experimental period and for a control period. For our experiments, the experimental period is usually a period in which an animal is making a movement, while the control period is a period of equal duration in which no movement is being made. Cross-correlation coefficients are calculated as follows using the algorithm of Gerstein and Kiang.[10]

First, the spike train of a cell is represented by $X(i)$, where i represents the time step ($i = 1, 2, \ldots N$ msec) from the beginning of the period of interest; $X(i)$ equals 1

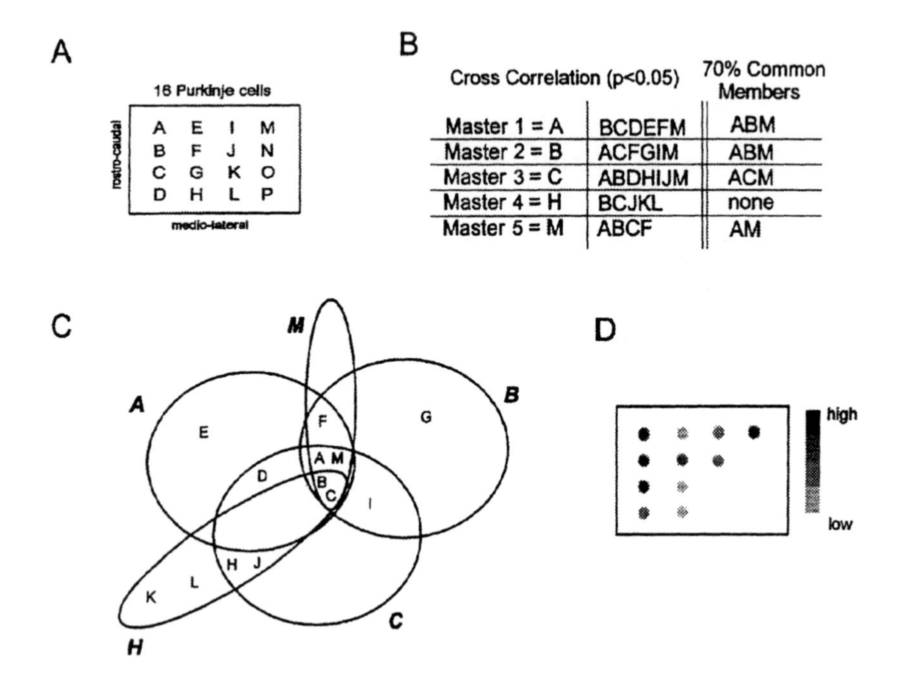

Cross Correlation ($p<0.05$)		70% Common Members
Master 1 = A	BCDEFM	ABM
Master 2 = B	ACFGIM	ABM
Master 3 = C	ABDHIJM	ACM
Master 4 = H	BCJKL	none
Master 5 = M	ABCF	AM

FIGURE 5.10
Schematic summary of a procedure for deriving spatial plots of synchronous neuronal firing. A. The spatial location of 16 Purkinje cells (A through P) in a 4-by-4 multielectrode recording array placed onto the surface of the cerebellar cortex. A zero-time cross-correlation matrix of spike firing is calculated for each cell in the array serving as a "master" cell, and significance is determined with a z-transform of each matrix using a control period to provide an estimate of the population mean and standard deviation (see text). B. In this example, the firing of five cells in the array (A, B, C, H, M) is significantly ($P < 0.05$) cross-correlated with the firing of at least four other cells. The left side of the table indicates each of the five master cells and their respective groups of significantly correlated cells. The right side of the table indicates families of neurons that have as their significantly correlated group at least 70% overlap with a particular master cell. C. A venn diagram illustrating the relationship in the groupings associated with the five master cells. Here, master cells A, B, C, and M have many significantly correlated members in common, while master cell H does not. The analysis forms families of cells that have many common relatives. D. A spatial plot of the mean degree of cross-correlation of each cell with a family of cells derived in B and placed in anatomical space. The gray scale provides a quantitative measure of the tendency of each cell to fire synchronously in the configuration shown.

if the onset of a spike occurs at the ith ms, otherwise $X(i)$ equals 0; $Y(i)$ is the same as $X(i)$, but for the "master" cell in the matrix. Second, the cross-correlation coefficient is represented by $\Sigma\{V(i)W(i)\}/\sqrt{(\Sigma V(i)^2\ \Sigma(W(i)^2))}$, where $V(i)$ and $W(i)$ are the normalized forms of $X(i)$ and $Y(i)$, respectively, such that $V(i) = X(i) - \Sigma X(i)/N$ and $W(i) = Y(i) - \Sigma Y(i)/N$. The normalization of the spike trains is essential to correct for differences in firing rate between the two neurons.

Third, a single mean ($\bar{x}_c$) and standard deviation (s_c) is calculated for the cross-correlation matrix of the control period for every master cell. These values are then used to convert each element in the same master cell's cross-correlation matrix from the experimental period into standard deviation units using the formula $z = (\bar{x} - \bar{x}_c)/s_c$. The elements of each z-transformed matrix can then be assigned a significance value. We typically use a one-tailed cutoff of 1.65 for a significance level of 5%. Significance in this analysis indicates that a given value of cross-correlation during the experimental period is statistically greater than the mean degree of cross-correlation during the control period. The end result of this step is that each master cell is associated with a group of cells whose tendency to fire synchronously during the experimental period was statistically greater than that observed during the control period. Such groups are typically small subsets of the entire population and are likely to be different among the master cells.

The fourth and most critical step in the analysis extracts groups of master cells whose population of significantly correlated cells is similar. This step requires assigning a coefficient of similarity, and we have found that a conservative criterion of 0.7 is useful for identifying the most reliable patterns of synchronous firing. Essentially, this step groups master cells that tend to have common "relatives." The result is that groups of relatives are formed into assemblies, or "families," that are a subset of the entire array and that the "master" cell is no longer a variable in further analysis.

The fifth step is to determine the mean cross-correlation of every member of the recorded population with the members of the assembly. Once this is determined, a plot of these values in anatomical space depicting the surface of the folium can be made. The final result is a plot whose ordinate and abscissa are anatomical distance representing the medio-lateral and rostro-caudal extent of the array space and whose elements are neurons, each represented by a value that reflects the degree to which it fired synchronously with the members of the derived family. The range of values can be assigned to a gray scale and a visual plot of the folial surface can be produced. A useful characteristic of this analysis is that it allows the experimenter to graphically represent the dynamic rearrangement of synchronous firing that occurs in time during a sensory, cognitive, or motor event.[20]

Figure 5.11 shows the results of the analysis for Purkinje cell recordings during a complex skilled movement in a rat. The subject was operantly trained to make synergistic movements of its left forepaw and its tongue in order to receive a drop of water as a reinforcer. In this experiment, 19 electrodes were implanted into 3 mm^2 of a cerebellar folium, and complex spikes were isolated and discriminated from single Purkinje cells. The arm movements were very rapid (about 150 msec), and during this time the subject extended its tongue toward a target in front of the mouth. Figure 5.11 plots the movement as five "frames" of a movie taken at 50 msec intervals. Below the cartoons of the movement, the spatial patterns of synchronous complex spike firing within the cerebellar folium are plotted, also in 50 msec frames in register with the cartoons. The analysis reveals a dynamic rearrangement of synchronous firing within the cerebellum during the execution of the movement. Such an analysis could be applied to a wide range of brain regions and behavioral paradigms in order to understand the functional significance of dynamic repatterning of synchrony with regard to motor and cognitive function.

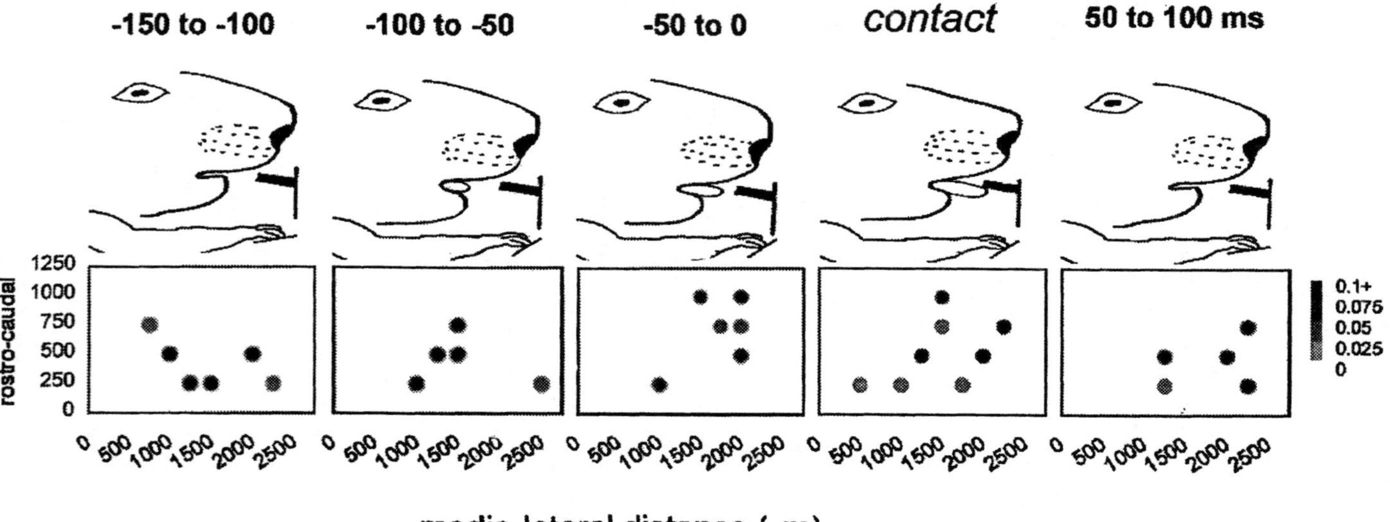

FIGURE 5.11
The spatial distribution of synchronous complex spike firing in a cerebellar folium during a complex movement synergy involving the forepaw and tongue. The top row shows the trajectory of the tongue and forepaw taken from video images obtained during the movement. The bottom row shows spatial plots of synchronously firing Purkinje cells in a matrix of 19 Purkinje cells recorded from the left Crus IIa folium of the rat. The spatial plots of synchronous firing are calculated from 5 epochs of 50 ms and are placed in temporal registration with the overlying video images. For the spatial plots, 0 reflects the postero-lateral limit of the recording array.

Acknowledgments

This work was supported by grants from the National Institute of Neurological Disorders and Stroke (NS–31224) and a biomedical engineering grant from the Whitaker Foundation. C.S. was supported by a fellowship of the Deutsche Forschungsgemeinschaft (Schw577/4–1).

References

1. Braitenberg, V. and Atwood, R. P., Morphological observations on the cerebellar cortex, *Journal of Comparative Neurology,* 109, 1, 1958.
2. Cajal, Ramon y, *Histology* (Fernan-Nunez, Transl), William Wood & Company, Baltimore, MD, 1933.
3. Eccles, J. C., Llinás, R. and Sasaki, K., The inhibitory interneurones within the cerebellar cortex, *Experimental Brain Research,* 1, 1, 1966.
4. Eccles, J. C., Llinás, R. and Sasaki, K., Parallel fibre stimulation and the responses induced thereby in the Purkinje cells of the cerebellum, *Experimental Brain Research,* 1, 17, 1966.
5. Eccles, J. C., Llinás, R. and Sasaki, K., The mossy fibre–granule cell relay of the cerebellum and its inhibitory control by Golgi cells, *Experimental Brain Research,* 1, 82, 1966.
6. Eccles, J. C., Llinás, R. and Sasaki, K., Intracellularly recorded responses of the cerebellar Purkinje cells, *Experimental Brain Research,* 1, 161, 1966.
7. Eccles, J. C., Llinás, R. and Sasaki, K., The excitatory synaptic action of climbing fibres on the Purkinje cells of the cerebellum, *Journal of Physiology,* 182, 268, 1966.
8. Fox, C. A. and Bernard, J. W., A quantitative study of the Purkinje cell dendritic branchlets and their relationship to afferent fibres, *Journal of Anatomy,* 91, 299, 1957.
9. Friedrich, V. L. and Brand, S., Density and relative number of granule and Purkinje cells in cerebellar cortex of cat, *Neuroscience,* 5, 349, 1980.
10. Gerstein, G. L. and Kiang, W. Y., An approach to the quantitative analysis of electrophysiological data from single neurons, *Biophysical Journal,* 1, 15, 1960.
11. Hillman, D. E. and Chen, S., Vulnerability of cerebellar development in malnutrition. I. Quantitation of layer volume and neuron numbers, *Neuroscience,* 6, 1249, 1981.
12. Lang, E. J., Sugihara, I. and Llinás, R. GABAergic modulation of complex spike activity by the cerebellar nucleoolivary pathway in the rat, *Journal of Neurophysiology,* 76, 255, 1996.
13. Llinás, R. and Sasaki, K., The functional organization of the olivo-cerebellar system as examined by multiple Purkinje cell recording, *European Journal of Neuroscience,* 1, 587, 1989.
14. Palkovits, M., Magyar, P. and Szentagothai, J., Quantitative histological analysis of the cerebellar cortex in the cat. I. Number and arrangement in space of the Purkinje cells, *Brain Research,* 32, 1, 1971.
15. Sasaki, K., Bower, J. M. and Llinás, R., Multiple Purkinje cell recording in rodent cerebellar cortex, *European Journal of Neuroscience,* 1, 572, 1989.

16. Schwarz, C. and Welsh, J. P., Temporal activity patterns within the cerebellar cortex evoked by the cerebral cortex: Studies of the lingual motor system, *Society for Neuroscience Abstracts,* 23, 18, 1997.
17. Sugihara, I., Lang, E. J. and Llinás, R. Uniform olivocerebellar conduction time underlies Purkinje cell complex spike synchronicity in the rat cerebellum, *Journal of Physiology,* 470, 243, 1993.
18. Sugihara, I., Lang, E. J. and Llinás, R. Serotonin modulation of inferior olivary oscillations and synchronicity: a multiple-electrode study in the rat cerebellum, *European Journal of Neuroscience,* 7, 521, 1995.
19. Szentagothai, J. and Rajkovits, K., Uber den Ursprung der Kletterfasern des Kleinhirns, *Z. Anat. Entwickl.,* 121, 130, 1959.
20. Welsh, J. P., Lang, E. J., Sugihara, I. and Llinás, R., Dynamic organization of motor control within the olivocerebellar system, *Nature,* 374, 453, 1993.
21. Welsh, J. P. and Llinás, R., Some organizing principles for the control of movement based on olivocerebellar physiology, in *The Cerebellum: From Structure to Control,* DeZeeuw, C. I., Strata, P. and Voogd, J. (Eds.), *Prog. Brain Res.,* 114, 449, 1997.

Chapter 6

Methods for Chronic Neuronal Ensemble Recordings in Singing Birds

Amish S. Dave, Albert C. Yu, John J. Gilpin, and Daniel Margoliash

Contents

6.1 Introduction

Techniques for recording muscular and neural activity from awake behaving animals are well established for larger vertebrates such as primates, and for invertebrates

0-8493-3351-2/99/$0.00+$.50

such as insects. Many species of primates will readily learn a behavioral task while working for a food reward under conditions of restricted movement, and primates can carry considerable weight, which facilitates preparation of implants for neurophysiological recordings. Behaviors under study such as locomotory behavior in insects can be elicited while animals are tethered, which effectively restricts movements and also facilitates deployment of recording devices. Such favorable conditions cannot always be achieved in all animal test systems, especially in small vertebrates producing behavior that is under volitional control. Yet these systems may offer unique advantages that are highly desirable for neurobiological research.

One such example is the study of song learning and its neural substrates in passerine birds (songbirds). Song learning in birds has been a central focus of ethological research. The study of the neurobiology of song learning, song production, and song perception has stimulated many seminal contributions to our understanding of brain/behavior relationships and how learning influences these patterns. The passerine birds commonly employed in these studies are small, however. For example, the most commonly used species, the zebra finch, weighs 13 g as an adult. Song learning occurs early in life when birds are even smaller and the skull may not be fully calcified. In addition, there is considerable movement associated with singing — the central behavior under study — especially of the head, beak, and tongue. Such movements can produce sizable electrical artifacts in the high impedance electrodes suitable for isolating individual neurons. Finally, birds will generally not sing when physically restrained. The combination of small size, movement, and volitional control have impeded progress in analysis of neuronal populations in songbirds under conditions of behavioral salience. This chapter reviews recent progress in achieving chronic neuronal recordings in singing birds. Descriptions of complementary techniques to record from peripheral structures during singing, including the muscles of the syrinx, can be found elsewhere.[1]

To the best of our knowledge, McCasland and Konishi[2] were the first to employ chronic recording techniques to record from "song system" nuclei (Figure 6.1A) while birds sang. In that study, relatively large coaxial electrodes were employed in a differential recording configuration to minimize movement-induced artifacts.* The electrodes were fixed in place and electronics were attached to the bird's head in a single surgery. Under these conditions, it was possible to record from many species of birds, ranging from the relatively large mockingbirds (approximately 70 g) to zebra finches. In all cases, however, the immobile coaxial electrodes achieved only relatively low signal-to-noise ratios (S/N) in the recordings, hence only multiple units that were not discriminated into single units were recorded. McCasland[3] later

* Movement-induced electrical artifacts will be similar for physically proximate sites. If electrode characteristics are matched, then a differential recording configuration will cancel out so-called common-mode signals present on both inputs to the amplifier. This is often the preferred configuration in practice, although in some cases single-ended configurations will give better performance. Thus it is best to test initially under both conditions. In addition, since the differential recording mixes two channels, brief periods of single-ended recordings are necessary in any case to determine which units are being recorded on which channel.

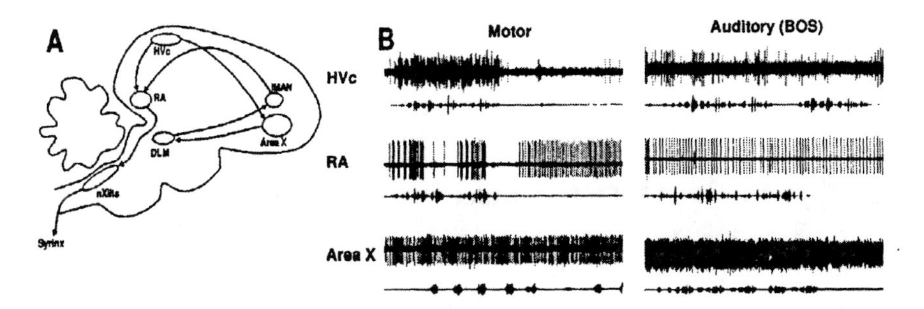

FIGURE 6.1
A. Schematic of the bird song system. The zebra finch brain is roughly 1.5 × 1.5 × 1.0 cm in dimension. B. Single unit recordings from HVc, RA, and Area X during singing, or while bird is presented with playback of his own song. Oscillograms of acoustics (stimuli/song) are below those of neuronal activity. Each trace spans 5 s, except Area X motor, which spans 1.5 s. Recordings are from different recording sites from 4 zebra finches.

reported an improvement on this approach, using a mechanical microdrive to advance standard Pt/Ir electrodes commonly used for single (individual) unit recordings in songbirds. In this case birds would sing when attached via a flexible wire to a commutator which passed the electrical signals outside the cage. The ability to move the electrodes permitted achieving the high S/N required for isolation of single units, and it was empirically observed that a differential recording configuration provided sufficient relief from movement artifacts, even between electrodes spaced several hundred microns apart. This approach was very difficult, and the success rate was very low. The difficulty of this approach is highlighted by the fact that although considerable effort was devoted to acquiring such interesting data, McCasland[3] only reported on unit recordings from a single bird.

Recently, Yu and Margoliash[4] were able to achieve considerable success in recording single units from singing birds. In different experiments, Yu and Margoliash[4] employed both fixed electrodes (bundles of fine wires) and a mechanical microdrive to move single-unit electrodes, and found much greater success with the latter approach. This chapter describes the techniques of Yu and Margoliash[4] and subsequent improvements, and the experience with these chronic recordings in our laboratory. The initial section describes the preparation of two custom-built devices, the mechanical microdrive used to drive the electrodes and the commutator that permits the bird freedom to move within the cage while being attached to a flexible cable. The subsequent sections are organized roughly along the lines of conducting an experiment: preparing the microdrive and electrodes, implanting the device, collecting data, and data analysis. Where appropriate, we also describe modifications of techniques that are required when recording from juvenile birds. Examples of the quality of recordings that can be achieved are shown in Figure 6.1B. These procedures are still very difficult to deploy, but nevertheless, recently there has been a resurgence of interest in chronic recordings in singing birds.[5-7] We hope that this chapter provides useful directions for laboratories embarking on such projects.

6.2 Mechanical Devices

6.2.1 Microdrive

Central to our approach is a lightweight mechanical microdrive that moves one or more electrodes through the brain when a central drive screw is turned. Starting with the general concept described by McCasland,[3] we developed a prototype device. Experience with that prototype resulted in a second-generation device, of which to date six have been manufactured and which is the device we describe here. The main parts and assembled microdrive are pictured in Figure 6.2A and schematized in Figure 6.2B. We have had approximately three years' experience with these six devices, and roughly four years with the prototype. The devices were developed from concept by one of us (JJG), an experienced machinist. The fine holes and tight tolerances required in the manufacturing process makes access to a professional machinist highly desirable when undertaking this project.

The body of the microdrive is made from 2024 T6 alloy aluminum rod of 3/16-inch diameter (Figure 6.2A). A central hole is drilled 0.345 inch in depth using a #31 drill. The side slots are cut to a depth of 0.300 inch using a slitting saw, 0.035 inch face width. The overall length of the body is 0.375 inch. The top is drilled and then tapped for a #000-120 machine screw (W. M. Berg, Inc., Part #Y7-S000-A5). The fine threads result in a movement of approximately 200 μm for each complete rotation of the drive screw. With practice, this system allows the experimenter to move electrodes probably as little as 2 to 5 μm, which is sufficient resolution to isolate single units. During operation, a locking nut fitted to the drive screw allows the screw to be locked into place. In addition, the microdrive body has four holes (two near the bottom and two near the top on each side) also tapped for #000-120 machine screws. These allow side-entry screws to securely lock the electrode holder. Originally we anticipated the side screws would be required during recordings to reduce movement artifact, but in practice we have not found that necessary. We use the side screws only to secure the electrode holder when loading the electrodes (see below), and to secure the microdrive body during implantation.

The electrode holder is made from white Delrin®* (McMaster Carr) (Figure 6.2B). The four wings are cut on a vertical milling machine using an index table that can rotate from a vertical to a horizontal position. The electrode holder wings are machined to a thickness of 0.034 inch, the outside overall width is 0.200 inch, the wing width is 0.085 inch, the length is 0.115 inch. A hole is drilled and tapped into the edge of each wing for the screw that secures and makes electrical contact with the electrode. Because the wings are thin compared to the electrode-securing screw, the Delrin tends to spread when machined. This necessitates the use of a very small parallel clamp to prevent deformation of the wing during this operation. The machine screw to secure the electrode is manufactured by Bestfet-Bergan (Switzerland). It is a #4 screw with thread diameter of 0.60 mm, length 2.70 mm, and 169.33 threads per inch. Four holes through which the electrodes will pass are drilled with a

* Registered trademark of E.I. du Pont de Nemours and Company, Inc., Wilmington, Delaware.

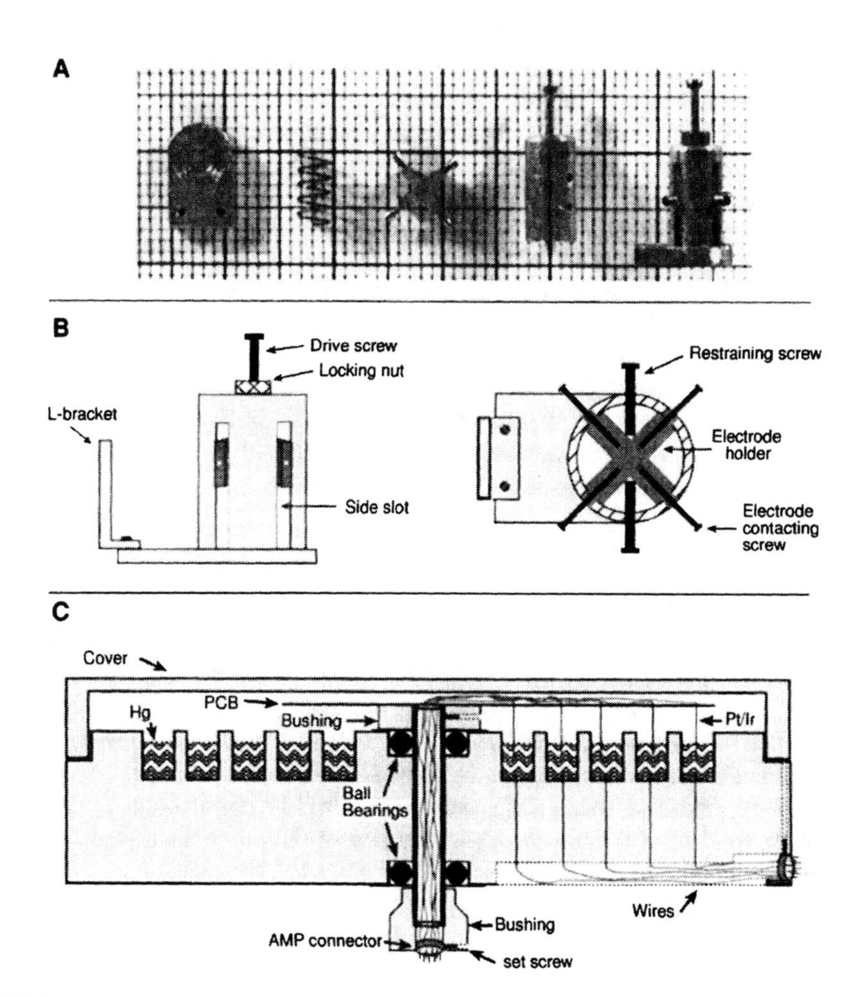

FIGURE 6.2
A. The 4 major components of the microdrive (base, spring, electrode-holder, and body, and a completed microdrive to the right. Small divisions are in mm. B. Schematic of microdrive, with attached L-bracket. C. Schematic of mercury-filled 5 channel commutator.

0.004-inch bit lengthwise along the wings 0.020 inch from the center of the electrode holder. After the hole is drilled it needs to be cleared by hand with the drill bit held in a pin vise. Note that this is a deep hole for such a fine diameter, which makes it difficult to machine. The hole spacings we use place the four electrodes at the corners of a square but other electrode spacings would be possible. In our application, it is possible to place all four electrodes in some of the structures we target (e.g., HVc, area X; see Figure 6.1A). By varying the relative depth of electrodes, it is also possible to place two electrodes in one structure and two in another (e.g., HVc and RA).

Prior to assembly, an interior open coil (compression) spring (Figure 6.2A) is placed between the base plate and electrode holder. The compression spring maintains pressure between the electrode holder and the drive screw. The spring is made from 0.008-inch spring wire, and has five open coils of 0.120-inch diameter, with an overall length of approximately 0.260 inch.

The base plate (3/8 inch × 1/4 inch × 1/16 inch, Figure 6.2A,B) has a semicircular end with a recess to hold the microdrive body, and a rectangular end with two holes drilled and tapped for #000-120 screws. To attach the microdrive body to the base plate, a pin point is used to pick up a small amount of epoxy. The epoxy is placed at four points on the bottom end of the body, which is then set into the recess of the base and clamped until set. Proper orientation of the body and base is essential for smooth operation. In general, some hand fitting of parts will be necessary to achieve optimal performance. Typically, a microdrive assembly can be reused multiple times, but eventually the epoxy breaks. If so, don't panic! After collecting all the parts, the microdrive can be reassembled. The screws of the base plate are used to secure a lightweight L-shaped bracket (approximately 0.017 inch) that holds the electronics and connector (Figure 6.2B), as described below.

6.2.2 Commutator

Ultimately, electrical signals from the bird need to be transmitted outside the cage. One attractive approach is the use of radio transmitters. Even the smallest of these, however, adds weight, limits the number of channels (one per transmitter), and requires that the bird carry a power source (battery), presumably on its back. When mounted on the head, we also found the simple, lightweight transmitters we tested unexpectedly exhibited considerable and erratic drift in transmission frequency. Eventually we determined that the high body heat of the zebra finch (42.5°C) was heating the components of the transmitter's tank circuit, hence changing its resonant properties, with the degree of heating dependent on whether the bird was stationary or active. This problem could be solved by potting the transmitter in foam (Great Stuff Sealant foam, Insta-Foam Products, Inc.), which achieves a hard surface when it cures while adding negligible weight. Nevertheless, in light of all the other problems we abandoned the transmitter approach.

In the current approach, signals from the bird are transmitted to the outside via a wire cable. This necessitates the use of a commutator. We use a custom-built, mercury-filled, five-channel commutator (Figure 6.2C). The body of the commutator is formed from Plexiglas and has five concentric 1/4-inch circular troughs. In the center, a precision-ground hollow 1/4-inch stainless steel shaft rotates on ball bearings. A fitting with a set screw is attached to the top end of the shaft, where it protrudes above the troughs. Epoxied to that fitting is a thin arm made from printed circuit board material that extends over all the troughs. Another fitting is pressed onto the bottom end of the shaft. Into a recess of that fitting is placed a circular connector (#8059-2G7 Augat transistor socket), which is fixed in place with a set screw. Five wires running inside the shaft connect the circular connector to 0.005 inch Pt/Ir wires that extend downward from the overhead arm into the troughs. Five other stationary Pt/Ir wires extend upward through holes drilled from a recess in the base into the troughs. The stationary wires are glued in the recess with dental cement to prevent leakage of mercury. When the troughs are filled with mercury, this system

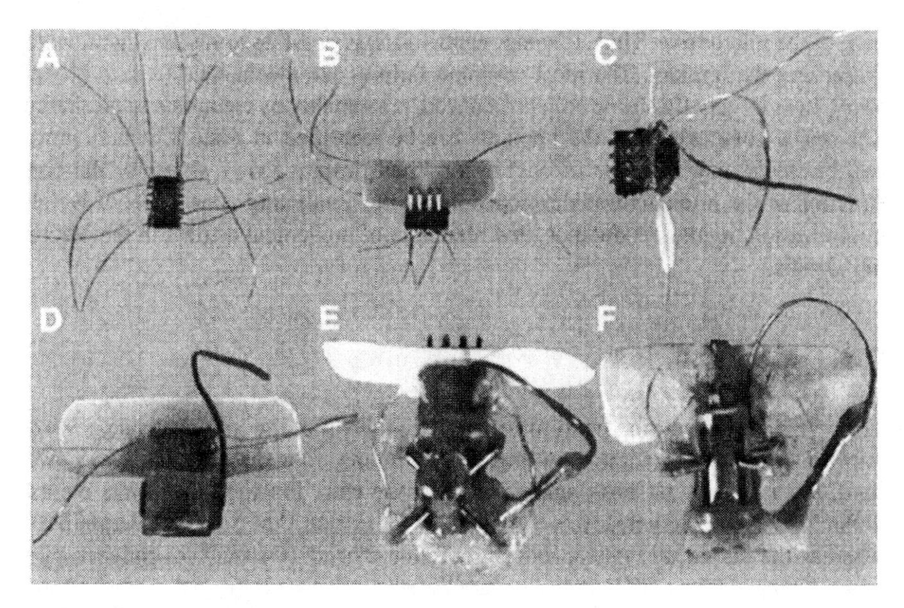

FIGURE 6.3
The implant. A. Quad channel op amp, with gold wires attached (bottom) to inputs, copper wires attached (top) to outputs and power. B. Plastic T-shaped piece and dual-row strip connector attached to op amp. C. Strip connector is wired to op amp and to a ground connection (bottom right). D. Implant is potted and attached to L-bracket. E&F. Two views of completed implant (taken post-mortem, with some of the skull and dental cement from implantation surgery still attached).

achieves continuous electrical contact for all five channels, while exhibiting sufficiently low friction and torque that zebra finches can easily activate it. Note that since mercury has a low vapor pressure and the vapor is poisonous, whenever the commutator is filled with mercury it is fitted with a tightly fitting cap, also made of Plexiglas. Filling and emptying the commutator should only be done under a hood. Commutators with similar functionality are available from commercial vendors, but we do not have experience with these.

6.3 Construction of the Implant

The microdrive loaded with electrodes forms the centerpiece of the implant that will be surgically attached to the bird. The other components include the L-bracket onto which components are glued, a low-power, four-channel op amp in a surface mount package (TLC25L4BC, Texas Instruments), a dual-row strip connector (which is connected during recording sessions to the commutator via a flexible cable), along with various wires and connectors (see Figure 6.3). In addition, a plastic shield surrounds all sides of the microdrive and ground wire, to protect the delicate wires and microdrive from becoming tangled with the bird's legs if it tries to scratch, and as protection from mechanical damage during collisions with cage walls, food and water containers, etc. The L-bracket is secured by two #000-120 screws onto the base of the microdrive. Thus, if components fail, it is possible to replace them, while preserving the implant. The most common failures are mechanical (e.g., a broken wire). Less frequently, the op amp is damaged, presumably by electrostatic discharge. The major components of the implants can be identified in Plate 3, which shows two birds carrying implants of slightly different design. Every aspect of the construction of the implants was developed under the conflicting constraints of weight minimization, implant endurance, and reduction of movement artifact in the recording signals.

6.3.1 Construction of the Electronics

The op amp is the starting point for construction. Each of the four channels is wired into follower configuration, resulting in four inputs, four outputs, and two power lines, for a total of 10 wires emanating from the chip. For the input wires we use Teflon®*-coated gold wires (0.003" bare, 0.0045" coated, AM Systems, Part #7510), whereas the output and power wires are Teflon-coated 10-strand 50 gauge copper wires (Cooner Wire, #CZ1103). To stabilize the power supply and protect the op amp a 1.5 pF ceramic chip capacitor is soldered between the positive and negative power pins. The op amp will have a "top" and "bottom" when part of the implant: channels #1 and 4 will be on the bottom, and channels #2 and 3 will be on the top. The input (gold) wires will emerge from the implant near the top, and the output and power input wires from the bottom, where they will eventually be attached to the strip connector. The attachment of wires, the lengths to which they are cut, and the orientation and path of the wires from the connection point are all prescribed so as to produce the spidery contraption shown in Figure 6.3A.

The four output (stranded copper) wires are attached first. Each is first stripped of enough insulation to allow soldering it to the output and noninverting input (+) pins of an op-amp channel, thus wiring the channel into follower configuration. The remaining wire will exit the op amp toward the bottom, for attachment to the strip connector. The power wires are attached next, also coming out near the bottom, although they exit between the bottom two pins of the op amp. Next, the four gold wires are carefully stripped of a small amount of insulation using a sharp blade under magnification, and soldered without pre-tinning to the inverting input (–) pins, using low-temperature solder. A temperature-regulated soldering station facilitates this procedure. As the gold wires are very delicate, it is arranged that they travel the length of the op amp and emerge on the opposite side of the op amp from the channel they supply. This provides increased length of wire to be potted with dental cement to the op amp, and decreases possible tension at the solder joint. Finally, the chip capacitor, with two very short lengths of stranded copper wires soldered to it, is cemented to the op amp, and the wires attached to the power supply pins (being careful not to detach the wires already soldered to these pins). At this point, acrylic dental cement is used to pot the entire package, focusing especially on protecting the solder joints at the op amp pins by embedding the emerging lengths of wires,

* Registered trademark of E.I. du Pont de Nemours and Company, Inc., Wilmington, Delaware.

and embedding the gold wires and firmly fixing the capacitor into place. However, the dental cement is used sparingly in achieving these aims. The amount of dental cement used for potting purposes constitutes the bulk of the variation in weight of the implants.

It should be noted that when using dental cement for potting purposes, the resulting package can be made much stronger (hence requiring less cement) and smoother by painting the drying surface with cyanoacrylate glue (e.g., Krazyglue®*, Borden, Inc.). This catalyzes the polymerization process of the cement. The resultant smooth surface is also translucent, allowing internal connections to be inspected, etc. The smoothness of the surface reduces the chance that the relatively thin layer of potting material will catch an edge and become detached. The mechanical properties of the dental cement-cyanoacrylate combination are attractive, and we use this material during both the construction of the implant and during the surgical attachment of the implant onto the skull.

At this point in the construction, the beginnings of the plastic shield are attached. A thin (0.35 mm) plastic piece is cut in the shape of a T, and attached with gel cyanoacrylate (Duro QuickGel, Loctite Corp.) to the bottom of the op amp, where the capacitor is potted with dental cement. Note this leaves a gap between the other end of the op amp (toward the top of the implant where the gold wires emerge) and the plastic T, which provides a space to prevent the plastic T from touching the gold wires. The arms of the T should protect the emerging gold wires from behind, while one leg of the T should be long and thin enough to allow for complete embedding onto the implant once the strip connector is attached to it and dental cement applied to the sides (Figure 6.3B). This back plate of the plastic shield is all that will be prepared until the last step in the surgery, when the sides and front of the shield will be attached to this back plate.

An 8-pin dual row strip connector is prepared from long single row strips (GM-12, GF-12, Microtech, Inc). The male strip connector is attached to the implant. (The female counterpart, which is more susceptible to contamination by dust, seed husks, etc., is used in the cable that will connect the bird to the commutator.) The strip connector is attached to the back of the plastic T near the bottom of the assembly. Care must be taken to ensure that it is properly centered and aligned, and that it does not protrude too far below the base of the op amp, or not close enough to this base. This is achieved by adjusting the position of the strip connector as the glue is drying. We have found that the gel cyanoacrylate glue works well for this task. Once the wires emerging from the bottom of the assembly have been soldered to the connector, the entire package will be potted with dental cement, so a strong attachment by the cyanoacrylate glue is not necessary at this time. We have also found that having the strip connector emerge at an angle to the plastic backing plate eases the connection and disconnection of the cable.

The wires emerging from the assembly must be shortened, stripped, tinned, and soldered to the tinned solder cups of the strip connector pins. The remaining wire should be folded into the available spaces, so as to protrude as little as possible. Finally, an insulated, stainless steel wire is soldered to one of the solder cups and

* Registered trademark of Borden, Inc., Columbus, Ohio.

bent to emerge from one side of the implant (Figure 6.3C). The other end of this wire will be attached to a female connector, to be connected to the ground wire (described below with the electrodes). In the final configuration, 7 of the 8 pins are in use: 4 output lines, 2 power supply lines, and 1 ground line. The implant now has 4 gold wire arms, in addition to the protruding stainless steel wire (Figure 6.3D). The solder joints and sides of the strip connector are then potted to the rest of the implant with dental cement, with the leg of the plastic T backing plate, the capacitor below it, and the stainless steel wire emerging along one side completely embedded.

The connectors to be attached to the gold wire are prepared by shortening female connectors (A-M Systems, #5201) with a diamond disk cutting wheel (Stoelting, #58605) attached to a rotary tool. These particular connectors were chosen for their snug fit onto the electrode contacting screws of the microdrive, because the gold wires and connectors constitute the parts requiring the greatest care if movement artifacts are to be avoided. The connectors must fit on the screws easily enough to be attached with little force, but firmly enough to be well-coupled to the bird's movements. The gold wire fulfills two objectives. First, the fineness of the relatively fine wire appears to prevent introduction of movement artifact. This obtains only if the wire does not touch anything between its two points of attachment. Second, the wire must be packed against the microdrive at the end of each recording session so as to avoid catching the bird's feet, and unpacked prior to the start of recordings. Gold is highly ductile, and the only solid metal wire of this fineness we found that does not break upon being manipulated across multiple recording sessions.

A similar female connector is attached to the free end of the stainless steel ground wire; the male counterpart pin will be soldered to an uninsulated electrode which will be implanted into the brain in the opposite hemisphere from the recording electrodes. Each of the connector attachments is protected with cyanoacrylate-hardened dental cement. The joints between gold wire and connectors, as well as the exit points of the gold wires from the TL25L4C package, are protected with a mass of Silastic®* medical adhesive (Dow Corning, Silicone Type A). Finally, the L-bracket is cemented onto the back of the op amp (as shown in Figure 6.3D). The resulting package is screwed onto the base of a microdrive. The connectors will be attached to the electrode contacting screws only after the electrodes are loaded.

6.3.2 Loading Electrodes into the Microdrive

For electrodes, we have used 0.003" platinum–iridium wire (A-M Systems, Inc., #7676) etched to a 10 to 15 μm bullet-shaped tip by repetitive dipping while applying 5 to 10 VAC. Insulation is applied by passing the upright etched wire through a platinum wire forge (2 to 3 VAC) that melts solder glass (Corning glass #7570). The insulation is removed at the tip by capacitive discharge. Note that it should be possible to adapt other electrodes, including multitrodes, to the microdrive mechanism.

The electrodes are prepared longer than necessary, but with the amount of insulation tailored to the desired target. The electrode contacting screws and drive

* Registered trademark of Dow Corning, Inc., Midland, Michigan.

screw are then withdrawn, and the microdrive's electrode holder is lowered as far as it will go and locked in place via the side-entry locking screws. The electrodes are then shortened to the proper length, and visibly guided, tail-end first, into the microdrive from the bottom. (We use a simple block of aluminum drilled for the top of the microdrive body with a set screw to secure the microdrive upside-down.) Each electrode is gently pushed into its shaft until it protrudes from the top end (viewed through the side slits of the microdrive body). Through the side slits, the electrode is pushed with small forceps until it is flush with the top of the drive, then the bare wire of the electrode shaft that protrudes from the bottom of the drive is grabbed with forceps to move the electrodes forward a bit, eventually resulting in all four electrodes at the correct relative position (level or staggered, depending on the experimental design). This complex procedure is necessary to assure the drive screw does not contact the top of the protruding electrodes, thereby causing an electrical short circuit. In addition, when loading a new microdrive the electrode shafts may become blocked with Delrin detritus, making it difficult to insert the electrode or even irrevocably damaging the electrode by bending it. Therefore, we typically clear the shaft of any debris by running either a drill bit or bare Pt-Ir wire down each shaft by hand before loading the electrodes.

Once all electrodes are in place, the electrode contacting screws are tightened. This procedure is also delicate, as too much tightening will break the electrode, and too little fails to secure it. The latter condition will result either in poor recordings with excessive movement artifacts or in an inability to drive the electrode. The forces exerted by the tightening and rotating of the contacting screw against the electrode shaft, however, result in visible movement of the electrode tip. These movements provide diagnostic information that can be interpreted with respect to the tightness of the contact. In practice, we have found that the contact between screw and electrode is secure when the electrode tip stops moving outward and begins to move in the direction of rotation. This amount of force will visibly crimp the electrode. Since it is possible that a debris particle lies between the contacting screw and the electrode shaft, preventing electrical contact, we test the resistance of this connection, and repeat the procedure if necessary.

At this point, the electrodes are visibly splayed outward. They are now carefully straightened with forceps, until they are parallel with each other and the microdrive. This procedure provides an opportunity to verify that the electrodes are firmly secured — any rotation during the straightening procedure should be impossible if the contact is secure. Finally, the resistance of the contact is checked again. The electrode holder may then be raised by removal of the side locking screw, and each electrode may be connected to an op-amp channel by gently pushing the connectors attached to the gold wire input lines onto the heads of the electrode contacting screws. This is a delicate procedure, especially given that the electrode tips are protruding below the base of the microdrive. As well, noticeable force is usually required, but nevertheless one must not torque the joint between the microdrive and its base. Once all connections are achieved, the implant is ready to be implanted. A completed assembly typically weighs approximately 1 g. This weight is increased by the dental cement used during implantation, and by the effective weight of the cable supported at one end by the commutator and the other end by the bird. We estimate the overall weight experienced by the zebra finch during recording sessions to be no more than 2 g.

6.4 Implantation Surgery

The implantation procedure begins with the selection of a subject. Obviously, the larger and stronger zebra finches, and those more inclined to sing, make for better subjects. Juvenile birds are smaller than adults; however, in some ways they are even better than adults in tolerating the weight of the implant. This is because the neck muscles of juveniles can grow quite large after only several days of carrying extra weight on their heads. In fact, we have occasionally prepared juveniles for the implant by carefully gluing a small screw with Flexible Collodion (J. T. Baker, Inc, #9204-04) upside-down to the feathers on top of the skull. We add one or more nuts to gradually increase the weight felt by the bird (from 0.3 g to 1.5 g). After several months of carrying an implant, even adults tend to become quite big and strong. Nonetheless, we stress that the preselection of appropriate recording subjects can greatly facilitate success.

There are many differences between the implantation surgery for juvenile and adult zebra finches, and it is likely these differences apply to most songbird species. Adult birds have a double-layered skull with the two layers separated by bony spicules, or trabeculae. This greatly facilitates the attachment of the implant to the skull, as dental cement can be flowed into the spaces between the two layers, providing a very strong attachment when it hardens. Juvenile birds have a single-layer skull, which is comparatively soft and wet. Songbirds have sagittal and horizontal sinuses separating the hemispheres and the cerebellum. It is above these sinuses that the double layer begins to develop in juveniles, first over the cerebellum, then spreading rostral and caudal. Thus, beyond approximately 50 days of age, even juveniles have some double-layer skull into which dental cement can be embedded. In many species, including the zebra finch, aspects of song development (e.g., sensory acquisition) occur at even earlier stages in life.[8,9] We have had some success implanting birds younger than 50 days of age, but those techniques are not yet reliable.

The implantation surgery requires the use of anesthesia lasting at least two hours. For this purpose, we have employed Equithesin, a chloral hydrate/sodium pentobarbital mixture, with 50 to 60 μL injected I.M. into the pectoral muscle. The scalp is plucked of feathers, cleaned with 1% iodine solution (EZ-Prep, Benton Dickinson, #372931), and prepared with thin layers of lidocaine and bacitracin. A longitudinal midline incision and two diagonal incisions (midline to caudolateral) are made so as to produce two triangular flaps of skin while avoiding cutting the large cutaneous blood vessels that enter these flaps caudally. The exposed skull is covered by a periosteal layer that provides the attachment for the occipital muscle just beyond the caudal edge of the field of exposure. The point at which the sagittal and horizontal sinuses meet (Y-zero) is used as a fiduciary stereotaxic point. Excessive tearing of the periosteum, as well as exposure of the occipital muscle, may reduce the bird's ability to hold his head up.

A 2 × 2-mm opening is made in the top layer of the skull, over the targeted region. Dental cement with a runny consistency should be applied to the edges of the large opening, in such a way that it is wicked several mm in each direction into the double-layer skull surrounding the opening, but without covering the exposed bottom layer. An opening is now made in this exposed lower layer of the large opening, rendering the dura and brain visible. It is extremely important to avoid bleeding or to stop any dural bleeding that may occur, as this blood may clot around the electrodes, preventing their movement. To prevent the dura from drying and hardening we apply a small amount of high-viscosity silicone (derived from mixing various viscosities of Dow Corning 200 fluid: dimethylpolysiloxane) to cover the exposed dura. The size of the exposure should allow at least 0.5 mm between any of the electrodes and the bone. Once the exposed dura is covered with silicone, a small opening is made in the top layer of the opposite hemisphere, approximately 1 mm × 1 mm, along with a smaller opening in the lower layer of the skull. The ground electrode is delicately lowered into the brain through this hole, held by forceps gripping the male pin to which the electrode is soldered. The ground wire must be long enough to temporarily anchor this pin, requiring at least 2 mm within the brain. Dental cement is gently applied to the base of the pin, and allowed to flow between the double layer, cementing the ground pin firmly in place.

Now attention can be returned to the microdrive assembly. The assembly is secured by attaching a hollow plastic tube that fits over the top of the microdrive, with slits for the electrode contacting screws, and locked in place with a set screw. The microdrive is lowered until the electrodes are just above the surface of the brain. A custom-built screwdriver that snugly fits through the plastic tube is used to turn the electrode drive screw, allowing the electrodes to be advanced and retracted within the microdrive. We have found it very useful to monitor the signals from the electrodes during the implantation procedure, allowing for the identification of structures based on electrophysiological characteristics. (One may also identify specific target locations with a traditional single-electrode approach, then use this information to orient the microdrive, but this may increase the duration of the surgery.) This procedure also requires a cable to be attached to the implant's strip connectors, and removal of this connector upon completion of the surgery must be delicately performed, as the dental cement mounting the implant to the skull may not have completely hardened by that time.

We find that it is necessary to be able to rotate the entire platform holding the zebra finch to overcome the visual obstruction created by the base of the microdrive, thereby permitting visual observation of the electrodes entering the brain from various angles. Furthermore, viewing from the front and the side facilitates confirmation of the medio-lateral and antero-posterior coordinates of the electrodes, judged relative to small marks that we make in the skull anterior and lateral to the targeted area. This crude stereotaxy has served sufficiently well for many of the large and/or superficial structures we have targeted, but obviously the procedure for stereotaxic implantation of the implants can be improved.

Once the position of the implant is confirmed, it is lowered until the base makes contact with the skull, and is cemented onto the skull. Dental cement is placed around the outer lips of the base so as to contact the cement layer already prepared on the skull. Dental cement should be used judiciously to minimize the weight of the implant. Nevertheless, we typically join the dental cement around the ground pin and essentially completely cover the exposed region of the skull with at least a 0.5 mm layer of dental cement, hardened with cyanoacrylate (Figure 6.3E,F). This can produce a very hard implant, which is important for longevity.

With the implantation completed, the tube securing the microdrive is removed, and construction of the plastic shield surrounding the implant is completed by gluing the side and front panels to the back panel (the arms of the plastic T piece). After allowing several minutes for the dental cement to dry, the cable attached to the strip connector is delicately removed, and the bird is removed from the surgical apparatus and allowed to recover. We find that within a few hours, immediately after a bird recovers from anesthesia, he is usually able to hold his head up straight against the weight of the implant. In some cases, the head may hang to one side at this time, but after a day or two of recovery, the bird is again able to hold his head upright.

6.5 Recording Procedure

Recordings only begin after a bird has had several days to recover from the anesthesia and implantation surgery. During this time, and from this point on, the environment of the bird must be carefully controlled to avoid the possibility of the bird catching or knocking the implant on perches, etc. Perches should ideally be placed very low to the ground, so that the bird will not attempt to go under them. The bird should also be discouraged from jumping on the sides of the walls by arranging the perches or covering a favored wall with a paper towel to remove footholds, not because this behavior poses a particular threat to the implant, but because during later recordings, this increases the possibility of the bird getting tangled in the cable or introducing movement artifacts.

6.5.1 Connecting Cable and Signal Conditioning Electronics

Recording begins with connecting the zebra finch's implant to signal amplifiers providing voltage gain, via a cable connected to the commutator. The cable joining the bird to the commutator is constructed by running several of the 10-strand, 50 gauge copper wires through Silastic medical-grade tubing (I.D. .062", O.D. .095", Dow Corning, #602-265). Although 7 channels (4 electrodes, ± power, and ground) must run through the cable, we have found it helpful to have 10 wires, with the positive, negative, and ground connections using two wires in parallel. This provides fault tolerance for these critical connections. The cable itself plugs into a battery pack, which plugs into the bottom of the commutator. The battery pack, for powering the op amp, employs two 3 V calculator batteries (Panasonic, CR2016) in dual-sided configuration. The midpoint between the batteries is connected to the ground pin of the bird. The cable exits from the battery pack at a right angle, therefore off-axis from the commutator, allowing forces exerted by lateral movements of the bird (e.g., perch hopping) to produce torque at the commutator instead of simply twisting the cable. It is important for the cable exiting the commutator to remain horizontal for a distance of at least several centimeters, and for the joint between the cable and battery pack to carry some of the weight of the cable. We accomplish this by stiffening this joint with Silastic medical adhesive (Dow Corning, Silicone Type A). The end of the cable to be connected to the bird's implant has a dual-row strip connector (GF-12, Microtech Inc.). The cable should extend from this connector at an angle such that when attached to the bird, and with the bird in a relaxed, perched position, the cable will point upward. The length of the cable should allow the bird free exploration of the cage, but not be so long as to form loops or be accessible to the bird's beak or feet when the bird is directly under the commutator. These combined restrictions restrict the size of the bird's recording environment to a maximum of approximately 32 × 32 cm, assuming a height of 20 cm.

The signals from the commutator include ground and four electrode signals. We almost always record the electrode signals differentially, but which electrodes make the best pairs can only be determined experimentally. Thus, it is very useful to be able to quickly check each channel and to try different pairings while the bird is still in hand. We accomplish this with a simple switch box. The signals from differential amplifiers in the switch box, together with that from a microphone inside the sound isolation booth, are connected to monitors, speakers, oscilloscopes, and to data acquisition hardware on a computer. Occasionally, we have also used a digital video camera in the sound box to broadcast images of the bird over an intranet. This permits remote monitoring of a bird in cases of long-duration recording sessions.

6.5.2 Preparation for Recording

When recording from freely-moving animals, there can be many issues that arise which affect the subsequent interpretation of the data. We cannot overly stress the importance of maintaining a detailed and accurate account of the conduct of each experiment. Whatever scheme the investigator adopts, this should be well thought out in advance and fully prepared prior to each recording session.

Before each recording session, the batteries are checked with a multimeter, the battery pack and cable are attached to the commutator, the low-torque rotation of the commutator is verified, and analog equipment and computer data acquisition is tested. The recording session begins with the transport of the zebra finch from his home cage in a sound attenuation booth to the commutator-equipped recording cage in another attenuation booth (Industrial Acoustics Corp., AC-1). The recording cage is electrically shielded with copper wire mesh. With the bird still in hand, the cable is carefully connected to the implant. The locking nut on the drive screw is unlocked with forceps, and an insulated screwdriver is used to advance or retract the electrodes. With the cable attached, the signals from the electrodes can be monitored on a nearby oscilloscope and audio monitor. By pairing the various wires against the ground channel, or against each other, an electrode with neuronal activity can be identified, and isolation of single units on that electrode can be accomplished. Although the electrodes can be moved a total of 2.8 mm in depth, it is advisable not to move them more than a few hundred microns per day.* Depth is recorded by noting the number of turns of the drive screw. Be patient! The period of time the bird is held captive while electrodes are advanced can range from 1 to 10 min.

Once suitable neuronal signals are obtained on one or more of the electrodes, and the desired electrode pairings chosen, the locking nut is gently tightened, and the bird is prepared for release. Care must be taken upon release of the bird to minimize violent jumping and hopping, which could jeopardize the neuronal isolation. We have found that individual birds react in predictable but different manners to release, and that this reaction can depend on how long they were held, etc. One useful trick is to release the bird in the dark on the cage floor near the middle of the cage, while slowly and quietly removing the hand. After several seconds to a minute, the lights can be turned on, usually with little associated jumping. Assuming that the signal quality is still acceptable, data collection can begin. Otherwise, the bird must be restrained again, and the electrodes once again moved, and the entire process repeated.

6.5.3 Data Collection

For most of our experiments to date, data collection has consisted of recording the neuronal activity from one or two pairs of the electrodes during identified behavioral conditions of the freely behaving animal. There are two general approaches to data collection: continuous data collection with simultaneous recording and categorization of the animal's behavior, and data collection of segments of activity during known behaviors. For the design of our experiments we use the latter option. We typically record neuronal activity in three different conditions: when the bird is singing, when the bird is quiet but listening to an auditory stimulus of our choosing, and when the bird is doing neither (spontaneous activity of these neurons).

Data collection is strongly affected by the unpredictability of an animal's reactions — patience is recommended here as well. For example, some birds are frequently stimulated to sing by many of the auditory stimuli (such as conspecific songs) we present to characterize auditory responses. Collection of spontaneous activity, which will provide a baseline reference of the recording site activity, must occur within a known behavioral context (for example, while the bird is quiescent and resting but alert, as opposed to during eating, beak wiping, other active movements, etc.). Thus, decisions regarding stimulus choice and whether or not to record neuronal signals at a given time must be made on-line by the investigator, and these must be documented. Nonetheless, some automation is possible: for example, our data acquisition software automatically identifies vocalizations and records the neuronal activity occurring before, during, and after the vocalizations.

* It is very important to never allow electrodes to retract from the brain. A thick tissue regrowth often forms on the surface of the brain after an implant. This can actually help stabilize the recordings, but if electrodes are withdrawn, they do not find the same entry points and are damaged trying to pierce the regrowth tissue. It is for this reason that the x-y stage of McCasland[3] proved of limited value.

In our experiments, we have been primarily concerned with characterizing the premotor activity of neurons in the song control system of male zebra finches. In accomplishing this, we have found that a variety of behavioral strategies can be used to elicit singing during chronic recordings. Birds tend to cease spontaneous singing immediately after being physically restrained, as we must do to isolate neurons. However, adult male zebra finches will often respond to the presence of another zebra finch. Thus, we have found it helpful to divide the cage into two compartments with a remotely controlled curtain in between. In the adjoining half-cage, we can place a mirror, a female, or a male conspecific. By raising or lowering the curtain, we can control — to some extent — the onset of the singing and calling behavior of the zebra finch being recorded, and prevent habituation to the singing-inducing stimulus. Typically, a female zebra finch is placed in the adjoining cage. In sexually mature males the presence of the female stimulates singing. The female zebra finch's acoustically simple calls (which can also be recorded in isolation for reference purposes) are comparatively easy to segment and identify from the microphone recordings. Additional strategies to induce singing include the playing of zebra finch calls and songs via a speaker located within the sound isolation booth, or even the investigator's producing similar-sounding calls vocally (which can be surprisingly effective with zebra finches).

In initial sessions, during the entire time that the bird is connected to the commutator, it is essential to visually monitor the bird surreptitiously. To this end, our sound isolation booths are equipped with one-way mirrors mounted in the doors. The investigator is immediately available to respond to the relatively rare emergency situation, such as the bird becoming entangled in the cable or catching his feet in the plastic shield surrounding the implant or in the gold wires. However, if the bird acclimates to this environment (most do after several recording sessions), the use of a video camera and remote monitoring is feasible. This is useful for very long-duration recording sessions such as is envisioned for recordings from juvenile birds.

The end of a recording session is determined largely by the investigator's assessment of the condition of the bird. We have found that most of our birds get visibly tired after 3 to 4 hours of carrying the extra weight of the cable. Signs of fatigue include the bird hanging his head to one side for longer and longer stretches of time, increase in breathing rate, and the beak remaining open for longer than usual durations (indicating panting). The duration of time that a bird can hold the implant, however, varies greatly across birds. The factors that affect the maximal recording duration are the quality of the implantation surgery, the strength and size of the bird, and the number and frequency of prior recording sessions. Recording sessions usually can occur as frequently as every other day or every day, depending on the individual bird. We have found that the size of the bird's neck muscles increases with how long the bird has carried the implant. Younger birds adapt more readily to the weight of the implant and the recording cable; within only several days they may quickly reach an endurance level that adult birds do not reach after months. Such birds exhibit visibly larger muscles than those of these adults. For some juveniles, we have succeeded in keeping the birds attached continuously for 48 h, with no visible sign of fatigue. The ability to monitor activity continuously over long periods of time can be useful for some experimental designs. For example, we wish to record during the sensorimotor stage of song development early in life, with the goal of following changes in neuronal activity during song learning.

Our implants can have surprisingly long lifetimes: in one case, we obtained recordings from an adult over a 12-month period. More typically, however, healing or growth processes cause the attachment of the implant to the skull to weaken over time, and eventually the implant falls off after a few months. (This problem is exacerbated in juveniles.) We have found that useful recordings are almost always possible throughout the period birds are carrying implants. In some cases, over time the quality of the recordings diminishes, probably as the result of excessive tissue damage around the electrode tracts. This can be partially controlled by "parking" electrodes above the target structure at the end of each recording session. At the end of all useful recording sessions, histological examination of the brain can be performed to reveal the location and path of the electrode tracts within the brain.

6.6 Data Analysis

Our recording system includes dual differential AC amplifiers (M. Walsh Electronics, Model μA244/2) each amplifying and filtering the signals from two pairs of electrodes, a microphone (Model #33-2011, Realistic) connected to a custom-built signal conditioning circuit that amplifies and filters the signal, and a PC-based multichannel data acquisition system with a 16-bit, 100 kHz A-D converter (AT-MIO-16X, National Instruments, Inc), with a custom driver and software operating under the Linux operating system. The microphone and two neuronal channels are sampled at a 20 kHz sampling rate. The resulting PCM data is stored to disk, and is also processed in real time to provide scrolling oscillograph, spectrograph, raster/histogram displays, as well as sound. The data can also be transmitted over a local computer network, with these same displays and audio monitoring capabilities, and sounds can be played to the bird from a remote terminal. This facilitates remote monitoring and long-duration recording sessions.

The real-time displays are useful for assessing the quality of recordings and improving on-line decision-making, but the PCM data that is stored to disk is analyzed off-line in a much more rigorous fashion. This has the disadvantage (compared to on-line processing followed by recording of spike times-of-events) that it is labor intensive, and that large amounts of data are recorded to computer disk. Nevertheless, we have found this approach essential. We find that the presence of movement artifacts and multiple units in the signal render on-line sorting of spikes into times-of-events inaccurate for single neuron recordings. Although many recordings are almost completely unaffected by the bird's movements, this is not always the case. Also, even recordings that appear to be single unit and have very high S/N can still be contaminated with a second unit. We have found the SpikeSort program (http://www.cnl.salk.edu/~lewicki)[10] to be very reliable and effective in detecting and sorting spike time-of-events from the raw PCM recordings. SpikeSort uses Bayesian statistical methods to formulate spike waveform models, and analyzes cases of overlapping spikes. Spike sorting is a very active field, and multiple approaches have been described.[11]

Another aspect of the analysis lies in segmenting and labeling the microphone recordings into occurrences of specific vocal behaviors, i.e., calls or syllables and songs. Custom software[12] is employed to assist in this task, making use of the dynamic time-warping (DTW) technique to identify vocalizations based on a library of templates created manually for each bird.

6.7 Future Directions for Improvements

The techniques described here can yield valuable data, but the reader will have appreciated that the procedures are extremely difficult to deploy. For example, even though these techniques are reasonably well established in our laboratory, an individual requires approximately two years to become proficient in them. It would probably take even longer for someone to develop this technology for the first time in another laboratory. It took one of us (ACY) approximately five years for the initial development. Thus, such a project should only be attempted by someone with considerable time to devote to the project, and a compelling need to record in small animals. We note that the difficulty of the techniques implies frequent failure. Even a skilled worker probably only achieves a success rate of 35 to 40%. In this regard, animals such as the zebra finch that can be acquired in quantity from commercial vendors have advantages over wild-caught animals. The project also requires significant resources in analog equipment, computers, and programming. In addition, access to a professional machinist (or proficiency therein) is probably essential.

Significant difficulties are encountered during fabrication of the microdrive. Because of the need for hand-crafting, it is difficult to mass produce. Our design can be improved, of course, probably greatly facilitating the construction. It should also be possible to miniaturize the electronics so that they are integrated directly with the microdrive. This would greatly facilitate construction of the implant and would eliminate a significant component of movement artifacts that we resolve only by using very fine, hard-to-manipulate gold wires. Finally, the commutator could be improved by remotely sensing the animal's position and driving the rotary shaft.

Recording from birds with these implants is complicated by the need for extensive manual manipulation of the bird. This results from the mechanical nature of the microdrive, which is a major limitation. One can envision piezoelectric or other mechanisms that could be activated remotely to move electrodes to their final recording positions. Such an advance would greatly increase the rate of data collection, and would greatly facilitate maintaining single unit recordings over long periods of time.

The limited ability to automatically process vast amounts of data off-line is also a significant limitation. Such problems can only be expected to increase in the future, as electrode counts in extracellular experiments rise. This issue is addressed elsewhere in this book.

Finally, the current system drives all electrodes simultaneously. In most cases this implies that only one site at a time can be recorded with good S/N. A significant improvement would be to be able to move electrodes independently. This would greatly facilitate simultaneous multisite recordings, as are described elsewhere in this book. Possibly the easiest way to achieve this would be to manufacture sufficiently small drives so that a bird could carry two or more simultaneously. Depending on the research question, multiunits may also be acceptable. These are relatively easy to obtain from multiple sites simultaneously.

References

1. Goller, F. and Suthers, R. A., Role of syringeal muscles in gating airflow and sound production in singing brown thrashers, *J. Neurophys.,* 75, 867, 1996.
2. McCasland, J. S. and Konishi, M., Interaction between auditory and motor activities in an avian song control nucleus, *Proc. Natl. Acad. Sci. U.S.A.,* 78, 7815, 1981.
3. McCasland, J. S., Neuronal control of bird song production, *J. Neurosci.,* 7, 23, 1987.
4. Yu, A. C. and Margoliash, D., Temporal hierarchical control of singing in birds, *Science,* 273(5283), 1871, 1996.
5. Plummer, T. K. and Striedter, G. F., Auditory and vocalization related activity in the vocal control system of budgerigars, *Soc. Neurosci. Abstr.,* 23, 244, 1997.
6. Hessler, N. A. and Doupe, A. J., Singing-related neural activity in anterior forebrain nuclei of adult zebra finch, *Soc. Neurosci. Abstr.,* 23, 245, 1997.
7. Schmidt, M. F., Interhemispheric synchronization of vocal premotor activity in songbirds, *Soc. Neurosci. Abstr.,* 23, 244, 1997.
8. Marler, P., A comparative approach to vocal learning: song development in white-crowned sparrows, *J. Comp. Physiol. Psychol.,* 71(2) Part 2, 1, 1970.
9. Price, P. H., Developmental determinants of structure in zebra finch song, *J. Comp. Physiol. Psychol.,* 93, 268, 1979.
10. Lewicki, M. S., Bayesian modeling and classification of neural signals, *Neural Computation,* 6, 1005, 1994.
11. Fee, M. S., Mitra, P. P. and Kleinfeld, D., Automatic sorting of multiple unit neuronal signals in the presence of anisotropic and non-Gaussian variability, *J. Neurosci. Methods,* 69(2), 175, 1996.
12. Anderson, S. E., Dave, A. S. and Margoliash, D., Template-based automatic recognition of birdsong syllables from continuous recordings, *J. Acoust. Soc. Am.,* 100(2.1), 1209, 1996.

Chapter 7

Methods for Simultaneous Multisite Neural Ensemble Recordings in Behaving Primates

Miguel A. L. Nicolelis, Christopher R. Stambaugh, Amy Brisben, and Mark Laubach

Contents

0-8493-3351-2/99/$0 00+$.50

7.1 Introduction

As described in previous chapters of this book, the development of new technologies for recording the extracellular activity of hundreds of single neurons simultaneously is having a profound impact in neuroscience. In this chapter, we describe the methodology used for yet another powerful application of this new experimental paradigm, the investigation of large-scale neuronal interactions in awake primates.

The importance of combining multichannel neurophysiological recordings and behavior studies in primates has been clearly recognized even before methods for single-unit recording in awake monkeys were introduced almost 40 years ago. In fact, the first attempts to obtain chronic simultaneous recordings from large populations of cortical neurons in awake primates were carried out by John C. Lilly in the late 1940s.[1] These experiments, which started in 1949, involved the implantation of multieletrode arrays, with up to 610 elements, on the pia surface of the monkey cortex. By recording simultaneously from up to 25 of these electrodes, Lilly reported the existence of "... traveling waves of activity ... that varied from one cortical area to the next ... and changed according to the animal's state..."[1]

Despite this auspicious start, the widespread acceptance of single neuron physiology combined with demanding technological requirements contributed to slow the development of methods for simultaneous multichannel recordings over the next three decades. This picture started to change rapidly in the late 1970s and early 1980s when microcomputers began to be introduced in neurophysiology labs and a handful of investigators started to develop new paradigms for simultaneous multichannel recordings.[2,3] Although the resurgence of multichannel physiology was primarily due to studies carried out in small mammals (see Reference 4 for a review), the tremendous impact caused by the introduction of these methods soon rekindled the goal of applying this experimental approach to behaving primates. With this goal in mind, in the last two years our laboratory has focused on the adaptation of most of our multichannel recording methods, which were originally developed for experiments involving rats,[5-7] for the investigation of large-scale corticocortical interactions in behaving primates.

Overall, our studies revealed that modern multichannel recording techniques could be used to monitor the activity of large populations of single neurons, distributed across multiple cortical areas, throughout the weeks or months required for primates to learn a new behavior task. These encouraging results suggest to us that chronic multisite neural ensemble recordings will become an important experimental tool for investigating the large-scale neuronal interactions that underlie the generation of high brain functions in primates. In this context, the final paragraph of Lilly's classic paper on primate multichannel recording seems to be as insightful today as it was 40 years ago.

> "... one of the large difficulties in correlating structure, behavior, and CNS activity is the spatial problem of getting enough electrodes, and small enough electrodes, in there with minimal injury. Still another difficulty is the temporal problem of getting enough samples from each electrode per unit of time, over a long enough time, to begin to see what goes on during conditioning or learning, especially when a monkey can learn with one exposure to a situation, as we see repeatedly. As for the problem of the investigator's absorbing the data — if he has adequate recording techniques, he has a lot of time to work on a very short recorded part of a given monkey's life."[1]

The goal of this chapter, therefore, is to provide the reader with a detailed description of some of the technical solutions that have been proposed to solve the main difficulties associated with chronic multichannel recordings in primates. For the purpose of clarity we grouped them in three major categories: (1) selection of electrodes and implantation surgery, (2) instrumentation for simultaneous multichannel recordings during behavioral tasks, and (3) design and implementation of data analysis techniques.

7.2 Electrodes and Surgical Procedures

7.2.1 Choosing Electrodes

The first fundamental decision one needs to make in order to perform multichannel neural ensemble recordings is choosing among the large variety of multielectrode configurations that have been described in the literature (see Chapter 1). Although there is no clear rule of thumb to guide this decision, there is already enough published information to help one through this process. In the particular example described here, the decision to employ microwire arrays (NBLabs, Denison, TX, see Plate 1 and Figure 2.3 in Chapter 2) for our primate recordings was guided by the extensive experience gained in several years of experimentation using these electrodes to record from populations of cortical and subcortical neurons in rats.[7] More recently, the same microwire arrays have allowed us to obtain simultaneous recordings from up to 135 single neurons, distributed across different somatosensory and motor cortical areas of owl monkeys, for periods of time ranging from 6 to 22 months.[8]

In its most common configuration, each of these microelectrode arrays is composed of 16, 50-μm-diameter Teflon®-coated stainless steel microwires, distributed in two rows of eight wires each. The distance between the microwire rows varies from 0.5 to 1 mm, while the distance between pairs of microwires in a row ranges from 100 to 300 μm. To make these arrays, the microwires are first glued together using epoxy. A coating of polyethylene glycol is then applied to provide further support to the matrix. If necessary, this final coating can be dissolved away with saline as the arrays enter the brain.

7.2.2 Surgical Procedures for Chronic Implantation of Microwire Arrays

Once a particular type of electrode is selected, the next step is to design a surgical procedure for electrode implantation. In our experience, the quality of the surgical procedure plays a decisive role in the outcome of recordings from chronically implanted electrodes. Fortunately, it is much easier to implant chronic electrodes in the cortex of owl or squirrel monkeys than in the rat cortex. This observation should not be taken as a suggestion that the surgical procedures required for performing a successful chronic implant in primates are trivial or not challenging. On the contrary, deficiencies in surgical procedures likely contributed to most of the difficulties experienced by several investigators who have tried to carry out chronic neural ensemble recordings in primates. In a demanding surgical procedure such as the one required in these experiments, there are no ways to substitute for good planning, experience, and patience. Thus, the first thing one needs to remember is that there is no need to finish the procedure at a record pace. Indeed, in more than 10 years of experience with these procedures, we have learned that the slower the microelectrode implantation surgery the higher the single neuron yield. Careful implantation also seems to increase the longevity and stability of neural ensemble recordings in both rodents and primates.

In our primate surgeries, an initial dose of ketamine (15 mg/kg i.m.) is used to produce a superficial sedation of the animal. Primates are then maintained under deep anesthesia with a 0.5 to 1.5% isofluorane–air mixture administered through an endotracheal tube with spontaneous respiration. Before intubation, the larynx is sprayed with 2% lidocaine to prevent laryngospasm. The animal's EKG, heart and respiratory rate, and temperature are continuously monitored during the surgery. Slow i.v. infusion of fluids containing the required electrolytes and glucose is used to avoid dehydration and hypoglycemia during the surgery. Arousal level is determined by continuously monitoring reflexes and the electrical activity of cortical neurons (EEG) through metal skews implanted in the skull.

The implantation of microelectrodes is carried out through a series of steps. First, prior to their implantation in the brain, microwire arrays are sterilized using gas or ultraviolet light to minimize the risk of infection. Several microwire arrays are implanted in each animal through small craniotomies (Figure 7.1), each of which can accommodate only a single 16-microwire array (2 mm^2 in area). A series of smaller holes are then drilled in the skull for placement of 6 to 12 metal screws which provide support for the microwire arrays and a common reference for our electrophysiological recordings (Figure 7.1). Once the craniotomies are completed, a small slit is made in the duramater just prior to slowly (approximately 100 μm/minute) lowering the electrode array through the pia surface into the cortex. Over the years, we have learned that very slow implantation is required to reduce the amount of cortical dimpling produced by the electrode array. Different from the cerebellar cortex (see Chapter 5), blunt tip microwires can be driven slowly into the primate cortex without causing major tissue depression. However, the experimenter has to be aware that cortical dimpling can occur in the early stages of implantation if the electrodes are moved too quickly.

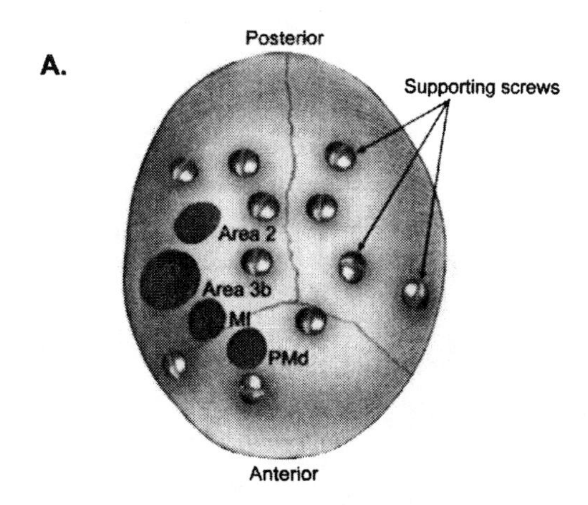

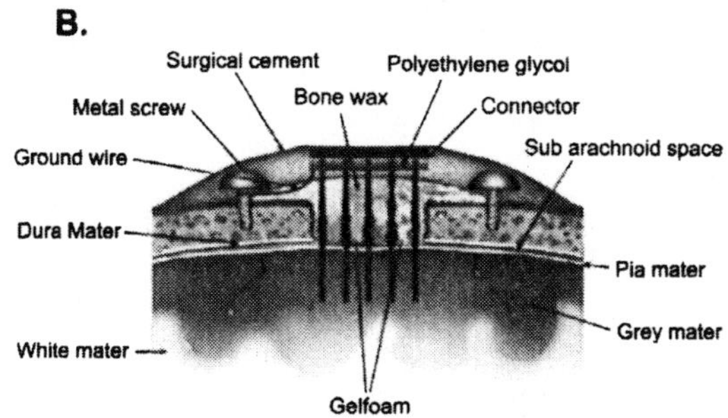

FIGURE 7.1
Schematic representation of the surgical approach used for chronic implantation of microelectrode arrays in the primate cortex. (A) Relationship between four small craniotomies and supporting metal screws. Craniotomies provided access for implantation of microwire arrays in the dorsal premotor cortex (PMd), the primary motor cortex (MI), the primary somatosensory cortex (area 3b) and area 2. (B) Cross section across one craniotomy illustrates the position of the implant and the three layers of materials (gelfoam, bone wax, and surgical cement) used to anchor the single microwire array implant around one metal screw.

Single and multiunit recordings are performed throughout the implantation procedure to ensure the correct placement of the microwires. The receptive fields of both single neurons and multiunits are qualitatively characterized to define the relative position of each individual microwire. Once the microwire array reaches the lower layers of the cortex, the craniotomy is filled with small pieces of gelfoam and then covered with either bone wax or 4% agar. Once the craniotomy is sealed, the entire microwire array is fixed in position using dental cement. This procedure is repeated several times until multiple microelectrode arrays have been implanted in different brain regions. In our hands, each microwire array implant requires about two hours to be completed. Therefore, 4 to 6 arrays (64 to 96 microwires) can be implanted during an 8- to 12-hour surgery. After the completion of the chronic experiments, we have found that individual microelectrode tracks can be readily identified in Nissl-stained sections months after the initial surgery.

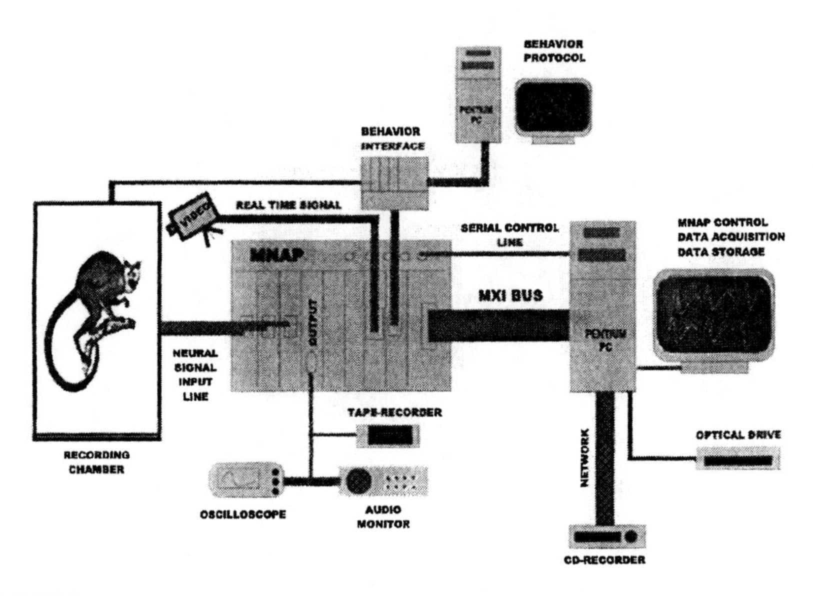

FIGURE 7.2
Schematic description of the complete experimental setup employed for chronic multichannel recordings in behaving primates. MNAP, many neuron acquisition processor.

7.3 Techniques for Simultaneous Multichannel Recording

Five to seven days after the electrode implantation surgery, the primates are transferred to a recording chamber containing a Many Neuron Acquisition Processor (MNAP, Plexon Inc., Dallas, TX). In its most complete configuration (Figure 7.2), the MNAP allows simultaneous sampling from 96 microwires and discrimination of up to four individual action potentials per electrode (Plate 4), for a maximum of 384 recorded neurons. Head stages (NBLABS, Dennison, TX), containing 16 field effect transistors (FETs, Motorola MMBF5459) arranged in two rows of eight at the end of insulated cables, are used to connect the 20-channel plastic connectors cemented in the animals' head to the MNAP preamplifiers. In all recordings, the FETs are set as voltage followers with unity gain. The MNAP preamplifiers contain differential OP-Amps (gain 100, bandpass 100 Hz to 16 KHz). The OP-Amp output signals are transmitted, through ribbon cables, to 96 A/C coupled differential amplifiers located on 6 input boards, each containing 16 amplifiers. Once in the input boards, the analog signals pass through a first level of amplification (jumperable gain of 1, 10, or 20), are filtered (bandpass 400 Hz to 8KHz), and reach the final stage of amplification (programmable multiplier stage, ranging from 1 to 30). These boards also include one 12-bit analog-to-digital (A/D) converter per channel which simultaneously digitizes the waveforms defining extracellular action potentials at 40KHz. After A/D conversion, the signals are routed to DSP boards, each of which contains four digital signal processors (DSP, Motorola 5602) running at 40 MHz (instruction read at 20 MHz). Each DSP handles data from eight input channels and contains 32 K 24-bits of SRAM and 4K 16-bit words of dual port SRAM memory.

An MNAP timing board is responsible for distributing timing stamps and synchronizing all input signals to the system. This board also provides a digital time output that can be used to synchronize external devices or to drive a video timer for a videotape recorder employed to document the animal's behavior. The timing board can also receive up to 16 TTL inputs generated by other devices which are responsible for controlling the behavioral cages used in our experiments. A single host Pentium microcomputer (100 to 200MHz, with 64 Mbytes of RAM, and 2.1 Gbytes of disk space), running a C++ software (SSCP, Plexon Inc., Dallas, TX) in the Windows 95 operating system (Microsoft, Seattle, WA) is responsible for controlling the MNAP over a serial line. Spike discrimination programs are downloaded from the PC host to the DSPs. Single spikes are discriminated by combining a modified version of a principal component algorithm,[7] running in real time, and a pair of time-voltage discriminator boxes per unit. The size and position of the time-voltage boxes (Plate 4) are defined by the experimenter to isolate the waveforms that belonged to a given unit. Spikes are only accepted as valid when they pass through both boxes. During the experiment, the time of occurrence of each of the valid spikes, for all 96 channels, is transferred to the hard disk of the PC host through a parallel bus (MXI-Bus, National Instruments, CA) which is capable of transferring 2MB of data per second. Digitized samples of the spike waveforms are also recorded periodically and stored for off-line analysis using a visualization program (SpikeWorks, Plexon, Dallas, TX). Optical drives (1.2 to 2.4 Gbyte cartridges, Pinnacle Inc., CA) are used for temporary backup of the data files. These files are then transferred, through an Ethernet-based network, to a computer server containing a CD-Recorder (Pinnacle, CA) which is used to produce permanent records of the data. The MNAP also supplies options for analog backup using 8 to 16 channel tape recorders.

In all experiments described in this chapter, discrete low-threshold tactile stimulation of the monkey fingertips was obtained by using a computer-monitored Grass S8800 stimulator to drive a small polygraph motor (MFE, Beverly, MA) in which a 10 cm wood probe was attached. The digital outputs of the stimulator were also transmitted to the MNAP to be stored together with the spike data.

7.3.1 Neural Ensemble Recordings in Behaving Primates

One of the most important advantages of carrying out chronic neural ensemble recordings in primates is that microwire arrays can be implanted in a naïve monkey and then be used to record the activity of large populations of single neurons throughout the several months required for the animal to master a variety of tasks. For example, in one of the behavioral tasks used in our laboratory, monkeys are required to execute visually guided arm movements. To execute this task, primates are placed in a restraining chair with a manipulandum and a Plexiglas® panel containing visual stimuli (LEDs) which can be placed in front of the animal. The manipulandum consists of a joystick-like, Plexiglas stylus, which is mounted on a linear motion slider (Small Parts, Miami Lakes, FL). The instantaneous position of the manipulandum is determined with a potentiometer and is sampled at 200 Hz. In addition, the instantaneous positions of the monkey's wrist, elbow, and shoulder are recorded via sensors attached to the monkey's arm (Flock of Birds, Ascension Technologies, Boston, MA). The sensors receive magnetic fields that are transmitted at 2000 Hz from a magnetic field generator located beneath the restraining chair and provide six degrees of freedom measurement of each joint's position and movement in space (x, y, and z position plus roll, pitch, and yaw). This behavioral protocol is controlled by a personal computer running the Tempo (Reflective Computing, St. Louis, MO) behavioral data collection system. Synchronization between Tempo and the MNAP software allows us to coordinate the timing of all behavioral events with the spiking time of large populations of neurons during the execution of the task.

In another task used in our behavioral studies, monkeys are required to perform visually guided arm movements while either a constant or time-varying external force perturbation is imposed on the stylus at different instances of the arm trajectory. This task is implemented to evaluate how the neural responses across the motor and somatosensory cortices respond to external perturbations represented by increased tactile and proprioceptive feedback. In addition, this experimental design leads to better understanding of how cortical neurons respond to proprioceptive and/or mechanoreceptive input during a perturbed movement, and how the brain processes this information to correct the motor output signal to compensate for the perturbation. To carry out these experiments, the monkey manipulates another type of stylus which is connected to the PHANToM (Sensable Technologies, Cambridge, MA), a haptic interface which can create a three-dimensional tactile feedback during visually guided arm movements. The virtual environment is created by reading in the robotic stylus' end point, then calculating the reactant forces that make up the objects in the virtual scene. The three-dimensional position, orientation, and reactant force data from the stylus point are collected by the Tempo system and saved with the other behavioral data. Finally, the animal's behavior is stored on videotape, and frames from the video systems are time-stamped with a resolution of 1 ms, using a video timer system controlled by an MNAP output timing signal.

To carry out these behavioral experiments with owl monkeys, we had to develop new procedures for restraining the animal during recording sessions and adapt a primate chair usually employed for squirrel monkeys. Our restraining device is an extensively modified version of a New World monkey chair (PlasLabs, Lansing, MI). The monkey is held in the restrainer by Plexiglas plates that are placed around its neck, middle torso, and waist. The animal's lower limbs are contained within a Plexiglas chamber that has a lower panel made of rows of delrin rods. An additional Plexiglas plate is attached to the plate that restrains the animal's neck. This plate extends 6 cm beyond the original plate and blocks the animal's upper limbs from accessing a metal tube that delivers fruit juice to the animal's mouth. The monkey's head is held in position by a bolt attached to the skull during the microwire implant surgery. This bolt is connected to a metal rod that is attached to the upper plate in the restraining device. Our restraining device allows the animal freedom of movement, especially of its upper limbs, and does not appear to cause the animal any undue stress from restraint.

7.4 Neural Ensemble Data Analysis

There are many challenges involved in performing simultaneous multichannel single neural recordings in behaving animals. The most daunting one, however, is analyzing the large data sets that are generated in each experimental session. The availability of new specialized electrodes and electronic hardware has increased the yield of multichannel recordings and, as a consequence, there is still a clear demand for the design of analytical tools dedicated to the quantitative characterization of firing patterns of large neural ensembles. Indeed, the development and implementation of such analytical methods have become one of the most difficult tasks for those working in this field.

Given the importance of this topic, several chapters in this book are entirely dedicated to introducing the reader to the most commonly used analytical techniques described in the literature. The methods described in these chapters range from the classical correlation-based methods, such as cross-correlation analysis (Chapter 8), to multivariate statistical methods, such as multivariate analysis of variance, principal component analysis, canonical correlation, and discriminant analysis (Chapters 10 and 11), artificial neural networks (see below), and some new techniques, such as direct coherence analysis (Chapter 9). It should be emphasized, however, that there is no definitive or optimal analytical procedure that can be universally used in any neural ensemble data set. In fact, since each method has advantages and disadvantages, different techniques should be applied to the same data set not only to examine different features of the data but also as a way to cross-validate results obtained with a given method.

The flow diagram depicted in Figure 7.3 illustrates a typical sequence of analytical steps employed in our laboratory to analyze neural ensemble data sets, including those obtained in behaving primates.

7.4.1 Single Neuron Analysis

The first step in any neural ensemble analysis routine involves the detailed characterization of the physiological properties of the individual neurons that form the simultaneously recorded population. Commonly, in this first stage poststimulus time histograms (PSTHs, Figure 7.4) are used to investigate the firing patterns of individual neurons. In our experiments, quantitative analysis of these histograms provides an important description of the range of firing patterns observed across the simultaneously recorded neural population during either passive sensory stimulation or the execution of different tactile exploratory tasks. PSTHs allow one to measure the average intensity of neuronal firing across several user-defined poststimulus time epochs (or bins). The "evoked unit response" or instantaneous firing rate of the neuron can then be expressed as: (1) spikes/per second, (2) average instantaneous firing rate over different poststimulus epochs, (3) firing rate minus background, or (4) firing rate divided by background (i.e., signal/noise ratio).

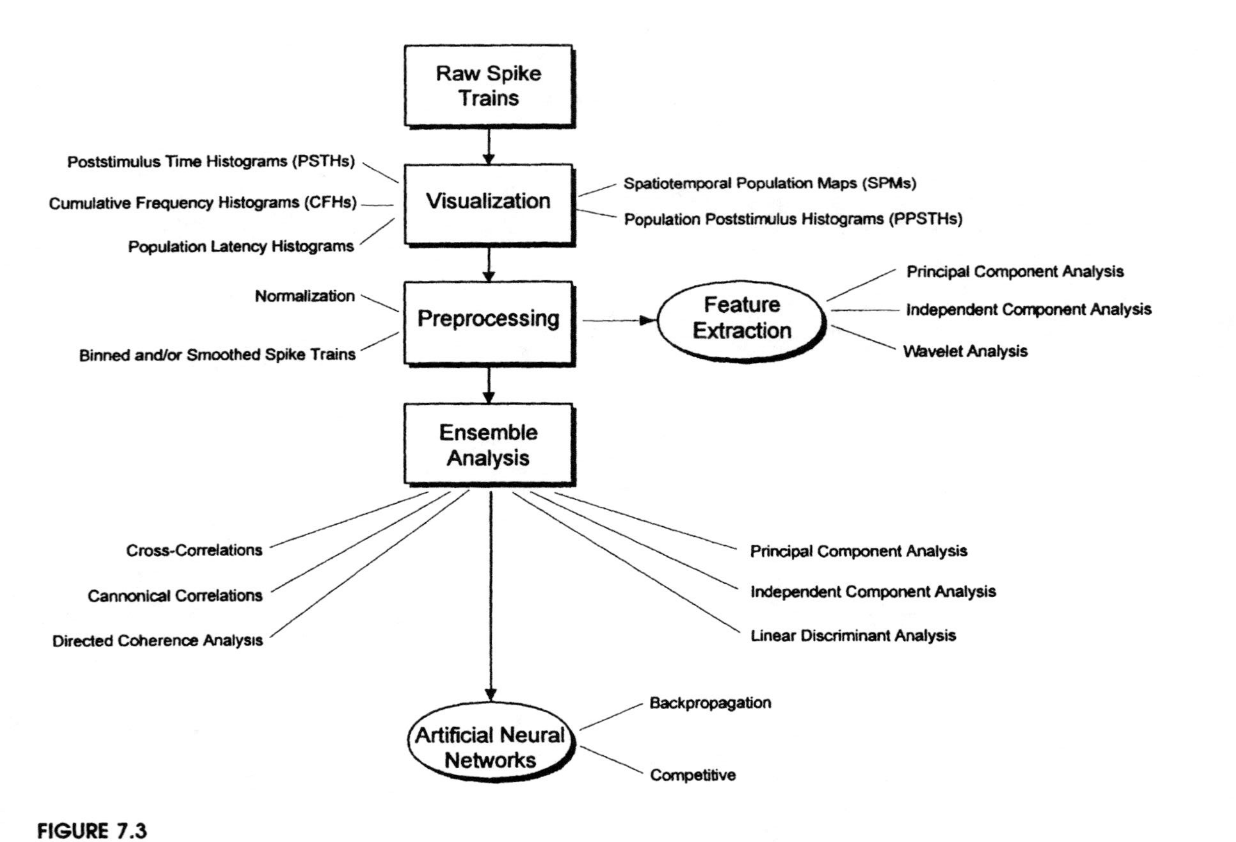

FIGURE 7.3
Flow diagram depicting the neural ensemble analysis routine employed in our laboratory.

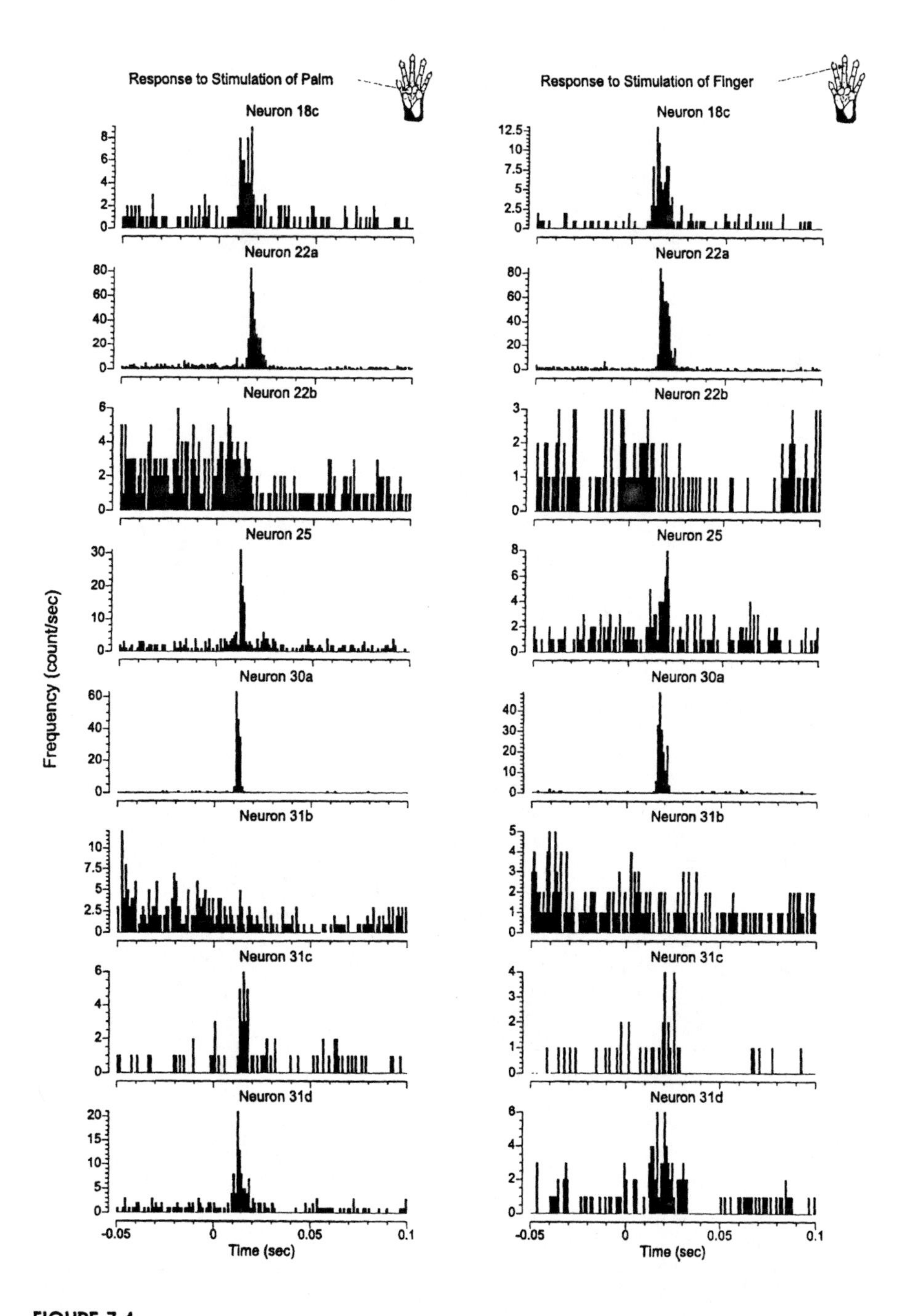

FIGURE 7.4

Poststimulus time histograms illustrate the sensory responses of eight simultaneously recorded neurons in the somatosensory cortex of an owl monkey following mechanical stimulation of two distinct regions of the animal's hand.

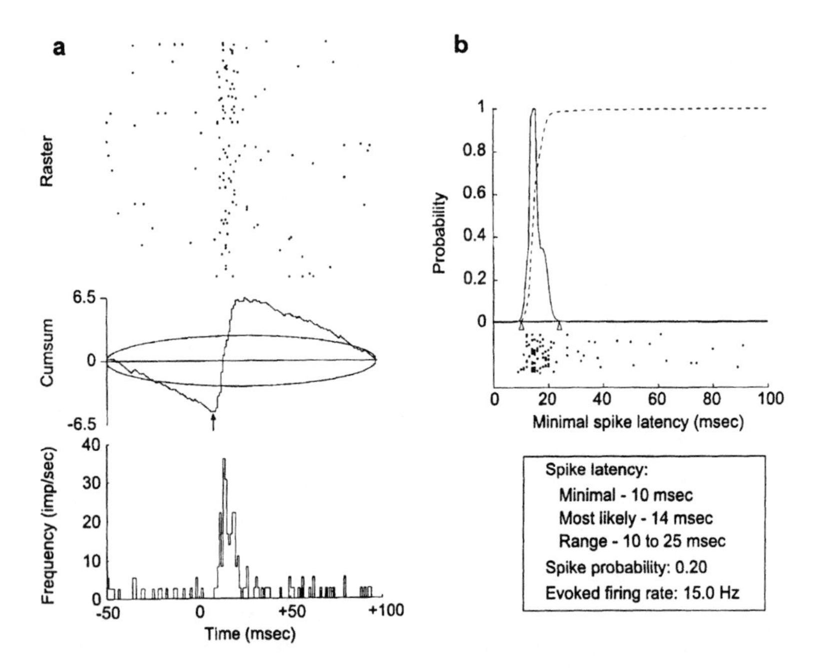

FIGURE 7.5
Calculation of neuronal sensory response latencies. (a) Cumulative frequency histograms (middle graph) are used to calculate the minimal response latency of individual neurons. The ellipse associated with this graph defines the confidence interval ($p < 0.01$) above which one can assume that the neuron sensory response differs from a random distribution. The solid arrow points to the inflexion point that defines the abrupt change in firing rate which defines the onset of the neuron's sensory response. The time of occurrence of this inflexion point defines the neuron's minimal response latency. A raster plot (top graph) and a PSTH (lower plot) are provided for comparison with the CFH. (b) Automatic quantification of neuronal response latency following peripheral tactile stimulation. The upper plot displays the probability of occurrence of the first spike after the stimulus, estimated using a kernel density method, as a function of poststimulus time. The arrowheads below the plot indicate the time associated with 1 and 99% of the total poststimulus spike density. This time interval is called the "minimal spike latency interval." The lower plot is a raster display of the poststimulus times of the first spike for each individual trial. The values given below the plots are for the minimal, expected (or mean), and range of spike latencies for the neuron. The spike probability per stimulus is defined as the number of stimuli with spikes in the minimal spike latency interval divided by the total number of stimuli. The evoked firing rate is the number of spikes in the minimal spike latency interval divided by the duration of the interval.

The statistical significance of sensory responses derived from PSTHs are routinely assessed in our laboratory through the combination of three techniques: (1) standard errors for all user-defined time epochs, (2) a Kolmogorov–Smirnov test which can be used to determine whether the sensory responses present in a histogram are significant (i.e., whether the distribution of counts across the histogram is nonrandom), (3) a Student's t-test which may be used to compare firing in control vs. experimental epochs.

When PSTHs are combined with cumulative frequency histograms (CFHs, Figure 7.5), the experimenter can also obtain precise measurements of the latencies of neuronal sensory responses. CFHs are preferable because they smooth the data without distorting their timing. Latencies in CFHs are determined with a 1 ms resolution by isolating the instant in which the firing distribution diverges from a

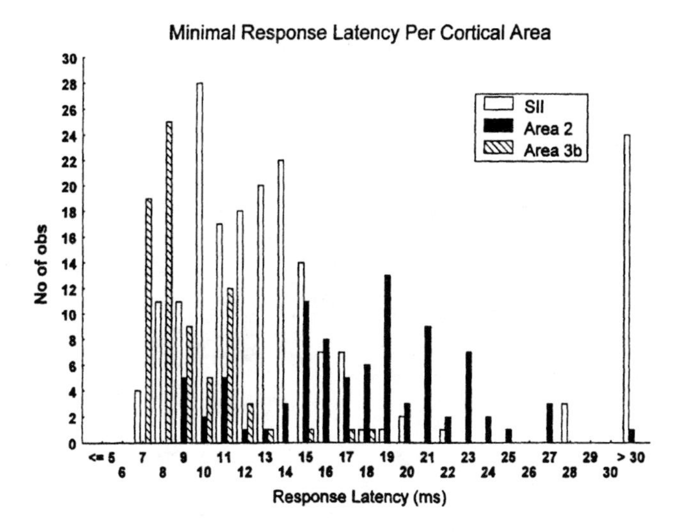

FIGURE 7.6
Population latency histogram depicting the distribution of sensory response latencies for cortical neurons located in areas 3b, SII, and 2. (From Nicolelis, M.A.L., Ghazanfar, A.A., Oliveira, L.M.O., Chapin, J.K., Nelson, R., and Kaas, J.H., Simultaneous encoding of tactile information by three primate cortical areas, *Nature Neurosci.*, in press. 1998. With permission.)

random distribution with $p < 0.01$. Usually, CFHs are used in conjunction with Kolmogorov–Smirnov tests to characterize the statistical significance of the sensory responses of individual neurons.[8] As a routine, only sensory responses that have a $p < 0.01$ probability to be generated from random firing are accepted for further analysis. Several commercially available software packages (Stranger, Biographics, Winston-Salem, NC; Nex, Plexon, Dallas, TX) provide users with PSTH and CFH routines for analyzing neural ensemble data. In addition to these packages, our laboratory has designed software that scans automatically all simultaneously recorded neurons in a data set to extract the PSTH and CFH measurements described above.

Latency data derived from individual CFHs can be lumped together to generate population latency histograms which depict the distribution of the minimal response latency of large samples of single neurons. These population histograms provide valuable information, particularly because they allow comparisons between neural populations located in different brain regions (e.g., different cortical regions or distinct subcortical nuclei) to be drawn. An example of such population latency histograms is illustrated in Figure 7.6 which describes the existence of a considerable degree of latency overlap between populations of neurons located in three distinct cortical areas.

PSTHs can also be used to characterize each unit's receptive field, a classical measurement used in single neuron physiology. An example of such analysis is depicted in Figure 7.7, which illustrates the multidigit receptive field (RF) of two cortical neurons located in the owl monkey secondary somatosensory cortex (SII). Notice that PSTHs were used to depict the response of each neuron to the stimulation of different fingertips. By measuring both the magnitude and the minimal latency of each neuron response, the center (i.e., maximum response magnitude and minimal latency) and excitatory surround of the RF can be clearly identified.

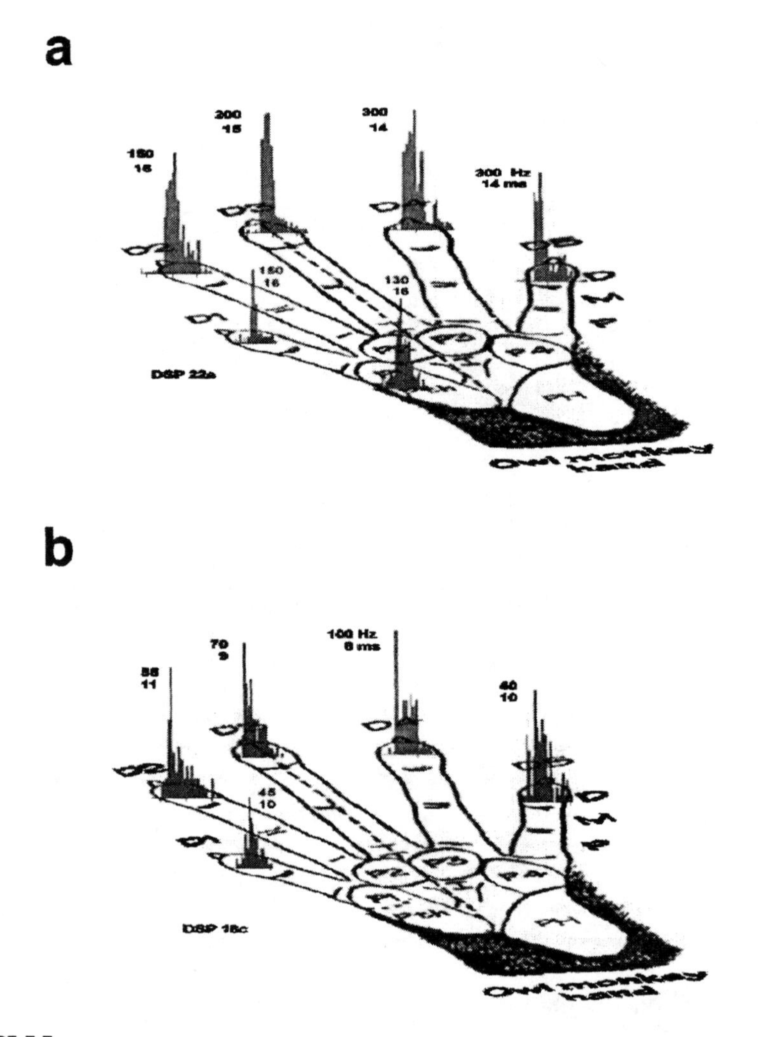

FIGURE 7.7
Multidigit receptive fields of two simultaneously recorded neurons located in the secondary somatosensory cortex of an owl monkey. PSTHs were used to define these RFs quantitatively. Two measurements were used to define these RFs: maximum firing rate (in spikes/sec, top number next to each PSTH) and minimal response latency (in ms, lower number).

Other classical techniques, such as cross-correlation analysis, can be used at this stage to characterize pairwise interactions between simultaneously recorded neurons. This technique is described in great detail in Chapter 8.

7.4.2 Population Poststimulus Time Histograms

Once the single neuron analysis is completed, the next step in our routine is to visualize the spatiotemporal patterns of neural ensemble firing. Software written in our laboratory is used for producing all 3D graphs and animation in our laboratory. This program also utilizes a series of MATLAB (Mathworks, Natick, MA) graphic routines to provide the user with a large variety of data fitting, 3D graphics, and animation options.

There have been many descriptions in the literature of methods to depict information from neural ensembles graphically. In this section we describe two of the graphics used in our laboratory to visualize such large data sets. The most commonly used graph in this phase of the analysis is the population poststimulus time histograms (PPSTHs). In our studies, PPSTHs are often employed to visualize how the entire neural population responds to a particular tactile stimulus or during a series of trials in a behavioral task. An example of a PPSTH is depicted in Plate 5. Basically, the PPSTHs is a 3D graph in which the X axis is used to depict postsimulus (or trial) time, the Y axis contains the simultaneously recorded neurons, ranked order according to a particular criterion (e.g., rostral–caudal position in the electrode array, firing rate, response latency, etc.), and the Z axis depicts instantaneous firing rate in spikes per second. The 3D matrix defined by this plot is filled with the average individual responses of the neurons, obtained over 100 to 300 trials. The final data set is then plotted and smoothed using a spline algorithm. The final values of the Z axis are then color coded using either absolute values or normalized z scores that depict the sensory evoked responses in terms of numbers of standard deviations above average spontaneous firing. Although these graphs are easily implemented in MATLAB, other graphical packages, such as CSS: Statistica, can be easily used for building these plots.

7.4.3 Spatiotemporal Population Maps (SPMs)

These 3D plots are used to depict the responses of several single neurons to stimulation of one or multiple peripheral sites as a function of poststimulus time. The difference between SPMs and PPSTHs is that, in the former, the actual spatial position of the neuron in the multielectrode array is preserved. Thus, SPMs are used to visualize how the spatial pattern of a given neural ensemble varies as a function of time. Although averages of multiple trials are usually used to build these plots, information derived from single trials can also be represented using this graphic format. In its simplest format (Plate 6, bottom scheme), SPMs are defined by a series of 3D matrices that represent the spatial pattern of firing of simultaneously recorded populations of neurons as a function of time. Data derived from single neuron PSTHs are usually used to generate these plots. Evidently, the spatial domain of these graphs reflects the type of multielectrode matrix configuration employed for neural ensemble recordings. In Plate 6, an SPM is used to represent the activity of a population of SII neurons recorded through a 2 × 8 electrode matrix. Data were plotted in consecutive 5 ms time epochs (depicted atop each graph). Each 3D graph (represented by 2 rectangular panels plotted side by side and separated by an empty space) in this plot is formed by a Y axis that represents the rostrocaudal position (top: rostral-most wires 1 and 9; down: caudal-most wires 8 and 16) of each electrode. In this case, the left-most panel represents the more medial row of electrodes, while the right panel represents the most lateral row of electrodes. In each panel, the X axis (horizontal axis) is formed by four different positions (labeled a to d). This reflects the fact that up to four single neurons per single electrode can be discriminated simultaneously using the MNAP system. Like PPSTHs, the Z-axis of the 3D matrices represents the variation in response magnitude of the neurons in spikes per second. Once SPMs like the one depicted in Plate 6 are generated, further numerical data analysis can be carried out to characterize the amount of tissue recruitment generated by a particular stimulus.

7.4.4 Multivariate Techniques to Analyze Neural Ensemble Data

The next step in our analysis routine is to obtain quantitative measurements to describe the spatiotemporal patterns of neural ensemble firing depicted in the different graphs described in the previous section. We have learned over the last 10 years that several multivariate statistical techniques, such as principal component (PCA) and discriminant (DA) analysis, can be employed in order to characterize these firing patterns and provide important insights into how information may be encoded by neural ensembles. A very detailed description of these methods is provided in Chapters 10 and 11.

7.4.4.1 Principal component analysis

By definition, PCA allows the reduction of large numbers of original variables into a smaller number of derived "components" which represent most of the variance observed in the original data set. PCA can be used to map the correlation patterns between large numbers of variables, because it provides a standardized way to represent the overall covariance structure of a large data set into a new and greatly reduced set of variables, known as principal components. Each of these components is formed by the weighted linear sum of the integrated firing of individual neurons and accounts for an independent source of variance of the data set. The contribution of each neuron, reflected by its component weight, varies from component to component. By definition components are orthogonal to each other, meaning that each of them accounts for an independent source of variance. Thus, different dimensions of information embedded in the firing pattern of the neural population can be extracted and examined independently by using PCA. PCA can be used to quantify the functional associations between neurons within neural ensembles and to measure how these associations are modified by either the delivery of a peripheral stimulus or the execution of a particular task.

In our PCA analysis, each neuron of the simultaneously recorded neural ensemble is considered as a variable. The firing rate of each neuron, integrated sequentially in 10 to 25 ms bins over 10- to 30-minute periods, is used as the raw data for this analysis. Neither the anatomical location of the neurons nor their receptive field properties are made available to the PCA algorithm. Time series describing the firing rates of each neuron are used to generate a correlation matrix involving all simultaneously recorded neurons in each animal. From this correlation matrix, a series of linear polynomia (known as principal components or eigenvectors) are extracted.

Several graphical displays can be used to describe the results of a principal component analysis. Two very useful representations are depicted in Figures 7.8 and 7.9. Figure 7.8 illustrates how reconstructed components can be used to reduce the original number of variables and capture particular sources of variances of the data set. The 3D scatterplot depicted in Figure 7.9, on the other hand, displays the distributions of neuronal weights obtained by calculating the first three components (PC1, 2, and 3) from a data set containing neurons distributed across two distinct cortical areas (SII and area 2). As one can see, neurons located in the same area tend to cluster together, while neurons from different areas tend to be clearly separated. This result has been reproduced many times in our laboratory and seems to indicate that within-structure correlations are much stronger than between-structure correlations. Similar cluster segregation has been found when thalamic and cortical neurons were analyzed together.

7.4.4.2 Discriminant analysis

Discriminant analysis (DA) is employed in our studies to statistically evaluate the differences between patterns of neural ensemble firing on a single trial basis (see Reference 8 and Chapters 10 and 11). DA provides a multivariate analysis of variance (MANOVA)-like treatment of the data set. It also derives classification functions for trial-by-trial discrimination between the different experimental groups (i.e., stimulus sites, different trial outcomes). When used to characterize neural ensemble responses to tactile stimuli, the multivariate data set consists of the measured spiking responses of each of the simultaneously recorded neurons, taking into account several poststimulus time epochs to depict each neuron's sensory response. DA generates a series of linear classification functions which optimally discriminate between the groups (i.e., stimulation sites). This allows one to calculate an estimation of how well neural ensemble activity can be used to predict the location of a tactile stimulus.

7.4.4.3 Independent component analysis (ICA)

Independent component analysis (ICA) has recently been introduced into the arsenal of analytical tools used in neural ensemble data analysis.[9] Methods for ICA find features that separate independent signals from linear mixtures of signals.[10] All methods for ICA work by maximizing higher-order statistical criteria that estimate deviations of the distribution of the variables from the Gaussian distribution. For example, one method, JADE,[11] finds features by projecting the data in multivariate space so as to maximize the kurtosis or "peakedness" of the multidimensional distribution. Other methods, which are based on neural network architectures, find features by maximizing information theoretic criteria for independence of the features.[12,13] The neural network based methods for ICA account for higher-order statistics by using nonlinear transfer functions, such as the logistic function, in the hidden layer of the networks.

To compare features identified by ICA and PCA, we applied several algorithms for ICA to neurophysiological data obtained in our simultaneous multielectrode

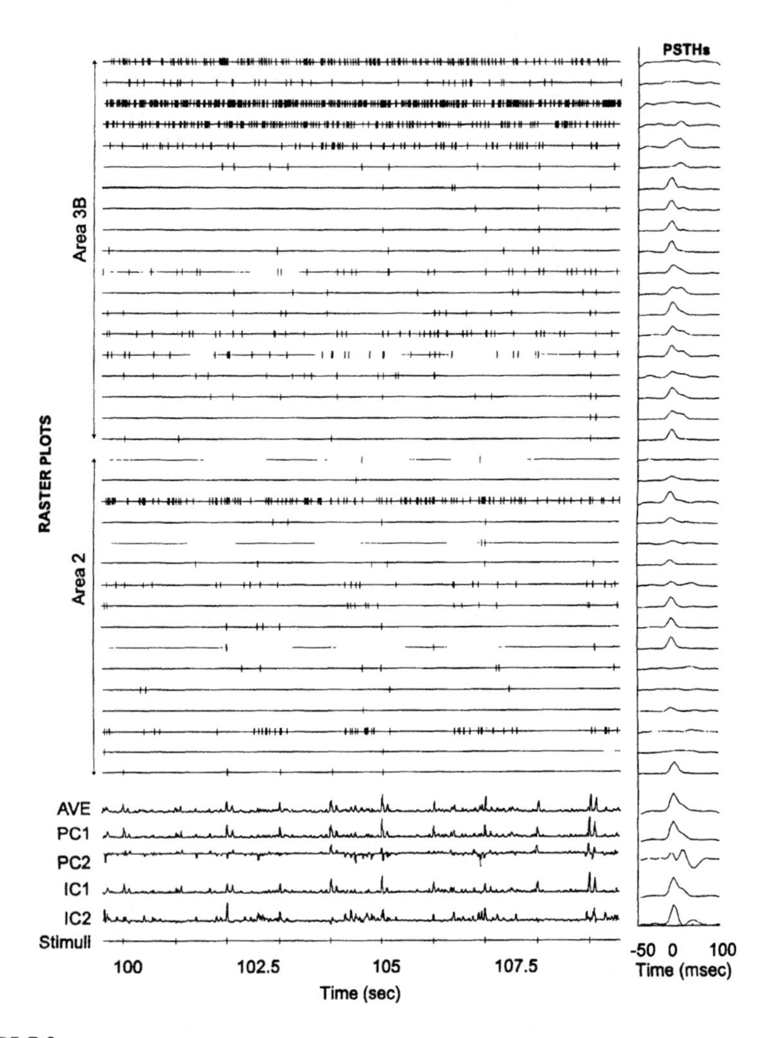

FIGURE 7.8
Extracting population vectors from simultaneously recorded neuronal ensembles. Spike trains from 35 neurons were recorded simultaneously in two different areas (area 3B [n = 19], area 2 [n = 16]) of the somatosensory cortex of an owl monkey during stimulation of hair above the monkey's eye (100 ms duration at 1 Hz). A population raster plot is shown in the left portion of the figure for illustrating the neuronal activity over 10 stimulus presentations. Each vertically oriented tick depicts the occurrence of an action potential from one of the simultaneously recorded neurons. At the bottom of the figure, strip charts depict the average response of the population of neurons (AVE), the first two principal components (PC1 and PC2), the first two independent components (IC1 and IC2), and the stimulus time. In the right portion of the figure, smoothed poststimulus time histograms (averaged for 331 trials) are shown for 50 ms before and 100 ms after the stimulus presentations (bin size 5 ms). Note that the time series for PC1 and IC1 were similar to the population average. Weights for PC1 (not shown) were broadly distributed over the neuronal ensemble; this mapping was to some extent reproduced by the weights for IC1, which were large and positive for the neurons in area 3B. By contrast, time series for PC2 and IC2 diverged from the population average response. Neurons in area 3B had positive coefficients for PC2, and those in area 2 had negative coefficients; this mapping of the neuronal ensemble gave a population vector that contrasted the activity of neurons in the two cortical areas. IC2 was based on sensory evoked response of the area 2 neurons that were not correlated with the sensory responses of area 3b neurons.

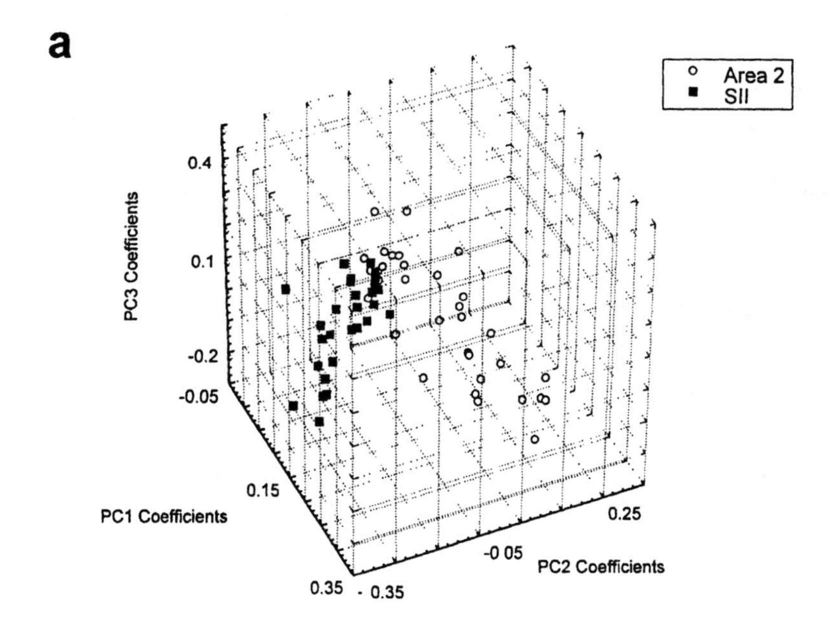

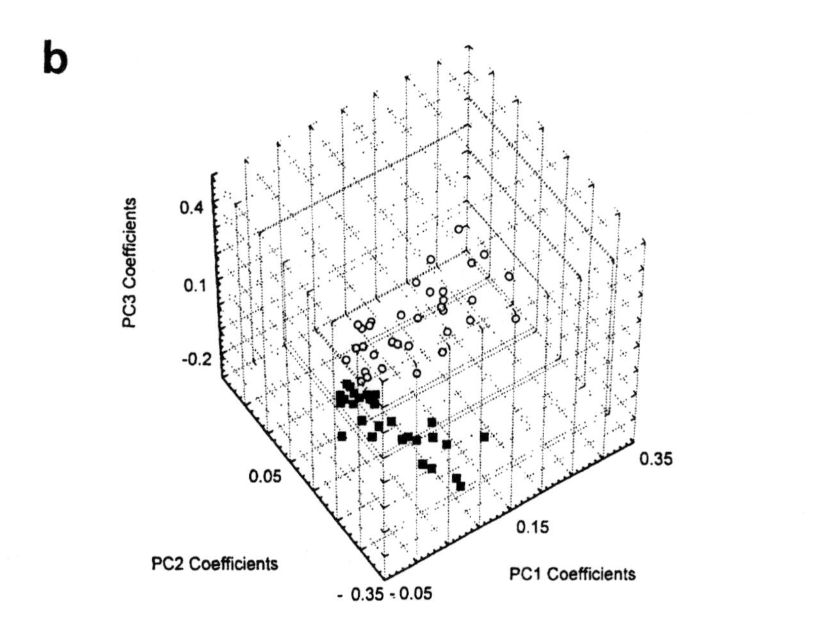

FIGURE 7.9
Three-dimensional scatter plot depicts the distribution of three component weights calculated for populations of neurons distributed across two cortical areas (SII, squares, and area 2, circles). Although the location of each neuron is not provided to the PCA algorithm, this analysis provides a clear clustering of neurons according to their cortical areal location.

recordings and to simulated spike train data.[9] Our implementation of ICA is as follows: a matrix of spike counts for all simultaneously recorded neurons is constructed using 5 to 25 millisecond bins. The mean spike count over all bins is subtracted for each neuron, and the residual zero-mean spike trains are normalized to have unit variance. This procedure is necessary to prevent neurons with higher overall spontaneous activity from masking interactions between neurons with low spontaneous activity. Next, we determine the number of principal dimensions for the data using a principal component analysis of the correlation matrix for the multineuron spike trains. Eigenvalues from this PCA are plotted in a scree plot, and the number of principal components above the break point eigenvalue (around 1) in the plot is chosen as the number of principal dimensions, or number of functional groups, in the neuronal ensemble. Then, we apply one of the methods for ICA to the multineuron spike train data and retain as many independent components as suggested by scree plot from PCA. (ICA is done using public domain code for MATLAB that is freely available over the Internet.) Scores for the independent components are computed over the collection of bins as the product of weights for the independent components and the multineuron spike trains. These procedures provide bin-by-bin estimates of the independent components over the time course of neuronal population recording experiments. These representations of the multineuron spike trains can later be analyzed with neural network methods to account for the effects of experimental manipulations on the activity of the neuronal population (see below).

Descriptions of neuronal population activity that are based on higher-order statistics (e.g., ICA) should be distinct from those based on low-order statistics (e.g., PCA). For example, the time-series for the first principal component (PC1) extracted from a neuronal population is usually very similar to the averaged response of the neuronal population (see Figure 7.8). The interaction between neurons that is usually accounted for by PC1 is distributed broadly over the population of neurons. ICA usually finds a similar feature for the neuronal population and represents this interaction as a sub-Gaussian independent component that is associated with specific neurons in the population. While coefficients for the first principal component and one of the independent components (such as PC1 and IC1 in Figure 7.8) can be similar, coefficients for the higher principal components and for the remaining independent components usually diverge considerably. For example, if neurons in more than one brain region are recorded during an experiment, then an analysis of these neurons typically finds that coefficients for the second principal component (PC2) contrast the activity of neurons in the different brain regions. ICA will represent these subsets of functionally related neurons not as a common component, such as PC2, but by a set of supra-Gaussian independent components, with one component for each functional group of neurons (see Figure 7.8).

The features of neuronal population activity identified by ICA may well be related to the degree of synchronization between subsets of neurons in the neuronal population. Computational experiments using simulated spike train data showed that each of the three algorithms for ICA that we use[11-13] are able to detect independent sources that synchronize the spike trains of multiple neurons.[9] Features from ICA can reconstruct the underlying sources of synchronization within a population of neurons with a high degree of accuracy. By contract, PCA represents sources for synchronization as linear combinations across the entire set of principal components. This representation is not able to account for the sources as independent functions. Therefore, our simulation studies show that ICA is a useful adjunct method to PCA.

7.4.5 Artificial Neural Networks

The final step in our data analysis routine involves the use of artificial neural networks (ANNs). ANNs are employed to investigate how well spatiotemporal patterns of neural ensemble activity reproduce the attributes of a tactile stimulus[8] or predict the outcome of a trial in a behavior task.[14] ANNs are very efficient pattern recognition devices which are particularly useful for large multivariate problems where the significance of each variable is not known.[15] Thus, unlike many other analytical techniques, no assumptions need be made about the structure of the data or its distribution to use ANNs for analyzing neural ensemble data. In addition, in our hands ANNs tend to perform as well as or better than other classification techniques used to analyze neural ensemble firing patterns.

There are many commercially available software packages and even public domain code that allow a fast and efficient implementation of different ANN-based analysis. In our analysis routine (Plate 7), ANNs are used to recognize and distinguish patterns of neural ensemble activity on a single trial basis. In the next few sections, we describe how supervised training paradigms applied to ANNs can be used to classify single trials that depict the sensory responses of a large population of neurons into one of a predetermined set of classes. In this analysis, classes can be used to group trials according to the nature of a sensory stimulus (e.g., the site of stimulation) or behavioral performance (e.g., right vs. wrong behavioral outcome).

Two classification methods are discussed in this chapter: feed-forward back-propagation and competitive ANNs. Because ANNs learn to solve problems from examples, a certain number of recorded trials must be used in teaching, or "training," the network. The remaining trials, known as "testing trials," are reserved to measure the performance of the discrimination system. Among other factors, ANNs' ability to learn and classify neural ensemble firing patterns greatly depends upon the robustness and consistency of the original data set. Thus, we also discuss methods to prepare the raw data so that maximum classification is achieved and valuable insights into the neural coding strategies can be attained.

7.4.5.1 Preprocessing of neural ensemble data for ANN analysis

In addition to a high signal-to-noise ratio and a large number of trials relative to the number of neurons, adequate preparation of the original data set is often required to obtain optimal performance in any procedure that involves ANNs. This initial phase of the analysis is known as data preprocessing, and it usually involves some type of data normalization and "feature extraction." Normalization is an easy way to improve classification and invariably consists of a simple linear transformation of the data to equalize the relative importance of each of the inputs. Feature extraction, on the other hand, allows one to constrain the number of elements from the original data set that

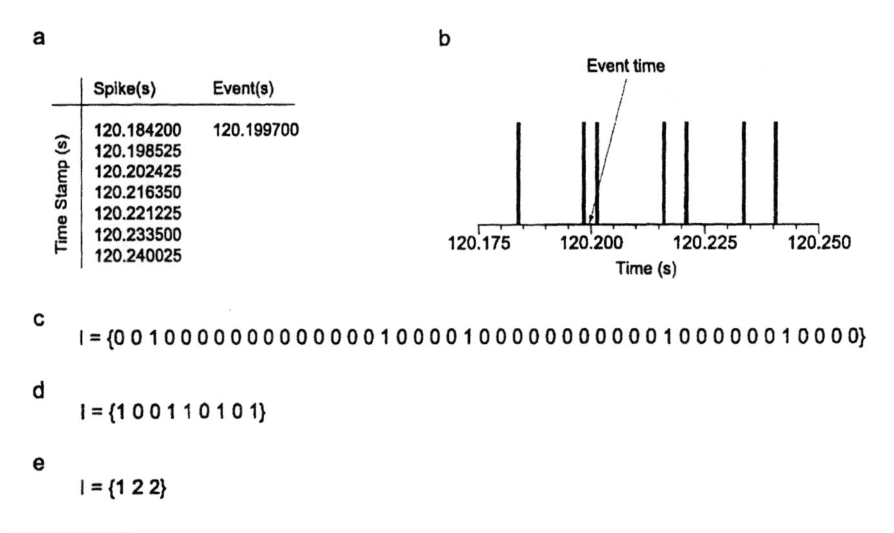

FIGURE 7.10
Five representations of a single neuronal spike train. (a) Initially, the spike train is stored on a computer disk as a series of time stamps. (b) Such a collection of time stamps can be more easily visualized in a strip chart. For numerical analysis, spike trains can be represented by vectors. The first element of each vector shown here represents the activity which occurred immediately following the event time (i.e., the stimulus). The next element contains the activity of the second time bin, and so on. Effect of varying the size of the time bin from 1 ms (c) to 5 ms (d), and 15 ms (e).

are important in making up a recognizable pattern. For instance, during data preprocessing we may take advantage of dimension-reducing algorithms, such as principal component analysis and independent component analysis, which tend to simplify the classification task and may improve discrimination ability. Normalization and feature extraction are discussed in detail in the next section.

Data preprocessing also aims at finding a simple and standard form for feeding the neural ensemble firing activity into the ANN, while preserving as much information as possible. Our laboratory employs an event timing system (see Chapter 3) to obtain a continuous record of the spiking time of all the single neurons that have been recorded simultaneously. Thus, our experiments yield multiple time series containing sequences of "time stamps" that represent the precise timing of each neuron's action potentials during the recording session. As depicted in Figure 7.10, the "time stamps" that represent the firing of a neuron may be transformed into many acceptable data structures. The time surrounding an event (e.g., sensory stimulus onset) can then be divided into discrete time bins (or epochs). Time bins of 1 ms, for example, form a binary code representation of a neuronal spike train (Figure 7.10c). This format is simple, consistent, and makes no *a priori* assumptions about the data set. In practice, however, time bins greater than 1 ms are usually employed to describe the time-varying nature of neural responses. One of the reasons for that is the existence of a refractory period, i.e., a time interval in which firing neurons are incapable or less likely to produce action potentials. Although this refractory period varies according to neuronal types, a 2 to 3 ms interval is usually considered appropriate.

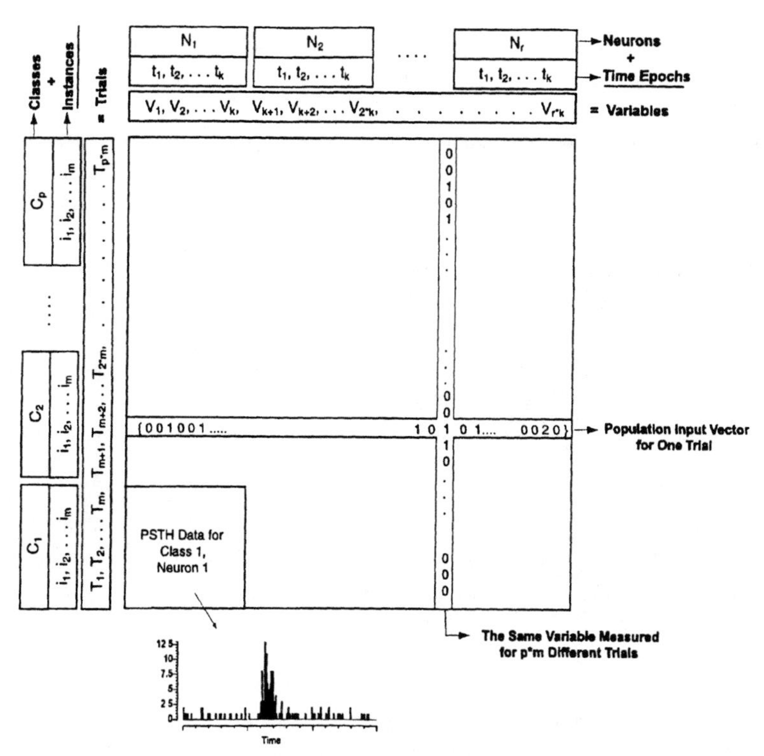

FIGURE 7.11
General organization of the data matrix used for neural ensemble analysis by ANNs.

In order to present information about the time-varying sensory response of neurons to the ANN, a sequence of bins is used to describe the changes in firing rate of each single neuron as a function of poststimulus time. Each of these bins is considered as a unique input variable to the ANN. The total number of input variables defines the size of the ANN input layer (see below). It follows that on a single trial, the firing of a population of *r* neurons (N), measured by using *k* time bins (t) (at a resolution of x ms), defines an input vector to the ANN:

Thus,

$$I = \{N_1t_1, N_1t_2, \ldots N_1t_k, N_2t_1 \ldots N_2t_K, \ldots. N_rt_k\} \text{ (see Figure 7.11)}$$

Routinely, large variable sets are obtained in neural ensemble recordings. This usually makes the discrimination of a single trial a very difficult task for most analysis methods. Then, it is always important to maintain the number of input variables to a minimum by achieving a compromise between the number of neurons and time bins used to describe the population response. Since one is usually interested in having as many neurons as possible in these analyses, the most common practice is to limit the number (or size) of time bins used to describe each neuron's firing pattern.

7.4.5.2 Normalization and feature extraction

Since different neurons exhibit different spontaneous and sensory-evoked firing rates, data normalization is usually necessary to equalize the general contribution from each neuron to the ANN. This is important because neurons with a large average firing rate are not necessarily more informative than those with a lower one. As a rule, we normalize the spike counts from each neural unit to zero mean and unit standard deviation before feeding the data into the ANN. Though theoretically an ANN should be able to incorporate normalization into its processing, simplifying the job of a classification system will generally improve its performance. When an ANN is used as a classification system, only the inputs to the network need to be normalized, since the outputs should always be on the same scale.

The next step in data preprocessing is feature extraction. The main goal of feature extraction is to reduce the number of input variables or dimensions that make up the input space. The input space is defined by all possible input vectors observed in the data set. Theoretically, a classification system should be able to handle massive amounts of information and find the important elements contained in the original data set. In practice, however, an overload of information will likely occur for large data sets and dramatically reduce the performance of the classification system.[16] This phenomenon, known as the "curse of dimensionality," results from the fact that each additional input dimension increases the size of the input space exponentially. Each input dimension also adds an exponentially increasing number of training trials required for proper mapping of the input space.[15] Therefore, the network's ability to perform an optimal classification of trials and the time it takes to achieve this classification are both adversely affected by overly large input spaces. Feature extraction can solve this problem by providing the ANN with just enough input variables to make the classification.

An effective way to implement feature extraction is to focus only on the data that are likely to contain the information on which the classification of trials is going to be based. For example, if in a set of trials a sensory stimulus is presented at a frequency of 1 Hz and significant neuronal responses are produced only during the first 50 ms after the stimulus onset, there is no need to include in the analysis the remaining 950 ms of recorded data in each trial. This decision can be guided by using data obtained in a previous analysis step in which poststimulus time histograms are used to characterize the time duration of neural sensory responses. This simple step would remove 95% of the recorded data from the input vectors and make the classification of trial much more efficient.

Selection of the neurons that make up the input vectors is extremely important as well. Neurons that are noisy, inactive, or fire only rarely should generally be removed from the input vectors. However, caution should be exercised every time the decision of removing a given neuron is made. To ensure that the network's classification decision is based on the information content rather than the amount or source of the information, all input vectors must be of the same size regardless of their class. Therefore, if a neuron is removed from the input vectors describing the ensemble response to one class of stimuli, it must be removed from all other input vectors describing all other stimulus classes. More commonly, specific neurons are retained or removed to explore the coding properties of different regions of the brain (see below).

Feature extraction, or dimension reduction, can also be carried out without using prior knowledge by employing one of the variety of unsupervised methods. In practice, most data sets can be reduced to a smaller number of variables (or dimensions) without significant loss of information. The immediate gain provided by data dimension reduction is a considerable decrease in analysis time, a factor that can be of great relevance when large data sets are used. Whether dimensional reduction leads to better discrimination varies from case to case. Classical examples of dimension reduction techniques are factor analysis, principal component analysis (PCA), independent component analysis (ICA), classification and regression trees (CART),[17] wavelet based selections,[18] or even another ANN.

It is important to remember that any dimension reduction technique used for ANN preprocessing should be applied to the set of training trials separately from the set of testing trials. If they are transformed together the testing set will have an effect, however small, on the transformation of the training set. This effect may allow the testing data to influence the training of the network and result in a false performance measure. Thus, for each training and testing session, the dimension reducing transformation matrix, such as the eigenvectors in PCA, should be found from the isolated training set but used to transform both training and testing sets. Then, all the data will undergo the same transformation, but influences of the testing set will not affect training of the network.

7.4.5.3 Different types of artificial neural networks

Once the neural ensemble data have been preprocessed, an ANN has to be selected. In this section, specific network architectures, learning functions, and training parameters are described which have been found to maximize performance. The goal here is to provide an introduction on how to employ two different ANNs for pattern recognition. For further details, the reader may consult several reference texts which describe the mathematics and theory behind the use of ANNs.[15,19] The basic unit of an ANN is the neuron model (Figure 7.12a). This artificial neural unit (ANU) accepts an input vector **I** and is initialized with an appropriate vector of weights **W** and a bias, b. The ANU output is produced by a transfer function, whose arguments are the input vector, the weight vector, and the bias. This output depends on the type of ANN used and the ANU's specific transfer function.

for backpropagation networks the output, $a = f_T(\mathbf{W} \cdot \mathbf{I} + b)$

for competitive networks the output, $a = f_T((\mathbf{W} - \mathbf{I}) + b)$

Various training processes and learning algorithms can be used to adjust the weights and bias of each ANU so that a given problem is most accurately solved. Complex patterns can be discriminated by connecting multiple ANUs and by using different transfer functions (e.g., log sigmoid, hyperbolic tangent sigmoid, etc.). Similar ANUs are generally grouped into layers which have particular connectivity. Multilayer ANNs are usually employed for large multivariate problems such as classification of biological images.

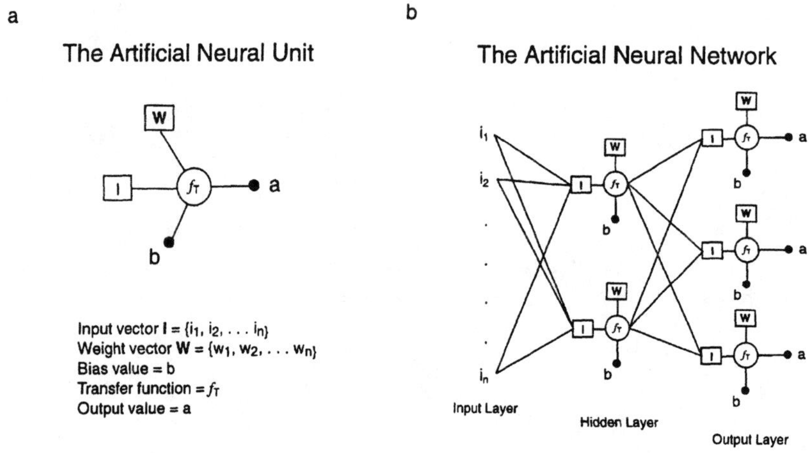

FIGURE 7.12
General description of an artificial neural unit (ANU, a) and the architecture of a multilayer artificial neural network (ANN, b).

In a multilayer configuration (Figure 7.12b), the first layer, also known as the input layer, is simply the input vector that is fed to the ANN during either the training or testing phases of the analysis. Every element of the input layer is directly connected to a group of ANUs that defines the next layer of the network. Usually this layer is known as a "hidden" layer because its outputs become the inputs of the next layer. The hidden layer must have enough units to handle the complexity of the data. However, too many hidden units may slow up training and actually overfit the data. Overall, the question of how many hidden units to use in an ANN can only be answered empirically, since the best value varies from problem to problem. Although one can design ANNs with several hidden layers, our experience is that efficient classification of neural ensemble patterns can be achieved by using only one.

The last layer of the ANN is the output layer. The units in this layer receive inputs from a hidden layer and provide the final output of the ANN. In our approach, the number of units in the output layer equals the number of classes to be discriminated. In this simple encoding scheme, each output unit has a value of 1 when an input vector is classified as belonging to its class, and a value of 0 at all other times. This is commonly known as 1-of-c encoding. Thus, a discrimination among three classes of trials (e.g., three stimulation sites), then, would have output vectors (1,0,0) for the first class, (0,1,0) for the second class, and (0,0,1) for the third.

To illustrate the potential application of ANNs in neural ensemble data analysis, we employed two different feedforward network designs, backpropagation and competitive networks, to classify single trials describing the sensory responses of neural populations located in different cortical areas of owl monkeys. By feedforward networks we mean ANNs that do not contain feedback loops connecting higher layers with lower ones.

The ANNs described here are representative of the two general approaches used to address the problem of pattern classification in multivariate systems. In a backpropagation network, the ANUs apply a mathematical transformation of the inputs in an attempt to produce the appropriate output vector. This process divides the input space and separates the classes with specific "hyperplanes." In a competitive network, on the other hand, each ANU's weight vector tries to match an input vector. The unit whose weight vector is closest to the input vector "wins" the competition. This process divides the input space into clusters. Thus, each hidden unit represents one cluster, and each cluster represents one class.

Backpropagation networks are generic problem solvers. These ANNs employ a learning algorithm which requires an error function and a method for evaluating the derivatives of the error function. The derivatives are used to compute the adjustments to be made to the weights.[15] There is a variety of powerful optimization schemes that can make these adjustments very fast, such as the Levenberg–Marquadt algorithm[20,21] and the resilient backpropagation algorithm.[22] The transfer functions that make up the backpropagation network must be able to handle the complexity of the input data. We have found that the nonlinear tangent–sigmoid transfer function works very well as an ANU transfer function. An output layer of log–sigmoid transfer functions is also advantageous for classification problems, since it produces values in the range of 0 to 1. The number of hidden units that is required to achieve maximum performance of a backpropagation network is unpredictable. One should begin with no more than five units and slowly increase the number of units until the performance ceases to improve.

Another important issue to consider when employing backpropagation networks is how to choose a training paradigm and determine the amount of network training that should be carried out. In this study, backpropagation training was carried out by using the resilient backpropagation method, and the number of training epochs, i.e., the number of examples presented to the ANN, was determined by "early stopping." Early stopping is an effective regularization technique for backpropagation networks that minimizes overfitting and automatically determines how many training epochs to use.[23] To use the early stopping method a test of the network's progress after each training epoch is required. When the performance of the network starts to decrease, training stops.

Competitive networks are ideal for classification problems because their output is always one clear "winner" (the output is either 0 or 1) that can be interpreted as the appropriate class for the input vector. This contrasts with the backpropagation networks which have multiple outputs which can assume intermediary values (between 0 and 1 in our case). The learning process in competitive networks matches the weights of each hidden unit to one reccurring input pattern. When testing the network, the hidden unit whose weight vector is closest in Euclidean distance to the input vector is the winner. The function of the output layer is simply to assign each hidden unit to one class. Note that the number of hidden units should match the number of pattern clusters there are to be recognized rather than the number of classes, as the number of clusters in the data set could be greater than the number of classes. This possibility depends on the noise level of the data and the consistency of the patterns, but the use of one to five hidden units per class is a good general rule to follow.

Like backpropagation, there are many competitive network algorithms that work very well. We have found that the optimized learning vector quantization (OLVQ) algorithm can be applied to a competitive ANN by using Kohonen's learning rule[24] and competitive transfer functions throughout the hidden layer. Competitive networks can suffer from overfitting because of too many hidden units, but not because of too much training. A safe number of training epochs to use for an OLVQ competitive network is the product of the number of hidden units and the number of training vectors as the number of training epochs.

Cross-validation should be used for every network training session. The performance of a network depends somewhat on the initial random weights and the randomly selected input vectors it is trained with. Cross-validation provides multiple training results so that a mean and a standard deviation can be found for one analysis. The training and testing sets should contain multiple examples of each class in a random sequence.

7.4.5.4 A case example

As an example for the analysis routine described above, this section details the sequence of steps involved in a typical analysis section. The question being addressed here is whether the sensory responses of a population of cortical neurons can be used to identify the location of the stimulus on a single trial basis. A full description of these results is provided in Reference 8. The neural network toolbox from MATLAB (The Mathworks, Cambridge, MA) was utilized for all ANN analysis described here. The MATLAB LVQ competitive network code was altered slightly by using the OLVQ rather than the default LVQ1. All routines were run on Intel-based Pentium workstations.

The first step in this analysis is to create a data set from the multiple time series that contain the time stamps of each cortical neuron's spiking obtained during the stimulation of different hand sites in adult owl monkeys. In these experiments, different regions of the monkey's hand were stimulated in random order by using a computer-controlled mechanical probe (stimulus duration = 100 ms, stimulus frequency 1 Hz) while the extracellular activity of neurons, located in somatosensory areas 3b, SII, and 2, was simultaneously recorded. A total of 360 single trials was obtained for each of the 25 stimulation sites. Since PSTHs revealed that the minimal latency of these neuronal sensory responses to tactile stimuli was around 10 ms and relevant information about the location of these stimuli was likely to be present in the subsequent 45ms (Figure 7.4), only the spiking activity from 5 to 50 ms post-stimulus time was used to create our input vectors. This time window (45 ms) was then divided into consecutive 3 ms time epochs (or bins), resulting in 15 variables per neuron per trial. A general description of the data matrix used in this analysis is illustrated in Figure 7.11.

Neuronal units from cortical areas 3b, SII, and 2 were included in this analysis. However, neurons that did not fire at least 100 spikes during one stimulus session (360 trials) were excluded from the analysis. For example, in a data set containing 39 SII and 40 area 2 neurons, a total of 58 neurons remained in the data set following feature extraction. Therefore, after this initial phase of feature extraction, the resulting data set in this example was defined by 58 neurons times 15 time epochs = 870 variables and 25 sites times 360 trials per site for a total of 9000 trials. Variables describing each neuron's firing pattern were normalized so that they had zero mean and unit variance (see above).

In this analysis (see Plate 7 for a step by step guide), each individual trial is associated with an expected output vector (using 1-of-c encoding) which defines the trial class (i.e., the location of the stimulus). All collected trials are then randomized, and separated into training and testing sets. The expected output vector of each trial is used to both teach the ANN to correctly associate training input vectors with their stimulus locations and later to quantify how well the ANN output matches the expected output vector for each individual testing trial.

Using the ANNs described above, we were able to demonstrate that neural ensemble firing patterns, derived from populations of 20 to 40 neurons located in areas 3b, SII, and 2, contain enough information to allow statistically significant identification of the location of a punctate tactile stimulus on a single trial basis[11] (see Figure 7.13). When normalized raw spike trains were used, backpropagation networks were always faster than competitive ANNs. However, back propagation nets never outperformed the classification accuracy exhibited by competitive networks. In fact, competitive networks sometimes outperformed the best backpropagation network by as much as 10% accuracy.

When PCA and ICA were used for data dimension reduction, performance was not increased but the computation power (e.g., memory) required for the analysis was significantly reduced. This is exemplified by the fact that in the example described above when principal components accounting for 99.9% of the total variance were calculated, the input vectors used in the analysis were reduced from 870 to just 391 variables (PCs).

Several controls were carried out to characterize the limits of the optimal ANN performance (see Plate 7). For example, during the analysis we measured how the discrimination performance was affected by gradually reducing the number of training trials presented to the ANN.[8] Overall, we observed that only 25% of the total trials were needed to properly train competitive networks. When 25% and 50% of the trials were used to train the network to distinguish the location of a punctate stimulus, considering many combinations of three different classes (chance level 33% of correct discrimination), proper trial classification occurred in 63 ± 0.02% and 66 ± 0.02% of the remaining testing trials, respectively. This was nearly equivalent to the 67 ± 0.02% accuracy realized when 90% of the trials were used in training, but far superior to the 55 ± 0.02% accuracy obtained when the ANN was trained on just 10% of the trials.

7.4.5.5 Using ANNs to characterize neural ensemble encoding strategies

In addition to classifying individual trials according to the location of sensory stimuli, ANNs can be used to investigate the potential encoding strategies employed by neural populations. To carry out this second step of our analysis, the performance of an OLVQ competitive network with two hidden units per class was measured

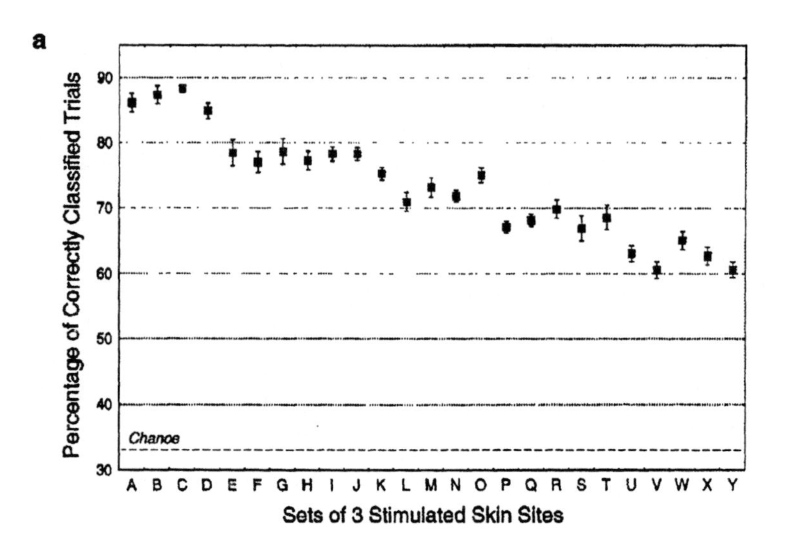

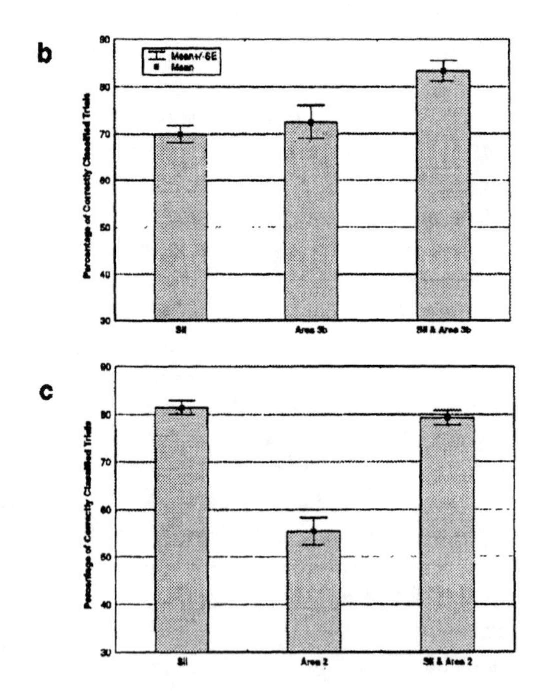

FIGURE 7.13
Populations of neurons in areas 3b, SII, and 2 were used for classification of the location of a punctate tactile stimulus. (a) Data from area SII allowed significantly greater than chance classification for 25 different random combinations of three hand locations. (b) Cortical areas 3b and SII from the same animal both contain discriminatory information. The combination of the two regions allows a significant increase in classification performance, suggesting that some of the information in these areas is unique. (c) Both cortical areas 2 and SII contain discriminatory information. The combination of the two regions, however, does not provide a significant increase in classification, suggesting that some of the information in these cortical areas is redundant.

after modifications in the original data set that included: (1) random shuffling of the trial number for each individual neuron (trial shuffling); (2) continuous increase in the size of the time epoch (bins), from 3 to 45 ms, that described each cortical neuron sensory response (bin clumping); (3) continuous reduction of the cortical ensemble size by sequential removal of individual neurons (neuron dropping); (4) testing different cortical populations either independently or in combination.

Each of these data manipulations aimed at testing different aspects of the encoding strategy employed by different populations of cortical neurons. For instance, trial shuffling provides a way of measuring the consistency of neuronal responses across trials and evaluating how robust a population pattern can be on a single trial basis. The procedure used here includes training the network with the original data set and then carrying out a random trial shuffling for each neuron independently so that each "testing trial" presented to the trained networks is actually formed by single neuron firing patterns obtained in different trials. This method can be used not only to test the performance of ANNs but also to measure the robustness of the neural correlation patterns observed in a single trial. In the test example described in this chapter, trial shuffling had little effect in the performance of ANNs. In some cases the ANN performance was enhanced after trial shuffling.

Bin clumping, the second data manipulation used in our laboratory provides an indirect way to investigate the relevance of temporal attributes of single neuron responses for the encoding of sensory or motor information. Although the strategy may vary from experiment to experiment, bin clumping for this study was obtained by varying the size of the time bin from 3 to 45 ms. Figure 7.14 describes the results obtained when such a manipulation was performed for two simultaneously recorded populations of neurons located in areas 3b and SII. Notice that while bin clumping had no effect on the ability of the ANN to correctly discriminate the location of a punctate tactile stimulus based on single trial responses of area 3b neurons, the same ANN experienced a significant reduction in discrimination capability when bin clumping was applied to SII neurons. One possible interpretation of these results is that while area 3b neurons may rely exclusively on variation in population firing rate to represent the location of a tactile stimulus on a single trial, SII neural ensembles may take advantage of the temporal structure of their sensory responses to accomplish the same task.

A third manipulation, neuron dropping, can be used to evaluate the contribution of individual neurons to the representation of tactile information. In addition, neuron dropping allows us to describe the discrimination capability of neural ensembles as a function of the number of neurons that form such ensembles. Since our multisite recordings allow multiple cortical and subcortical structures to be sampled simultaneously, we use this procedure to compare the resulting functions (discrimination vs. ensemble size) across brain regions for the same set of sensory stimuli or behavioral tasks.

Our studies of the owl monkey somatosensory cortex revealed that the areas surveyed (areas 3b, SII, and 2) tend to rely on a distributed encoding scheme to represent the location of a punctate tactile stimulus. Neuron dropping further supported this conclusion by demonstrating that a serial reduction of the neural ensemble leads to a graceful degradation of the single trial discrimination (Figure 7.15). The

a

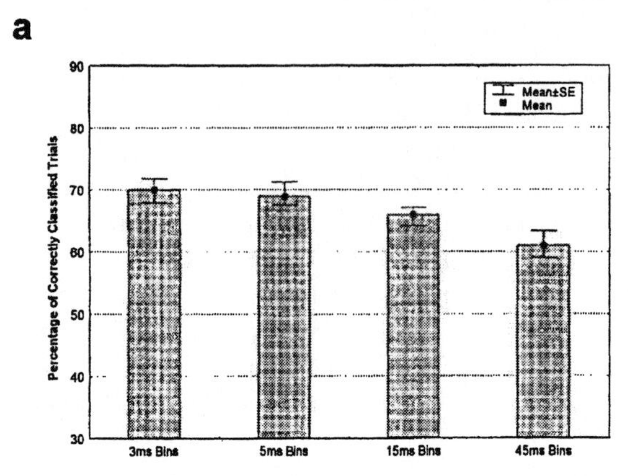

b

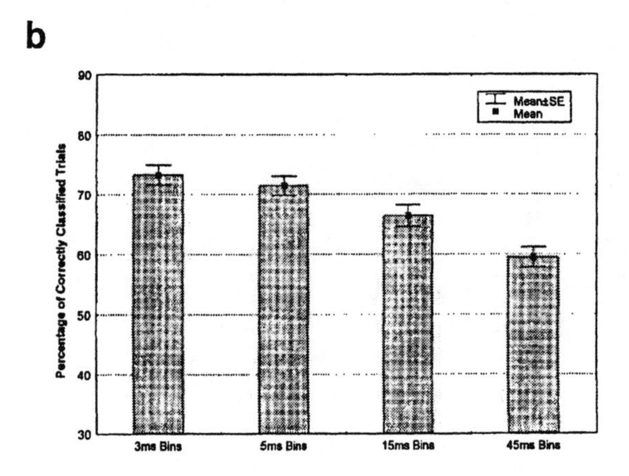

c

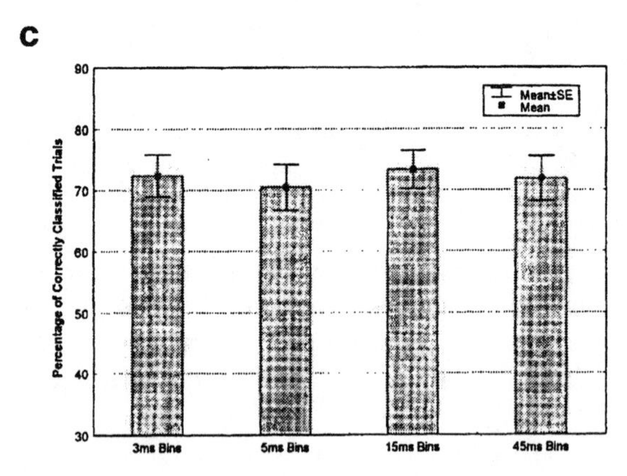

FIGURE 7.14
Effects of bin clumping on ANN classification performance. Bin sizes of 3, 5, 15, and 45 ms were used to represent the recorded data for the ANNs. Discrimination based on data from SII, shown in both (a) and (b) for two different animals, was adversely affected by large bin sizes, demonstrating that temporal precision in the order of 5 ms exists within SII. (c) Area 3b performs equally well for all bin sizes used, suggesting that information may be primarily encoded by variations in average firing rate. (From Nicolelis, M.A.L., Ghazanfar, A.A., Oliveira, L.M.O., Chapin, J.K., Nelson, R., and Kaas, J.H., Simultaneous encoding of tactile information by three primate cortical areas, *Nature Neurosci.*, in press. 1998. With permission.)

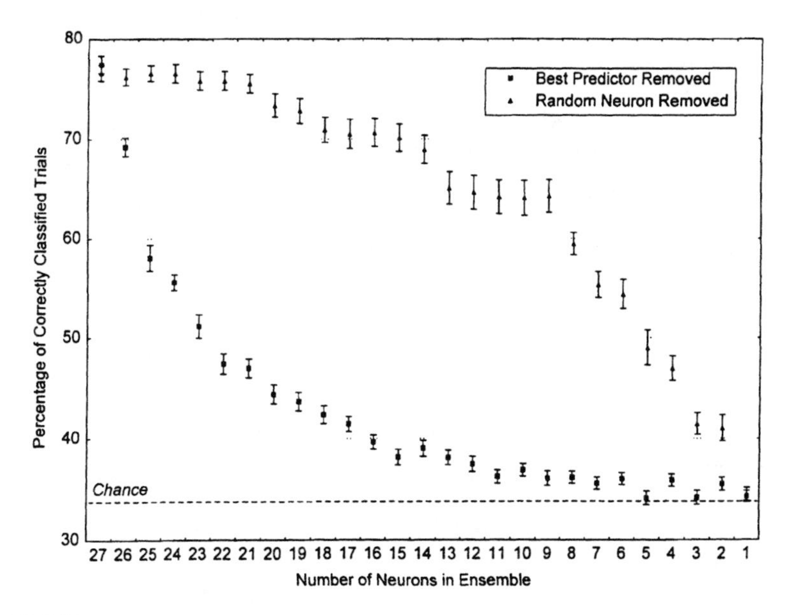

FIGURE 7.15
The effect of neural ensemble size on classification performance. The best predictor reduction method for neuron dropping (solid square) shows a markedly different trend than the random neuron removal method (solid triangle). The neural ensemble used for this analysis was located in the secondary somatosensory cortex. Five combinations of three stimulus locations were used to test the performance of each neural ensemble size.

procedure used for this analysis was very simple. First, the discrimination performance was measured with the entire ensemble. Next, the best single neuron predictor of the ensemble was removed and the classification of single trials was repeated. This procedure was repeated until only one neuron remained. As an alternative, random or removal of the least important predictor can be used as the main algorithm for neuron dropping.

The final data manipulation procedure used in our analysis is to use ANNs to test how interactions between neural ensembles located in different brain regions (e.g., cortical areas or subcortical nuclei) may affect the ability of discriminating the location of a stimulus on a single trial basis. To perform this analysis, single neurons are added to the ANN in a controlled manner such that the importance of cortical areas as well as ensemble size in the classification of single trials can be determined. Figures 7.13b and 7.13c describe an example in which an OLVQ competitive network was used to classify the stimulation sites based on pairwise combination of these cortical populations. Comparison of these results with the performance obtained for individual populations revealed that, whereas the combination of area SII and 2 neurons does not improve single trial discrimination, blending of area 3b and SII neurons does enhance the probability of correctly identifying the location of a tactile stimulus on a single trial basis. This analysis suggests that while SII and area 2 neural ensembles may share the same type of information used to represent the location of a stimulus, area 3b and SII neurons may encode different features involved in the genesis of these representations. Therefore, combination of these latter ensembles may enhance the overall discrimination performance at the single trial level.

As mentioned previously, the results obtained with all these data manipulations have to be considered so that one can reach meaningful conclusions about any data set. Moreover, it is also important to remember that none of the analytical techniques described in this section are known to reproduce brain-like mechanisms of information processing. Consequently, the results obtained by applying these methods have to be interpreted with great caution since they only indicate what potential computations could be carried out by the population firing patterns embedded in the neural ensemble data sets.

7.5 Conclusion and Future Directions

This chapter provides a description of some of the methods that are currently employed to record and analyze the extracellular activity of hundreds of single neurons distributed across different cortical areas of behaving primates. However, it is important to realize that this is just the beginning of a new era in neurophysiology. Several key technological developments, such as the production of reliable 3D multielectrode cubes, automatic multichannel real-time spike sorting, and the application of modern very large scale integration (VLSI) circuits that incorporate multichannel telemetry to acquire and transmit brain signals, are likely to increase the yield of these experiments to several thousand single units per animal and allow such recordings to be performed in freely behaving primates. To cope with this data flood, new analytical tools will have to be continuously developed not only to visualize the information available in these data sets but also to characterize the parallel computations performed by such vast neural ensembles.

In conclusion, the implementation of these and other techniques promises to further develop the field of neural ensemble physiology while allowing neuroscientists to venture into fundamental questions in modern cognitive neuroscience. We believe that among other applications, chronic multichannel recordings will likely provide a unique way to quantify the effects of training and learning upon the physiological properties of cortical and subcortical neural populations that define large brain systems. These two examples alone justify our contention that the development of methods for chronic multichannel single unit recordings in behaving primates is at the frontier of the emergent field of neural ensemble physiology.

Acknowledgments

The preparation of this chapter and the work discussed in it were supported by grants from DARPA-ONR, The McDonnell Pew Foundation, The Duke Sandoz Program, and The NINDS Neuroprosthetic Program to M.A.L.N.

References

1. Lilly, J.C., Correlations between neurophysiological activity in the cortex and short-term behavior in monkey, *Biological and Biochemical Bases of Behavior.* Harlow and Woolsey (Eds.), The University of Wisconsin Press, Madison, pp. 83-100, 1958.
2. Chapin, J.K. and Woodward, D.J., Modulation of sensory responsiveness of single somatosensory cortical cells during movement and arousal behaviors, *Exp. Neurol.* 72, 164, 1981.
3. Kruger, J., Simultaneous individual recordings from many cerebral neurons: techniques and results, *Rev. Physiol. Biochem. Pharmacol.*, 98, 177, 1983.
4. Deadwyler, S.A. and Hampson, R.E., The significance of neural ensemble codes during behavior and cognition, *Annu. Rev. Neurosci.*, 20, 217, 1997.
5. Nicolelis, M.A.L., Lin, R.C.S., Woodward, D.J., and Chapin, J.K., Dynamic and distributed properties of many-neuron ensembles in the ventral posterior medial (VPM) thalamus of awake rats, *PNAS U.S.A.*, 90, 2212, 1993.
6. Nicolelis, M.A.L., Baccalá, L.A., Lin, R.C.S., and Chapin, J.K., Sensorimotor encoding by synchronous neural ensemble activity at multiple levels of the somatosensory system, *Science,* 268, 1353, 1995.
7. Nicolelis, M.A.L., Ghazanfar, A.A., Faggin, B., Votaw, S., and Oliveira, L.M.O., Reconstructing the engram: simultaneous, multiple site, many single neuron recordings, *Neuron*, 18, 529, 1997.
8. Nicolelis, M.A.L., Ghazanfar, A.A., Oliveira, L.M.O., Chapin, J.K., Nelson, R., and Kaas, J.H., Simultaneous encoding of tactile information by three primate cortical areas, *Nature Neurosci.*, in press. 1998.
9. Laubach, M., Shuler, M., and Nicolelis, M.A.L., Principal and independent component analysis for multi-site investigations of neural ensemble interactions, submitted 1998.
10. Comon, P., Independent component analysis — a new concept?, *Signal Processing*, 36, 287, 1994.
11. Cardoso, J.F. and Souloumiac, A., Blind beamforming for non gaussian signal, *IEEE — Proceedings-F,* 140, 362, 1993.
12. Bell, A.J. and Sejnowski, T.J., An information maximization approach to blind separation and blind deconvolution, *Neural Computation*, 7, 1129, 1995.
13. Hyvarinen, A. and Oja, E., A fast fixed-point algorithm for independent component analysis, *Neural Computation*, 9, 1483, 1997.
14. Laubach, M. and Nicolelis M.A.L., Interactions between sensorimotor cortical and thalamic neuronal ensembles are altered during the acquisition of a reaction-time task, *Society for Neurosci. Abstract*, 1, 132, 1998.
15. Bishop, Christopher M., *Neural Networks for Pattern Recognition,* Clarendon Press, Oxford, 1995.
16. Bellman, R., *Adaptive Control Processes: A Guided Tour.* Princeton University Press, Princeton, 1961.
17. Breiman, L., Freidman, J.H., Olshen, R.A., and Stone, C.J., *Classification and Regression Trees*, Wadsworth International Group, Belmont, 1984.
18. Buckheit, J. and Donoho, D., Improved linear discrimination using time-frequency dictionaries, *Proceedings SPIE,* 2569, 540, 1995.

19. Ripley, B.D., *Pattern Recognition and Neural Networks,* Univ. of Cambridge Press, Cambridge, 1996.
20. Levenberg, K., A method for the solution of certain non-linear problems in least squares, *Quarterly Journal of Applied Mathematics* II, 2, 164, 1944.
21. Marquardt, D.W., An algorithm for least-squares estimation of non-linear parameters, *Journal of the Society of Industrial and Applied Mathematics*, 11, 431, 1963.
22. Riedmiller, M. and Braun, H., A direct adaptive method for faster backpropagation learning: The RPROP algorithm, *Proceedings of the IEEE Int. Conf. on Neural Networks,* 586, 1993.
23. Sarle, W.S., Stopped training and other remedies for overfitting, *Proceedings of the 27th Symposium on the Interface of Computer Science and Statistics*, 352, 1995.
24. Kohonen, T., *Self-Organizing Maps,* Springer, New York, 1997.

Chapter 8

Correlation-Based Analysis Methods for Neural Ensemble Data

George L. Gerstein

Contents

8.1 Introduction

Whenever neurobiological data contains two or more streams of events in time (such as action potentials, postsynaptic potentials, discrete stimuli), it is appropriate to seek temporal relationships between the streams. Basically we search for coincidence or near coincidence of events in the several streams; the appropriate mathematical tools are related to correlation. Direct application to the data produces what we will call the "raw" correlation or coincidence structure. However, even completely random and unrelated streams will show some degree of "accidental" coincidence; this level will be modified if two or more streams are influenced by the same modulating

0-8493-3351-2/99/$0.00+$.50

process (stimuli, for example). Thus it is necessary to develop a measure that extracts the excess or deficit coincidence structure, i.e., the deviation from the amount expected by chance and from all known sources of shared influence. Finally, we need a significance measure for this measured deviation.

Coincidence structure among spike trains (or among spike trains and other point events) can be interpreted in two different ways:

1. Excess coincidence above the expected amount suggests a functional connectivity among the observed neurons; this is not the anatomical connectivity, but a minimal logically equivalent circuit. Functional connectivity is one way to define membership in neuronal assemblies, and its dynamics can be interpreted as dynamics of assembly organization. It is important to note that the calculation of excess or deficit coincidence is a process that is only available in the laboratory and cannot be accomplished by the brain itself.
2. The "raw" correlation is, however, directly available to the brain. The raw coincidence structure is potentially significant in representation and coding because synchrony among convergent neurons greatly increases influence on the postsynaptic neuron. Thus synchrony likely plays a role in the transmission and transformation of information, and has been implicated in higher order processes, such as binding in a visual image.

In this chapter we will examine several forms of correlation measures as tools to delineate the temporal relationship between event streams. We will start with ordinary cross-correlation between two trains of action potentials. This is a measure of relationship that is averaged over the entire length of available data, thus losing any dynamics in the relationship. Whenever the neuron firing patterns are not stationary, or if they are modulated by context like stimulation or behavior, other methods are needed. Here we will examine two of these, the joint peri-stimulus time histogram (JPSTH) and the gravity representation. The first of these delineates the average stimulus modulation of coincidence structure between two spike trains; the second gives a nonaveraged dynamic measure of near coincidence structure among any number of spike trains. The latter is particularly appropriate for dealing with the newer methods of dense recording when spike trains from up to 100 individual neurons are simultaneously observed.

8.2 Cross-Correlation

The cross-correlogram of spike trains from neurons A and B is defined by the number of spikes from neuron B at time lag $\tau = k\,\Lambda$ relative to spikes from neuron A. Λ is the binwidth, and the index k runs from $-W \ldots 0 \ldots W$, where $2W+1$ is the total number of bins in the correlogram. Spikes from neuron A are called the reference event, spikes from B, the target event. With the assumption of independence of the two spike trains, the expected value in each bin is $E = N_A N_B \Lambda / T$, where N_A and N_B are the numbers of spikes in the two trains, and T is the total duration of the data. Assuming Poisson counts in each bin, the standard deviation of $E = E^{1/2}$. A cross-correlogram is significant to $P < .001$ if some bins differ from E by 3 standard deviations.

If there is periodic stimulus and related comodulation of firing of the two neurons, the expected correlogram can be computed by (a) the simple shift predictor, (b) the average of all possible shift predictors, or (c) the correlation of the two PST histograms. The simple shift predictor is obtained by shifting one of the two spike trains by exactly one stimulus period, and again computing the correlogram. This destroys any possible physiological spike for spike correlation, but preserves the effects of the comodulation. The average of shift predictors of all orders or the PST predictor are identical, differing only by the sequence of summing and correlating, and are statistically the better choice.

Considerably more detail on this and related topics can be found in two books,[1,2] and in the methods sections of several papers.[3,4]

8.3 Joint Peri-Stimulus Histogram (JPSTH)

The ordinary cross-correlation between two spike trains and the shift or PST-based predictor are averages over the entire length of data, and hence ignore whatever dynamics there may be. This means we can only infer the static average interaction between the neurons. A portion of the dynamics may be associated with the repeated presentation of stimulus, and it is this portion that is delineated by the JPSTH (as an average over the presentations). The JPSTH also gives a better approach to extracting excess correlation than do the traditional shift or PST predictors described above.[5,6]

Construction of the JPSTH begins with creation of the scatter diagram shown in Figure 8.1. For each presentation of the stimulus, we take the stimulus time as origin and lay off the activity of the two neurons, each along its own axis as indicated. Points are then plotted on the plane at all logical ANDs of the several action potentials on each axis. The process is repeated for subsequent stimuli, building up a scatter of points on the plane. Various features of the neural activity will produce regions of higher point density. If, for example, one neuron responds to the stimulus with a brief increase in firing rate, there will be a bar of high point density parallel to the appropriate axis at a location corresponding to the response latency. If there is a high probability of near coincidence in the two trains of action potentials, there will be a bar of high point density parallel to the principal diagonal, and at a distance from it that represents the latency of the favored near coincidence. There may be modulations of density along this bar, indicating the averaged stimulus locked modulations of the probability for the near coincidences.

The scatter diagram described in the previous paragraph is constructed at the maximum temporal resolution available in the data, i.e., corresponding to a single clock tick. In order to allow statistical manipulation of the point densities we need to make coarser bins within which to count points. The appropriate types of bins are shown in Figure 8.2. For the JPSTH matrix, we use square bins at the same values that are used for the ordinary PST histograms of each spike train (Figure 8.2A). For the region near the principal diagonal we use the bins indicated

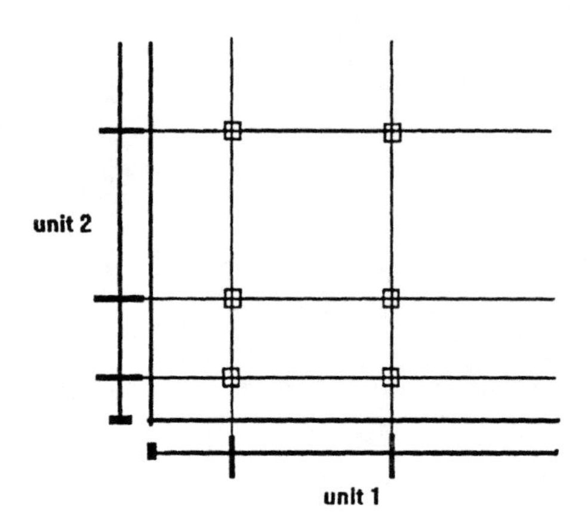

FIGURE 8.1
Construction of JPSTH.

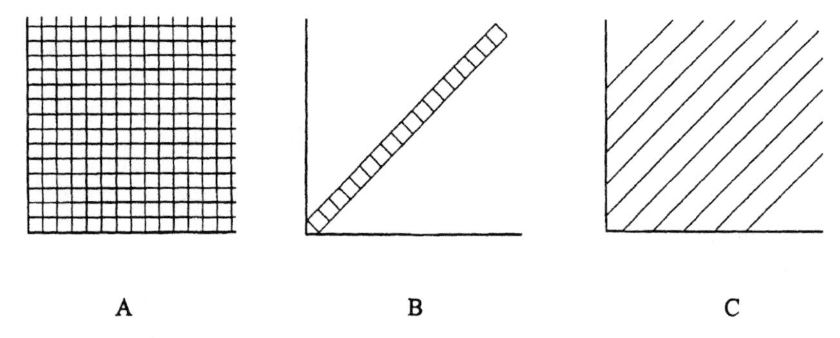

FIGURE 8.2
Bin arrangements for: (A) JPSTH matrix, (B) Peristimulus coincidence histogram, (C) cross correlogram.

in Figure 8.2B, producing the PST coincidence histogram. This measures the averaged stimulus locked near coincidence probability in the same sense as the ordinary PST histogram measures the averaged stimulus locked firing probability of each neuron individually. Finally, by summing in paradiagonal bins shown in Figure 8.2C we produce the ordinary cross-correlogram of the two spike trains. Note that since these bins are not of equal length (and hence area), we must either normalize by the bin length, or compute a larger matrix so as to allow making all bins of equal length.

In the following sections we will apply the JPSTH measurement to a set of spike trains produced by a simulator. The simulated circuit arrangement is shown in Figure 8.3. The observed neurons "10" and "12" are each "spontaneously" active because of independent and stationary inputs from networks G10 and G12. In addition, a periodic stimulus input briefly elevates firing probabilities for the observed neurons. Finally, there is an excitatory synaptic connection from neuron "10" to neuron "12" whose strength is itself modulated by the stimulus; this is indicated in Figure 8.3 by the small square box in the path. The temporal profile of this modulation is an initial rapid rise followed by a slower decay.

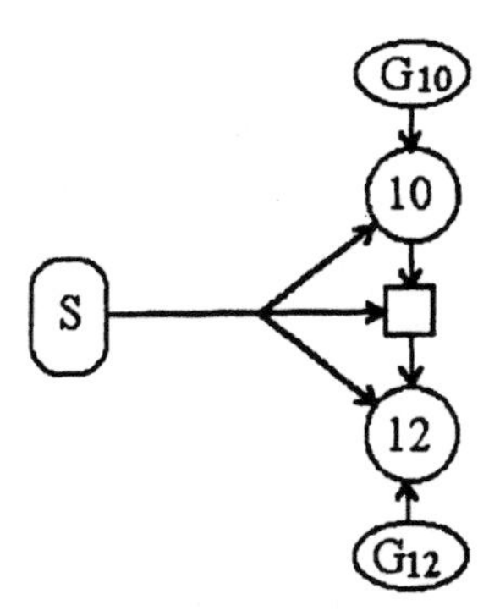

FIGURE 8.3
Circuit used to generate spike trains with simulator program.

The basic JPSTH analysis of these "raw" data is shown in the upper left quarter of Figure 8.4. At left and below the JPSTH matrix are the ordinary PST histograms of the two units. The abscissa PST (for unit 10) is repeated below the histogram grouping at the right of the matrix. Rising toward the upper right is the PST coincidence histogram, and at the extreme right is the ordinary cross-correlogram. The PST histograms show the time course of the stimulus-driven increase in firing. As expected from the descriptions of the scatter diagram construction and the simulated circuit used to generate the data, the JPSTH matrix shows a hill, a fairly broad stripe parallel to each axis, and a narrow stripe parallel (just above) the principal diagonal. The PST coincidence histogram shows a broad peak at the times when both neurons have stimulus-induced higher firing rates, and is otherwise fairly small. Finally, the cross-correlogram shows a narrow peak near the origin riding on a flat background.

Because the observed neurons are being comodulated by the direct stimulus input, their firing rates increase and decrease in unison, thus creating a large part of the "raw" correlation structure. In order to infer "effective connectivity" or organization among our observed neurons we need to compare the "raw" correlation to that expected for two independent neurons with precisely the observed PST histograms. This is accomplished with the corrected JPSTH as shown in the upper right quarter of Figure 8.4. Here we subtract the matrix produced by the bin-by-bin product of the two PST histograms (divided by the number of stimuli) from the "raw" JPSTH matrix, and then divide, again bin by bin, by the product of the standard deviations in each PST histogram. Each bin in the matrix has now become a correlation coefficient. PST histograms are shown as before, but now the two diagonal binnings on the corrected matrix produce, respectively, the corrected PST coincidence histogram and the corrected cross-correlation. Note that the hill and bars parallel to the axes have disappeared from the corrected JPSTH matrix, while the diagonal stripe has persisted. The corrected PST coincidence histogram shows an initial rapid rise and subsequent slow fall, as expected from the stimulus modulation of the synaptic connection in the simulator. The corrected cross-correlogram shows only a central peak, with no "background" correlation. We note that this matrix-based correction

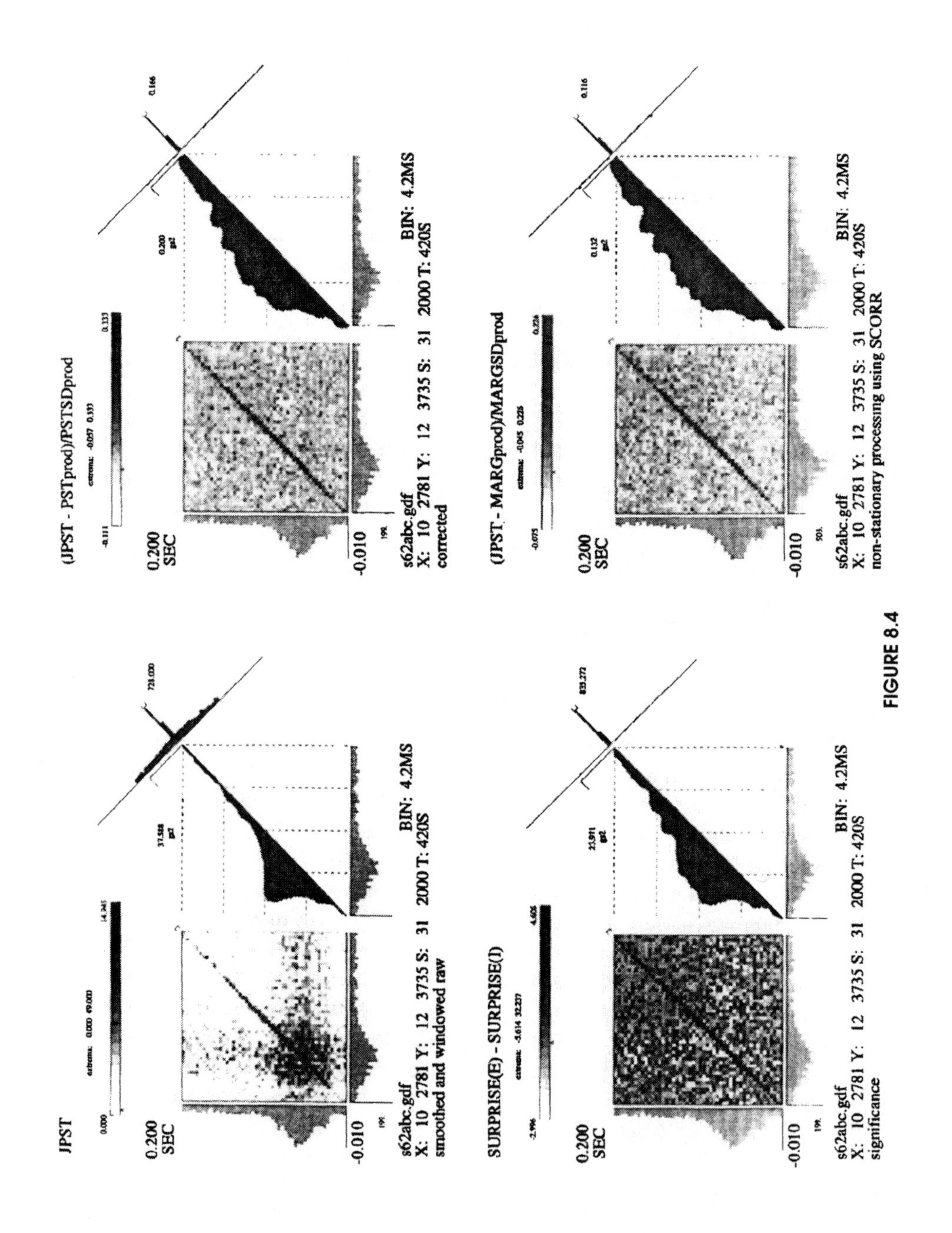
JPST
0.200
SEC
-0.010
s62abc.gdf
X: 10 2781 Y: 12 3735 S: 31 2000 T: 420S
BIN: 4.2MS
smoothed and windowed raw
(JPST - PSTprod)/PSTSDprod
0.200
SEC
-0.010
s62abc.gdf
X: 10 2781 Y: 12 3735 S: 31 2000 T: 420S
BIN: 4.2MS
corrected
SURPRISE(E) - SURPRISE(I)
0.200
SEC
-0.010
s62abc.gdf
X: 10 2781 Y: 12 3735 S: 31 2000 T: 420S
BIN: 4.2MS
significance
(JPST - MARGprod)/MARGSDprod
0.200
SEC
-0.010
s62abc.gdf
X: 10 2781 Y: 12 3735 S: 31 2000 T: 420S
BIN: 4.2MS
non-stationary processing using SCORR

FIGURE 8.4

of the cross-correlogram is somewhat different but more accurate than correction based on the more usual shift or PST-based predictor. Thus, this correction has extracted the (known) stimulus time-locked modulation of the "effective correlation" between the two neurons, and has removed the direct correlation effects of the stimulus-related comodulation of firing rates.

It is obviously necessary to establish significance for the features shown in the corrected JPSTH and its associated histograms. This can be done by assuming Poisson counts in each bin of the PST product (divided by the number of stimuli) matrix that is subtracted from the raw matrix. The standard deviation of this count is its square root, so that we can express the matrix in terms of standard deviation units. Values over 3 or 4 can be taken as highly significant. A second method depends on combinatorial manipulation of the observed bin counts. If we assume independence of the two spike trains, we can calculate the actual expected distribution of possible counts in a bin of the JPSTH matrix given the number of stimuli and the number of counts in each PST histogram bin (see Reference 6 for more detail). This allows us to evaluate the probability of observing the actual x,y bin count. The log of this probability is called "surprise."[7] The lower left quarter of Figure 8.4 shows the "surprise" calculation on the JPSTH matrix. The gray-level bar is set to indicate 99% confidence in the direction of excess counts (an excitatory interaction) and at 95% confidence in the direction of deficit counts (an inhibitory interaction).

The entire treatment described above assumes stationarity in the large, i.e., that the responses to any presentation of a particular stimulus are statistically identical. (The JPSTH deals with the nonstationarity *within* a stimulus presentation which are expressed by the stimulus time-locked modulations.) In real physiological experiments, stationarity is of course frequently violated as the state of the brain varies. The general effect on the JPSTH calculations is an underestimation of the predicted matrix, so that the observed counts appear to be more significant than they "should." Thus, it is appropriate to seek modifications of the JPSTH calculations which deal somewhat more gracefully with the slow changes along an experiment. One possible approach is to use the marginal distributions of the raw JPSTH matrix instead of the PST histograms in creating the corrected JPSTH. This is shown in the lower right quarter of Figure 8.4; this calculation considerably reduces the effects of response nonstationarity while producing essentially the same results as the ordinary

FIGURE 8.4 (opposite)
JPSTH analysis of spike trains generated by a simulator according to the connections shown in Figure 8.3. Top left: The raw data. The matrix shows the JPSTH; to the left and repeated twice underneath are the ordinary PSTs of the two spike trains. Rising diagonally to the right is the PST coincidence histogram taken over the bins shown by the scoop in the upper right corner of the matrix. The remaining diagonal is the cross-correlogram. Top right: Correction for stimulus-locked firing rate modulations as described in text. The computation has removed "Background" correlation in the matrix and each diagonal histogram. The PST coincidence histogram now shows the (known) time course of effective connection between the neurons, while the cross-correlogram has only two non-zero bins. Bottom left: The "surprise" measure of significance. See text. Bottom right: Correction for stimulus-locked firing rate modulations using the marginal distributions instead of the PSTs. For these data this method produces results almost identical to the PST correction, but is considerably less sensitive to possible nonstationarities that are not locked to stimulus.

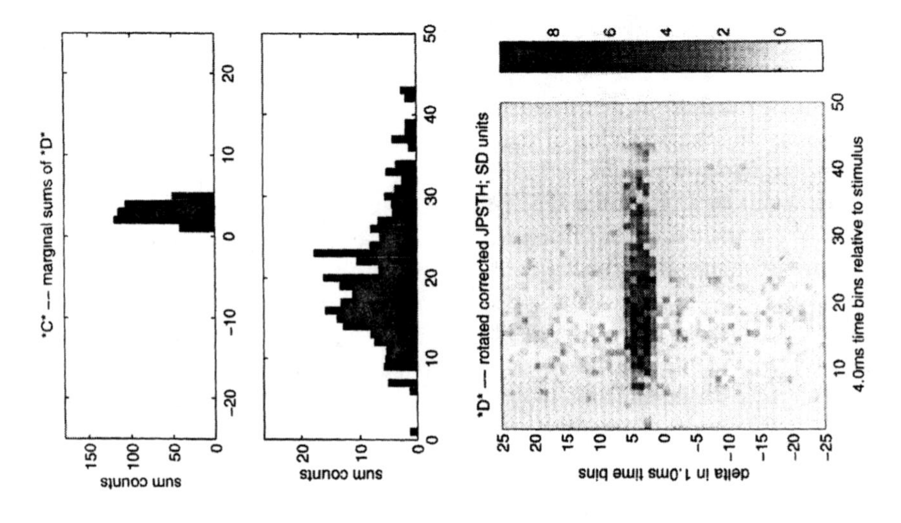
"A" -- raw JPSTH
1.0ms time bins relative to stimulus
"C" -- marginal sums of "D"
sum counts
sum counts

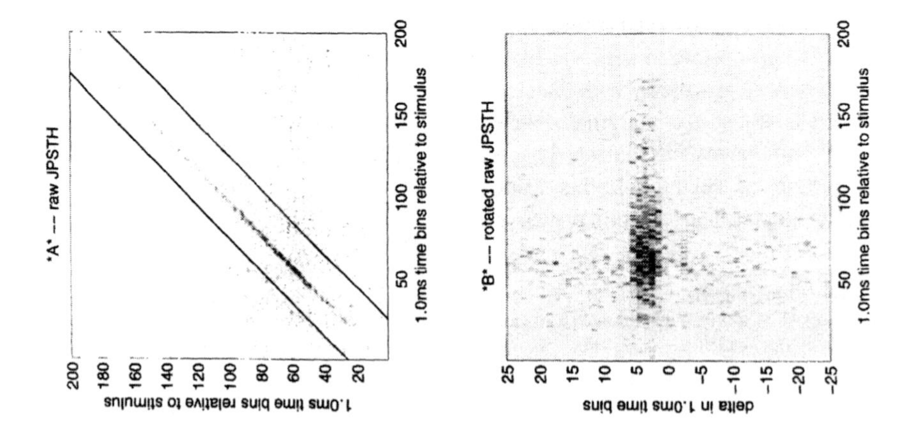
"B" -- rotated raw JPSTH
delta in 1.0ms time bins
1.0ms time bins relative to stimulus
"D" -- rotated corrected JPSTH; SD units
delta in 1.0ms time bins
4.0ms time bins relative to stimulus

FIGURE 8.5

PST histogram-based correction for stationary data. A different approach based on the number of spikes from each neuron for each presentation of the stimulus has been worked out[8] and is similarly effective. More development work needs to be done on this important issue.

A final problem with the use of the JPSTH is time resolution and how it affects the available bin counts and statistical treatment. As presented above, we use square bins of delay relative to the moment of stimulus presentation for each unit's response spikes. Our goal might be to achieve finer time resolution along the diagonal "relative" delay axis (upper left to lower right), which is used to delineate the cross-correlogram, and to have coarser time resolution along the other diagonal axis, which is used to delineate the PST coincidence histogram. This is easily done by computing the original square bin JPSTH matrix with the maximum possible time resolution, and then making appropriate rectangular combinations of such bins. This process is shown in Figure 8.5. The upper left is a standard JPSTH at maximum time resolution for the same type of data used for Figure 8.4. Figure 8.5B (lower left) shows the same data on a 45 degree rotated coordinate system. Note that the axes are scaled differently, so that the bins which appear rectangular actually represent the same time in both dimensions. We can now sum in four bin groups along only the abscissa to obtain usable bin counts while still maintaining high temporal resolution in the "relative" timing. Figure 8.5D shows the corrected JPSTH at these resolutions, now showing square bins in the display which actually represent different time along the two dimensions as desired. It is not possible to use the "surprise" calculation in bin summations of this sort, so that Figure 8.5D is expressed in units of standard deviation of the appropriate PST histogram bin products.

8.4 Gravity Representation

The JPSTH delineates the average stimulus time-locked portion of the coincidence structure of two spike trains, allowing the type of inferences described above. Dynamics which are not time locked to the stimulus are lost. In addition, if many neurons have simultaneously and separably been recorded as the technology now permits, analysis in pairs is a combinatorial disaster. For example, 20 neurons require analysis of 190 pairs. Because computers are fast, this is not an excessively time-consuming set of calculations, but it is an impossible load for the experimenter to simultaneously assimilate and intelligently sort so many pictures.

FIGURE 8.5 (opposite)

Rotation of the JPSTH matrix to allow individual control of time axes. A: Raw JPSTH matrix as in Figure 8.4, but at maximum time resolution. B: The raw matrix rotated. Shown are 200 bins relative to stimulus, but only 50 of delta time (corresponding to the region between the two para-diagonal lines in A). Time resolution along both axes is the same at 1 ms. D: Corrected JPSTH treated in the same way as in B, but with lumping of bins along the abscissa. The bins now have 1 ms resolution along delta time, but only 4 ms resolution along the stimulus time axis. This panel is shown in units determined by the standard deviation of the PST products used in the correction process, with a gray-level bar at the right. Values higher than 3 are considered significant signatures of excess correlation.

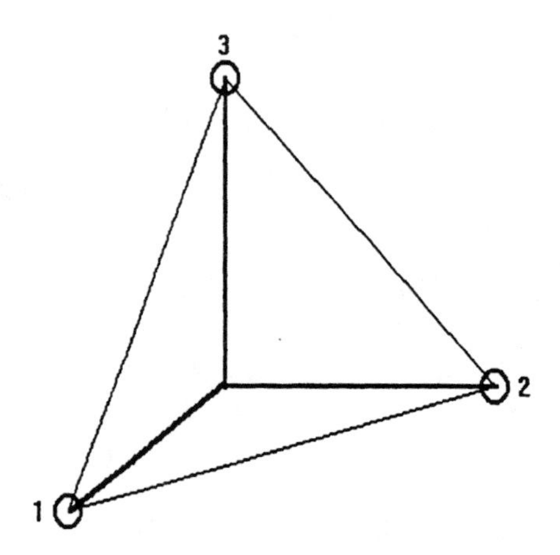

FIGURE 8.6
Gravity representation for three neurons. These are the initial positions of the three corresponding particles in a 3-space.

The gravity representation deals with both the above difficulties. It makes simultaneous analysis of all available spike trains, and it shows all dynamics of the raw or excess coincidence structure, whether stimulus related or not. An important application is to use it as a filter, to select those neurons that are worthy of more detailed examination with ordinary cross-correlation or with the JPSTH.

8.4.1 Basic Arrangements

The basic gravity calculation represents each of the N recorded neurons by a (massless) particle, all located initially at equal distances from each other in an N-space. This initial arrangement puts each of the N particles at an equal arbitrary distance (say 100 units) from the origin along its own cartesian axis. For three neurons and the corresponding three particles, we can directly show these starting positions in the three-dimensional N-space, as in Figure 8.6.

We give each particle a time-varying charge (similar to an electrical charge) that is a low pass filtered version of the corresponding neuron's spike train. The nature of this filter is quite arbitrary and can be chosen in various interesting ways for special purposes.

However, in the basic version of gravity, we can use a simple exponential filter. Thus, every time there is a spike, the corresponding particle receives an increment of charge which subsequently decays back toward zero with the time constant of the filter. The process is reminiscent of the summation of PSPs by a dendritic membrane, although much simplified in detail. Figure 8.7 shows a cartoon of the transformation of a spike train to charge: the upper panel is the (simulated) spike train; the lower panel is the resulting charge. The time scale here is arbitrary, but in a typical real recording situation a time constant of 10 ms is a good choice.

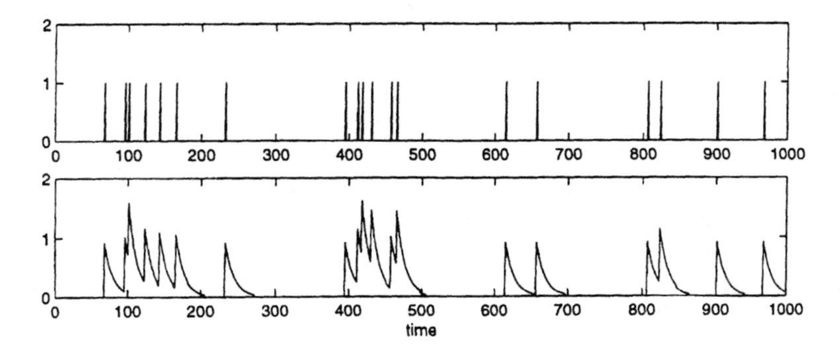

FIGURE 8.7
The spike train to particle charge transformation. The elemental charge here is a simple exponential. Upper panel: spike train; Lower panel: charge. Time is in arbitrary units, but scales have been chosen to show how charge builds up if spikes come frequently enough relative to the time constant of the exponential.

Now assume that there will be a force between two charged particles along the line joining them, with a magnitude proportional to the product of the two charges. Unlike the analogous interaction of two electrically charged particles, the force here is not a function of the distance between the two particles, and is attractive for identical sign of the charges. Thus the force between particles i and j at time t is a vector (indicated throughout by bold type):

$$\mathbf{f}_{ij}(t) = Q_i(t)\ Q_j(t)\ \hat{\mathbf{r}}_{ij} \qquad (1)$$

where $\hat{\mathbf{r}}_{ij}$ is the unit vector along the line joining the two particles, and Q is the charges. The total force on particle i is a vector which is the resultant of a vector sum of the individual pair forces over all particles j not equal to i:

$$\mathbf{f}_i^{tot}(t) = Q_i(t)\ \Sigma_{j \neq i}[Q_j(t)\ \hat{\mathbf{r}}_{ij}] \qquad (2)$$

Suppose that the particles are moving in a viscous medium rather than a vacuum. In this situation the velocity (rather than the acceleration) of a particle is proportional to the total force on it. The constant of proportionality is related to the inverse of the viscosity σ.

$$\mathbf{v}_i(t) = (1/\sigma)\ \mathbf{f}_i^{tot}(t) \qquad (3)$$

Finally, the distance $\Delta\mathbf{s}_i$ moved by particle i in time Δt is proportional to this velocity, so that

$$\Delta\mathbf{s}_i = \mathbf{v}_i(t)\ \Delta t = (1/\sigma)\ \mathbf{f}_i^{tot}(t)\ \Delta t \qquad (4)$$

Thus, at each time step of the computation we can incrementally move each particle a small step whose magnitude and direction are determined by the resultant of interactions with all other particles.

Note that the charge on particles has a non-zero mean, so that on average there will always be an attractive force between any two particles; the final result of this computation will be an aggregation of all particles. However, those particles which represent neurons which fire with appreciable amounts of synchrony will simultaneously have high values of charge. Such particles will have stronger than average attractive force between them and will aggregate more rapidly, with the speed of aggregation directly related to the amount of synchrony in the firing. In order to eliminate the universal background aggregation from the system, we can make each particle charge have a zero mean. Thus in all the above equations we will use an effective charge:

$$Q_i^{eff}(t) = [Q_i(t) - Q_i^m] \tag{5}$$

where Q_i^m is the mean charge on particle i. The mean can be calculated globally over the entire available data, but it is much better to calculate it locally over some window defined either by a fixed time increment or by a fixed number of spikes. Both latter strategies give an effective charge which better deals with dynamics in the data. In the case of a periodic stimulus presentation it is also possible to use a local mean charge that is proportional to the PST histogram of the spike train; this would repeat at the time of each stimulus presentation. Basically the use of effective charge reduces the dependence of the particle aggregation process on differences between or temporal variations in spike rates.

A final rule that helps ensure stability of this calculation is to turn off the force between any two particles if they approach each other to within (say) 10% of the initial distance. It is also necessary to choose a viscosity constant that is optimal for the situation, and this may involve some trial and error. Too low a viscosity coefficient increases the "random" noise in the particle movements and tends to mask any systematic aggregations. Too high a viscosity coefficient reduces all movement, systematic or otherwise.

The gravity representation we have defined above converts the time structure of the original set of observed spike trains from N neurons into the spatiotemporal evolution of a mechanical system. Aggregation of particles is the signature of firing synchrony that is above that expected from the firing rates; such synchrony is one means to define the membership in a neuronal assembly. The computation can easily deal with 100 or so neurons (given a modern, fast computer), and shows the dynamics of the inferred assembly organization. See References 9 and 10 for additional detail.

In order to demonstrate the application of the gravity representation we have used a simple simulator program to produce 10 spike trains with the following characteristics: neuron 1 drives neurons 2 and 3; neuron 4 drives neurons 5 and 6; neuron 7 drives neurons 8 and 9; neuron 10 is independent. Thus we have created three neuronal assemblies (a) neurons 1, 2, 3; (b) neurons 4, 5, 6; (c) neurons 7, 8, 9; and neuron 10 unrelated to any of the three assemblies. All neurons fire at randomly chosen rates near 10/sec, and have approximately Poisson spike trains. Analysis of these data are shown in the next section.

Direct observation of the particle system in the N-space is obviously impossible. One way to detect aggregating particles is to plot the evolution of all possible particle

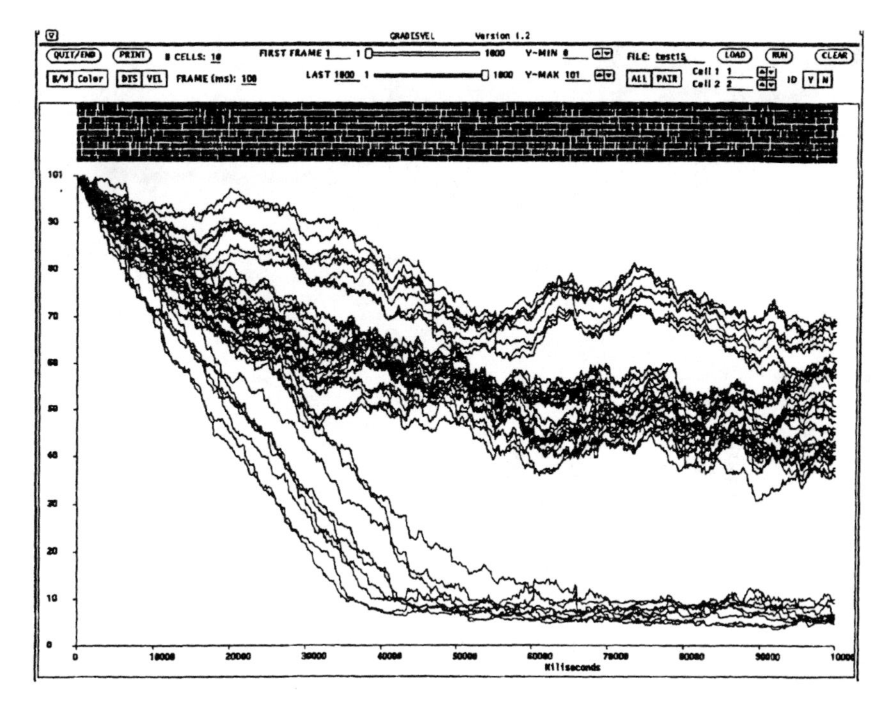

FIGURE 8.8
Trajectories of particle pair distances as a function of time. The group of rapidly decreasing distance curves and the particle aggregations inferred from them correspond to the three sets of (known) interacting neurons. See text.

pair distances as a function of time along the data stream. All pair distances have the starting value of 100 arbitrary units (the initial equidistant configuration of the particles in the N-space), and then variously evolve, as shown in Figure 8.8. The data run and whole abscissa cover 100 seconds, and on this time scale the activity of each of the 10 neurons shows continuous activity (black) with only occasional short pauses (white). The particle pair distance trajectories roughly fall into three groups. The top group (9 trajectories) is for pairs where one member is particle 10 (which represents the independent neuron). Pair distances in this group vary up and down in time, but do not diminish much from the starting configuration. The bottom group (9 trajectories) represent pairs within each of the three neuronal assemblies. Since each assembly has three members, there are three pairings in each. All these trajectories diminish rapidly and reach the minimum distance of 10 at which the pair forces are turned off; these trajectories represent the three aggregations corresponding to the three assemblies of the data. Finally, the middle group of trajectories corresponds to all the inter-assembly pairings. These trajectories diminish to about half the original distance value, but then remain noisily constant. This corresponds to the three separate aggregations of three particles each.

An analysis as shown in Figure 8.8 allows rapid identification of those pairs in the data set that are showing signs of interaction and incidentally determines the relative strength of the interaction (trajectory slope) and whether the interactions are

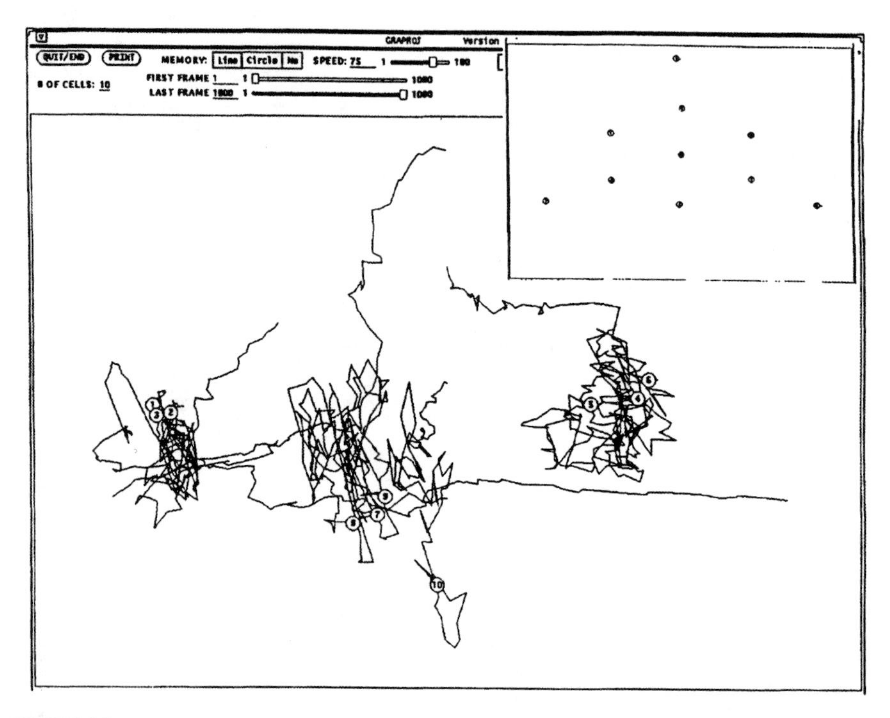

FIGURE 8.9
Projections of the individual particle trajectories from the 10-space to a 2-space. The same data as in Figure 8.8. Projections of the initial particle positions are shown in the inset. Final particle positions are indicated by the numbered circles; each trails behind it the trajectory from the initial position. Three groups of particles have aggregated, corresponding to the (known) three sets of interacting neurons. See text.

constant in time (i.e., produce a constant trajectory slope). It is then possible to concentrate on the data from those neurons using all appropriate tools.

Another way to examine the particle aggregation process in the N-space is to make a projection down to a 2-space. This obviously loses information and can show spurious aggregations. It is vital to check results in such projections against the pair distances as shown in Figure 8.8. There is also the problem of choice of projection method. We have found that the most useful geometric projections for up to 10 particles are based on the centers of gravity of various combinations of the particles. In this situation the plane to which the projection is made itself evolves with time instead of being a constant in the N-space. There are also various nonlinear projection methods that try to preserve distances.[11] Much more computation is involved, and it is not obvious that better results are obtained. An example of geometric projection of aggregation for the same data analyzed in Figure 8.8 is shown in Figure 8.9. The coordinates represent distance in arbitrary units; each circle contains the identifying number of the particle it represents and uses a thin line to show its past trajectory. The end stage of the aggregation process is shown here and indeed demonstrates the three known neuronal assemblies and the independent neuron 10. It is usually instructive to watch this picture dynamically as the projected particle positions move around on the plane. Again strength of interaction will be expressed by the rapidity of the aggregations, and dynamics by changes in the speed of the aggregation.

Now that we have demonstrated the application of the gravity representation to a data set that simulates neuronal assemblies and seen that this organization can be recovered in the analysis, it is appropriate to examine sensitivity to variations in the underlying firing rates. Such longitudinal nonstationarities are all too common in real experiments and can cause problems in calculating the expected amounts of firing synchrony. Again we use a simulator to provide 10 spike trains with appropriate characteristics for the present test. In these data simulated neurons 1 to 9 fire in a continuous Poisson manner at a base rate which switches between approximately 3/sec and 15/sec every thirty seconds; neuron 10 remains at the 3/sec base rate throughout. Neuron 10 drives neurons 8 and 9. The pair distance evolution for these data is shown in Figure 8.10. At the top there is a large group of trajectories which correspond to all pairs except combinations of 8, 9, and 10. These are the noninteracting particles (neurons); their trajectories remain at the initial pair distance, but with a gradual "fan-out" over the data run. This is the result of the residual background noise movement of the particles even when the neurons they represent are completely independent. In order to be interpreted as signature of neuronal interaction, a trajectory with progressive decrease of pair distance must fall outside this envelope, a requirement that obviously sets a lower limit to the strength of neuronal interaction that can be detected in a given situation. Note also that these trajectories for independent pairs show much larger variance (noise) during the periods of high firing rate than during the periods of low firing rate. The lower group of trajectories which show a progressive and quite smooth decrease in pair distance all correspond to pairings within 8, 9, and 10, the interacting neurons. In these trajectories there is also an increase in local variance during the periods of high firing rate.

This demonstration shows that the rate normalization based on local average of the particle charge, as described above for the basic gravity representation, is adequate even with fivefold variations in firing rates. Even in these rather extreme nonstationary conditions, the gravity representation allows the extraction of those neurons in the data which show signs of interaction. We note parenthetically that similar, but generally more brief, firing rate changes of many of the observed neurons may accompany presentations of stimuli. The local average charge normalization suffices also to largely eliminate particle aggregation simply because of stimulus related comodulation of firing.

Finally, we should examine the problem of significance: when do we accept an aggregation as real? It is obvious in all the examples above that particles which represent even completely independent neurons execute something like a Brownian motion in the N-space, and that this can lead to various degrees of spurious aggregation. No simple mathematical way has yet been developed for significance tests in the gravity representation. However, a Monte Carlo method is easily applied.[12,13] The data are shifted (shuffled) in the following way: the first spike train is left as in the original; the second is shifted by one stimulus interval, the third by two stimulus intervals, etc., (data from the stimulus intervals at the end of each train are rotated around to the beginning, so that no data are lost). With this shuffle we have destroyed any real physiological interactions, although we preserve stimulus-related firing rate

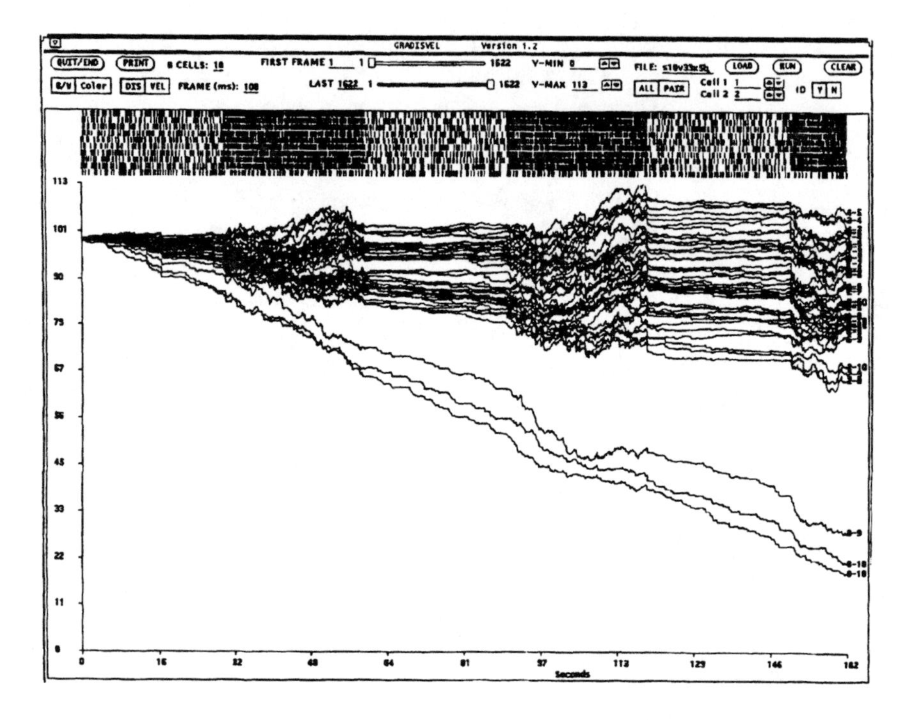

FIGURE 8.10
Demonstration that the firing rate compensation used in the gravity representation allows detection of particle aggregation even with considerable nonstationarities of rates. See text.

modulations. The gravity computation is rerun. Now a new series of shifts are prepared: first spike train as original, second shifted by two stimulus intervals, third by four stimulus intervals, etc. The gravity computation is again rerun. Additional shifts with similar rules are carried out, and each time used for a rerun. After 100 such runs, each with a different rotation of the data, we can examine the envelope of all trajectories in the pair distance plots. This envelope gives us a 99% confidence limit. Any trajectory in the computation with the original, unshuffled data that falls outside this envelope can be considered significant at that level.

8.4.2 Variations of the Basic Representation

8.4.2.1 Aggregation velocity

We have already noted that the trajectory slope in the pair distance plot reflects the strength of neuron interaction. Although we have not shown it here, analysis of real recordings frequently shows changing slopes in some trajectories, presumably reflecting modulations of neuronal interaction during the recording. Such modulations may represent changes of context in the experimental situation if they are temporally associated with known external events or stimuli, or they may occur when apparently nothing is happening in the laboratory. In the latter situation we can only assume that there are some internal state changes which we are not directly manipulating with the experiment.

It is obviously convenient to show the trajectory slope values directly. This is simply accomplished by taking the derivative of values in the plot of distance trajectories, and produces the velocity plot. A given trajectory may have both positive and negative values in the course of the data, corresponding to variations in the aggregation process; periods of disaggregation are often observed in real data.

In practice, several velocity trajectories may simultaneously show high values. We can identify moments during the recording when a criterion number of trajectories exceed a criterion value of velocity. (In terms of physiology, this means that there is a criterion number of simultaneous pair interactions that exceed a strength criterion.) More than one such repeating pattern may occur (i.e., the criterion number can be made up of different membership). Activity patterns defined in this way are often temporally related to laboratory events; for more detail see Reference 14.

The activity patterns detected by aggregation velocity criteria reflect short periods of excess pair correlation. Other definitions of firing pattern have been expressed more directly in terms of a specific pattern of several intervals in the firing of one or more neurons. Such a "melody" of any length among one or more neurons can be detected and tested for significant repetition by several types of algorithm.[15,16] A specific method for patterns in one, two, or three trains made up of three spikes (two intervals) depends on a modification of the joint PSTH. Here one of the spike trains is used for the triggering events, the others for the x and y coordinates. Specific repeating patterns appear as a significantly large bin in the matrix; they occur in approximate temporal vicinity to various behavioral events.[17]

8.4.2.2 Elemental charge functions

In the basic gravity representation we converted spike trains to charge functions using a single exponential that decayed with an appropriate time constant after the occurrence of the spike. Many other choices are possible for various special purposes. One variation allows differentiating between the presynaptic and postsynaptic neurons in an interaction. Here we require each particle to carry two independent charges, one for use when it is the particle being moved (particle i in Equations 2 through 4), and one for use when it is one of the particles causing another to move (particle j in Equations 2 through 4). We call the first of these the "acceptor" charge and the second the "effector" charge. If both charges are made identically with the standard exponential, the situation is no different than the basic representation, and the particles representing the members of a presynaptic to postsynaptic neuron pair will simply move toward each other and meet at the halfway position. On the other hand, if the "acceptor" charges are computed with a time reversed exponential function, while the "effector " charges are computed with the ordinary exponential, the particle representing the postsynaptic neuron will move toward the particle representing the presynaptic neuron. That particle will be approximately stationary. If instead only the "effector" charges are computed with the time-reversed exponential (and "acceptor" charges with the standard exponential), the particle representing a presynaptic neuron will move toward a stationary particle which represents the postsynaptic neuron. With these rules we can also easily distinguish neurons which receive shared input from those which are synaptically coupled. The situation becomes more complicated when there is a chain of neurons. For more detail see References 10 and 18.

8.4.2.3 Overlapping assemblies

Individual neurons can belong to more than one assembly, or may switch allegiance between several assemblies at different times. The consequences of this situation on the aggregation process can, at first glance, be confusing.[19] Suppose there are two assemblies. Some neurons are members exclusively of one or the other assembly, but there are some neurons which belong to both assemblies. (Membership here, as throughout this chapter, implies effective connectivity that can be recognized by excess correlation and hence by aggregation in the N-space.) The gravity representation for this situation will produce two clusters corresponding to the neurons which belong exclusively to one or the other assembly, but will put the particles representing the "shared" neurons more or less between these two clusters. Suppose we now "erase" temporarily from the data all neurons that have at this stage been identified as belonging exclusively to one of the assemblies. A rerun of the computation will now allow the particles representing the "shared" neurons to aggregate with the remaining cluster, since there is now no competition from the cluster representing the deleted neurons. Thus with unknown data it is appropriate to make several repeated computations, removing aggregated clusters singly and in combinations. Obviously, this combinatorial process can get out of hand if too many separate clusters have shown up in the original run, but that is unlikely in our experience. For more detail on this procedure see Reference 18.

8.5 Conclusion

We have examined two types of analyses for simultaneous separable spike train recordings of many neurons. Both methods use measure of near synchrony (correlation) above that expected from the firing rates in order to identify interactions among the neurons. Both methods address the dynamics of such interactions, so that we are going beyond an average measure over the entire data.

The JPSTH evaluates the stimulus time-locked modulation of interaction between two neurons, as averaged over many presentations, and also produces a cross-correlogram. The procedure that corrects for the direct modulation of individual firing rates also gives a better correction to the raw cross-correlogram than does the ordinary shift or shuffle predictor. Note that the JPSTH does not deal with other, nonstimulus-locked, modulations or nonstationarities. In fact, such general nonstationarities, which are unrelated to the times of stimulus presentation, can cause problems, particularly with the correction process and the calculation of significance. Slow nonstationarities are all too common in neuron observation. If the recording length and spike count statistics allow, it is worthwhile to divide the data into two or more pieces for separate analysis;

if the results are similar, they can be lumped with confidence. It is also a good idea as a preliminary to have examined raster plots of the entire data set to identify any large variations in response properties and firing rates.

The gravity representation simultaneously analyzes the activities and interactions of all neurons in the recorded data, so that it is not limited to analysis of neurons in pairs. Basically it converts the analysis of time relations in many spike trains into a spatiotemporal clustering problem. As in the JPSTH, we have procedures that correct for local, possibly stimulus induced, changes of firing rates, so that the particle aggregations can represent the excess correlation, above that expected for independent neurons with the same (ordinary) PSTs and interval structures. The membership, speed, and dynamics of particle aggregations can be interpreted in terms of the dynamics of neuronal assembly organization. A number of variations of the basic gravity representation allow further analysis of the details in the neuronal interactions. Nonstationarities in the data appear simply as part of the dynamics in this representation, and do not represent any difficulties in computation.

Both methods of analysis described in this chapter are useful with an appropriate experimental design to demonstrate changes in neuronal assembly organization with changes in the laboratory context. For example, if two (or more) flavors of stimuli are presented in a random interleaved sequence, the data can be segmented according to stimulus, and each such group of segments concatenated. This gives "continuous" runs of each flavor of stimulus with the obvious benefit that whatever nonstationarities there are in the data are distributed equally among all stimulus types. Analysis of such data from different stimulus flavors with either JPSTH or gravity shows that there are large changes in organization between the two contexts. It requires the concatenated data to make this inference of state switching. But the actual switching is occurring on the time scale of the interleaved stimulus presentation — usually well under one second. Application of such data analysis to various experimental situations can be found in References 13 and 20–27.

The methods discussed here are an approach to problems of neuronal organization in different circumstances. Much additional work is needed, however, to develop comparable tools for extracting how the recorded neurons are representing information. It is likely that the time course of both firing rates and synchronies or coincidence rates (pair and higher) will play an important role in such methods. In their initial calculation of the raw data both JPSTH and gravity can be used as a good starting point. But strategies aimed at the representation problem, and differing completely from the present "correction" procedures, must be invented.

References

1. Glaser, E. and Ruchkin, D.S. (1976). *Principles of Neurobiological Signal Analysis,* Academic Press, New York, 1976.
2. Eggermont, J., Neural interaction in cat cortex, *J. Neurophysiol.,* 71, 246, 1994.
3. Eggermont, J., *The Correlative Brain.* Springer, Berlin, 1990.
4. Eggermont, J., Neural interaction in cat primary auditory cortex, *J. Neurophysiol.,* 68, 1216, 1992.

5. Aertsen, A.M.H.J., Gerstein, G.L., Habib, M., and Palm, G., with the collaboration of P. Gochin and J. Kruger, Dynamics of neuronal firing correlation: modulation of "effective connectivity," *J. Neurophysiol.,* 61, 900, 1989.
6. Palm, G., Aertsen, A.M.H.J., and Gerstein, G.L., On the significance of correlations among neuronal spike trains, *Biological Cybernetics,* 59, 1, 1988.
7. Legendy, C.R. and Salcman, M., Bursts and recurrences of bursts in the spike trains of spontaneously active striate cortex neurons, *J. Neurophysiol.,* 53, 926, 1985.
8. Aertsen, A.M.H.J., private communication, 1992.
9. Gerstein, G. L., Perkel, D. H., and Dayhoff, J. E., Cooperative firing activity in simultaneously recorded populations of neurons: Detection and measurement, *J. Neuroscience,* 5, 881, 1985.
10. Gerstein, G.L. and Aertsen, A.M.H.J., Representation of cooperative firing activity among simultaneously recorded neurons, *J. Neurophysiol.,* 54, 1513, 1985.
11. Sammon, J.W., *IEEE Trans. Comput.,* 18, 401, 1969.
12. Lindsey, B.G., Hernandez, Y.M., Morris, K., Shannon, R., and Gerstein, G.L., Respiratory related neural assemblies in the brain stem midline, *J. Neurophysiol.* 67, 905, 1992.
13. Lindsey, B.G., Hernandez, Y.M., Morris, K., Shannon, R., and Gerstein, G.L., Dynamic reconfiguration of brain stem neural assemblies: Respiratory phase-dependent synchrony vs. modulation of firing rates, *J. Neurophysiol.,* 67, 923, 1992.
14. Lindsey, B.G., Morris, K.F., Shannon, R., and Gerstein, G.L., Repeated patterns of distributed synchrony in neuronal assemblies, *J. Neurophysiol.,* 78, 1714, 1997.
15. Dayhoff, J. E. and Gerstein, G.L., Favored patterns in spike trains. I. Detection, *J. Neurophysiol.,* 49, 1334, 1983.
16. Abeles, M. and Gerstein, G.L., Detecting spatio-temporal firing patterns among simultaneously recorded single neurons, *J. Neurophysiol.,* 60, 909, 1988.
17. Prut, Y., Vaadia, E., Bergman, H., Halman, I., Slovin, H., and Abeles, M., Spatiotemporal structure of cortical activity: properties and behavioral relevance, *J. Neurophysiol.,* 79, 2857, 1998.
18. Gerstein, G.L. and Ramachandran, P., Neuronal assemblies: extensions to the gravity representation, submitted, 1998.
19. Strangman, G., Detecting synchronous cell assemblies with limited data and overlapping assemblies, *Neural Computation,* 9, 51, 1997.
20. Aertsen, A.M.H.J. and Gerstein, G.L., Dynamic aspects of neuronal cooperativity: fast stimulus-locked modulations of "effective connectivity," in *Neuronal Cooperativity,* Ed. J. Kruger, Springer, New York, 1991.
21. Aertsen, A.M.H.J., Vaadia, E., Abeles, E., Ahissar, M., Bergman, H., Karmon, B., Lavner, V., Margalit, E., Nelken, I., and Rotter, S., Neural interactions in the frontal cortex of a behaving monkey — signs of dependence on stimulus context and behavioral state, *J. für Hirnforschung,* 32, 735, 1991.
22. Gerstein, G.L., Bedenbaugh, P., and Aertsen, A.M.H.J., Neuronal assemblies, *IEEE Trans. Biomed. Eng.,* 36, 4, 1989.
23. Gerstein, G.L., Bloom, M.J., and Maldonado, P.E., Organization and perturbation of neuronal assemblies, in *Central Auditory Processing and Neural Modeling,* Eds. J. Brugge and P. Poon, Plenum, New York, 1998.

24. Lindsey, B.G., Shannon, R., and Gerstein, G.L., Gravitational representation of simultaneously recorded brain stem respiratory neuron spike trains, *Brain Research,* 483, 373, 1989.
25. Maldonado, P.E. and Gerstein, G.L., Neuronal assembly dynamics in the rat auditory cortex during reorganization induced by intracortical microstimulation, *Experimental Brain Research,* 112, 431, 1996.
26. Sillito, A.M., Jones, H.E., Gerstein, G.L., and West, D.C., Corticofugal feedback in the visual system produces stimulus dependent correlation, *Nature,* 369, 479, 1994.
27. Vaadia, E., Halman, I., Abeles, M., Bergman, H., Prut, Y., Slovin, H., and Aertsen, A., Dynamics of neuronal interactions in monkey cortex in relation to behavioral events, *Nature,* 373, 515, 1995.

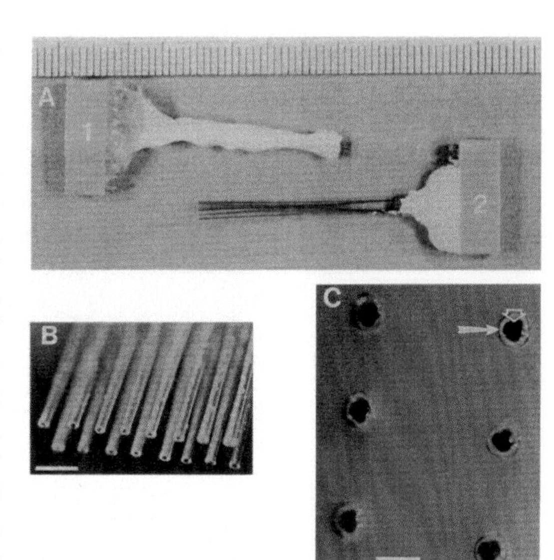

Plate 1 – Examples of custom wire microelectrode arrays available from NB Labs, Denison, Texas. (A) Parallel microwire arrays (1) designed for cortical implantation. Wire bundles (2) designed for brain stem implants. (B) High-power view of parallel microwire array. (C) End view of cut-off wires showing Teflon® insulation (solid arrow) surrounding stainless steel microwire (open arrow). Scale: (A) 1 mm per division; (B) 250 μm; and (C) 60 μm. (From Nicolelis, M.A.L., Ghazanfar, A.A., Faggin, B.M., Votaw, S., and Oliveira, L.M.O., Reconstructing the engram: Simultaneous, multisite, many single neuron recordings, *Neuron*, 18, 529, 1997. With kind permission of Cell Press, Cambridge, Massachusetts.)

Plate 2 – Reitbock-type seven-microelectrode drive system manufactured by Uwe Thomas Recording (Marburg, Germany). Individual electromagnetic brakes maintain the position of the microelectrodes. When an electromagnetic clutch is engaged and the brake released, the microelectrode can be moved with a stepper motor that is coupled by a flexible shaft, in 2 μm increments.

Plate 3 – (Left) Close-up photograph of an implanted male zebra finch. The plastic pieces are from an older, less protective design. (Right) Photograph of another experimental subject connected in a recording environment.

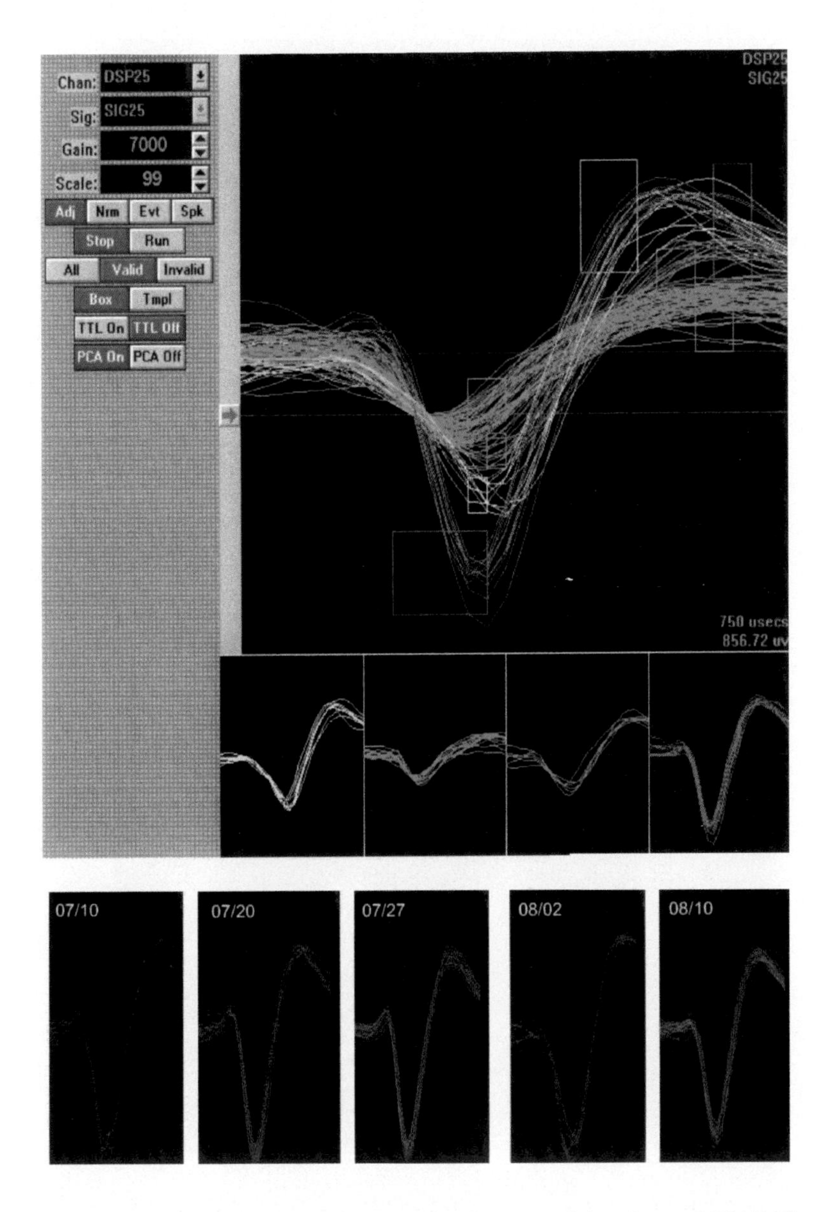

Plate 4 – Real-time discrimination of neuronal action potentials using an MNAP. The top panel illustrates how pairs of time-voltage boxes can be used to discriminate up to four distinct types of action potentials (illustrated here by different colors) per implanted microwire. The lower panel illustrates the fact that these chronic microwire recordings are very stable and that some single neurons can be maintained for several weeks. In this particular example, a single neuron (red action potentials) located in the primary motor cortex was recorded for over five weeks (from 07/10 to 08/10).

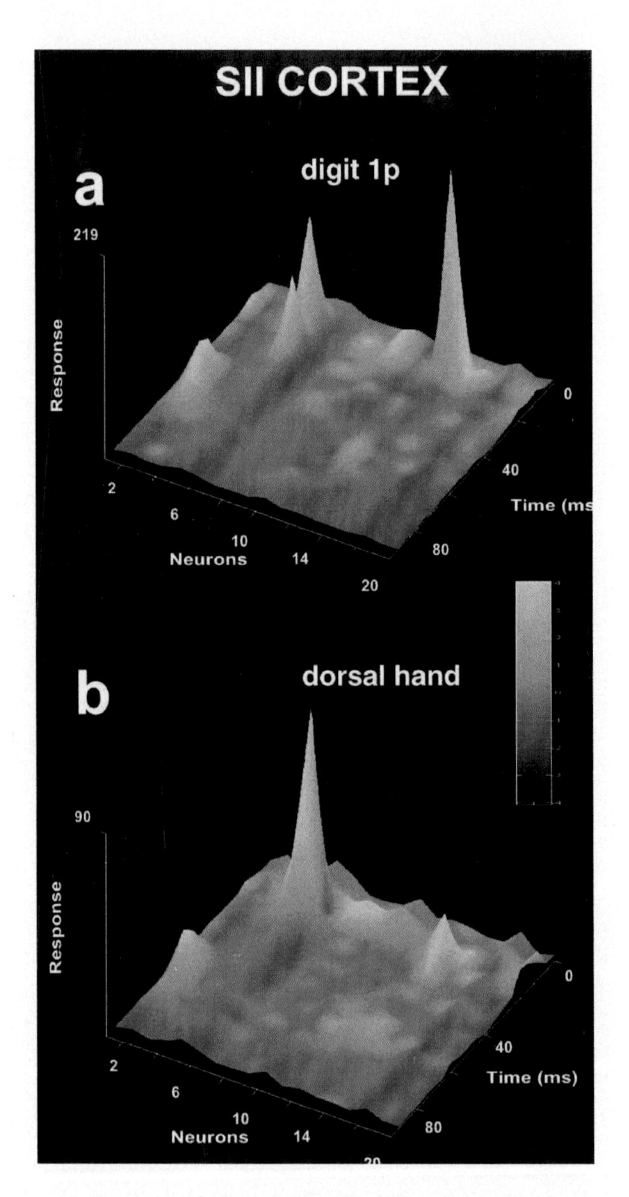

Plate 5 – Population post-stimulus time histograms illustrate the spatiotemporal pattern of sensory responses observed across a population of 20 SII cortical neurons, following the stimulation of two distinct hand areas: the proximal phalanx of digit 1 (top graph) and the hairy skin of the dorsal surface of the hand (lower graph). The color-coded Z axis depicts the magnitude of neuronal firing in number of standard deviations above (+4SD, light brown) and below (-4SD, black) spontaneous firing rate. (From Nicolelis, M.A.L., Ghazanfar, A.A., Oliveira, L.M.O., Chapin, J.K., Nelson, R., and Kaas, J.H., Simultaneous encoding of tactile information by three primate cortical areas, *Nature Neurosci.*, in press, 1998. With permission.)

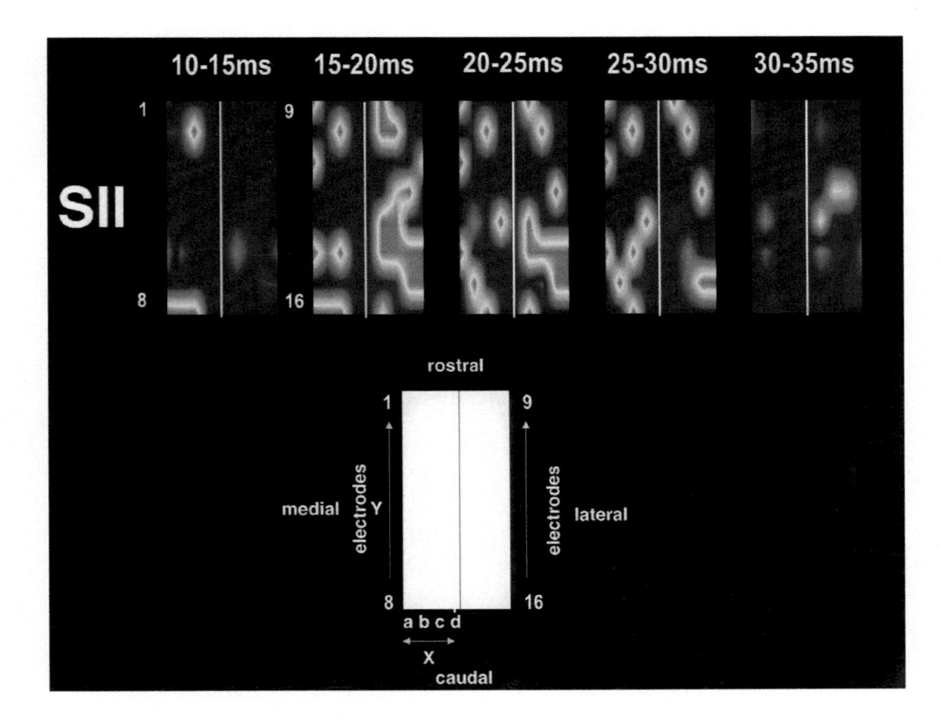

Plate 6 – Spatiotemporal population map depicting the sensory responses of a population of SII neurons following the mechanical stimulation of a finger tip. The scheme in the lower half of the figure illustrates how SPMs are built. In this example, the Z represents firing rates in z scores (light red equals 4 SDs above spontaneous firing rate, which is depicted in dark blue).

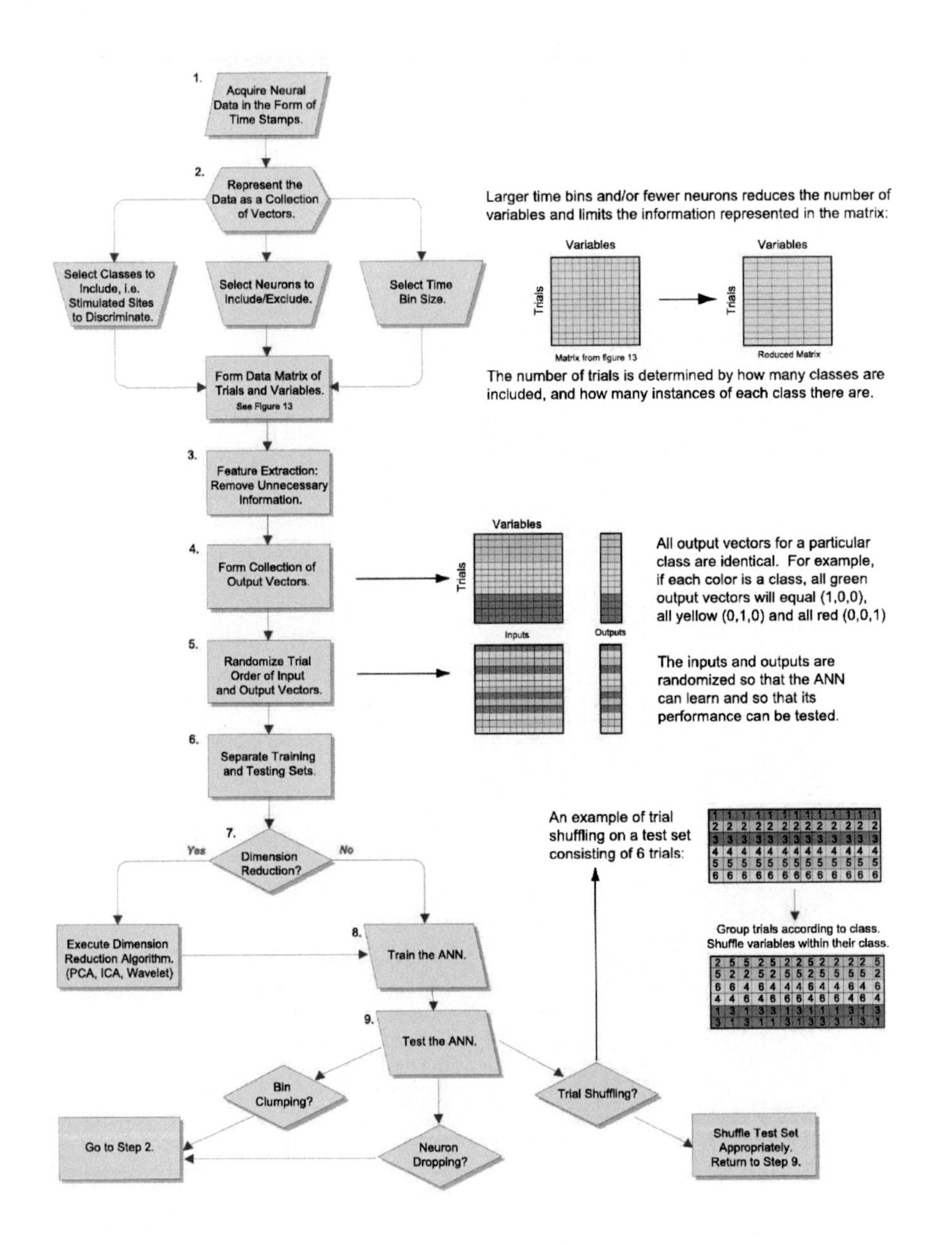

Plate 7 – Routine for applying ANNs in the analysis of neural ensemble data. Notice that modifications to the original data set (bin clumping, neuron dropping, and trial shuffling) are performed after a control performance has been attained.

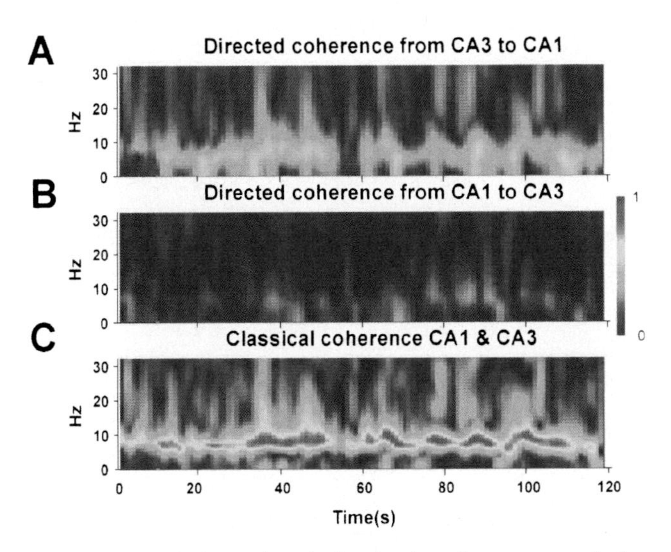

Plate 8 – A color-coded scale is used to depict the time-frequency evolution of Directed and Classical Coherences between CA1 and CA3. All colors except dark blue (γ_{ij} (f) < 0.1) constitute significant values. The activity of CA3 drives that of CA1 because the directed coherence in (A) is always larger than that in (B), from which one may infer short-lived episodes when information feeds back from CA1 to CA3 (light blue). Only coactivation and synchronicity can be inferred from the classical coherence plot (C).

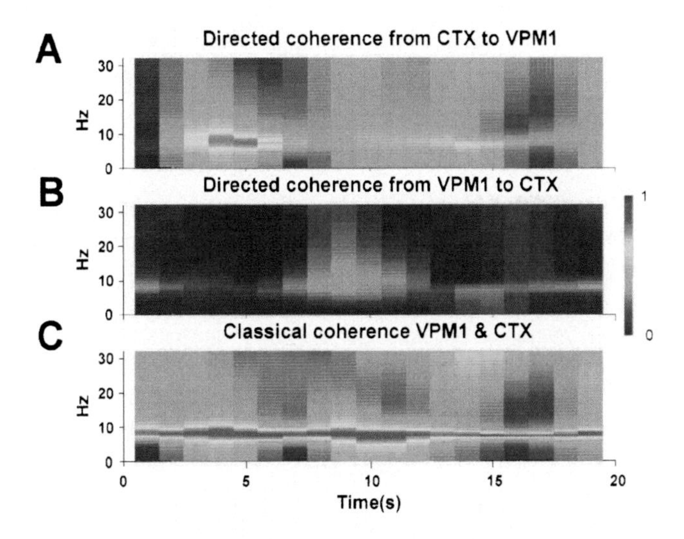

Plate 9 – Time-frequency evolution of Directed and Classical Coherences between VPM1 and CTX of Figure 9.3 (Chapter 9) represented by a color scale (all colors but dark blue are significant for γ_{ij} (f) > 0.1). Most information originates in the cortex (A), with an extended episode of bidirectionality where substantial mutual feedback was present (B). Significant synchronous coactivation is portrayed by classical coherence (C).

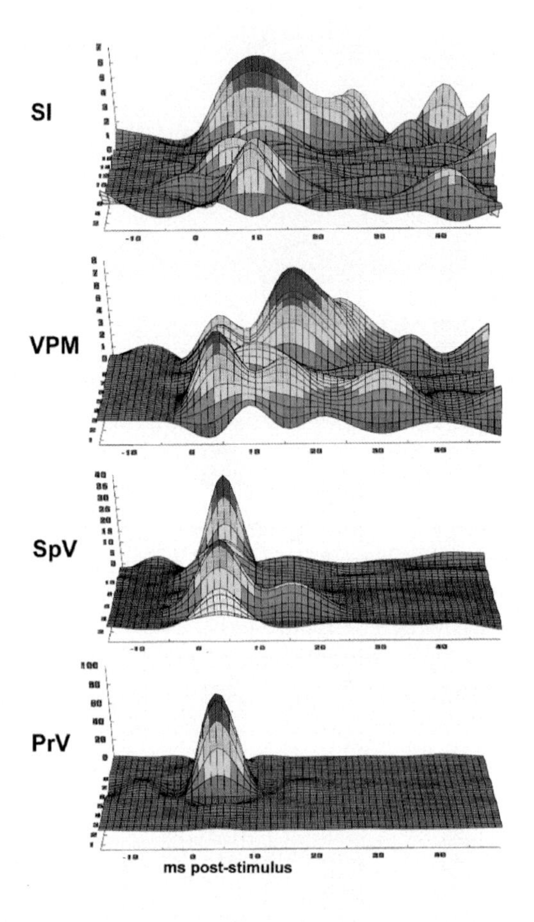

Plate 10 – Four three-dimensional surfaces define Population Poststimulus Time Histograms (PPSTHs) depicting the averaged responses of neuronal ensembles simultaneously recorded at four levels of the rat somatosensory system to 309 discrete mechanical deflections of a single whisker. The PPSTHs are arranged according to processing level within the system, including the principal trigeminal nucleus (PrV), spinal trigeminal nucleus (SpV), ventral posterior medial thalamus (VPM), and somatosensory cortex (SI). Each surface was built from peri-event histograms for each of the ensemble neurons, covering post-stimulus time from -10 to 57 ms (3 ms bins). The histograms were rank ordered by their response amplitudes and then reconstructed into a continuous surface using a spline smoothing algorithm. Stimuli consisted of 3° amplitude step deflections delivered at a rate of 1 Hz through a fine probe placed 10 mm from the whisker follicle. Probe movement was computer controlled via a vibromechanical actuator. Vertical axis: average instantaneous firing rates, in spikes/sec. (From Nicolelis, M.A.L., Ghazanfar, A.A., Faggin, B.M., Votaw, S., and Oliveira, L.M.O., Reconstructing the engram: Simultaneous, multisite, many single neuron recordings, *Neuron*, 18, 529, 1997. With kind permission of Cell Press, Cambridge, Massachusetts.)

Chapter 9

Directed Coherence: A Tool for Exploring Functional Interactions Among Brain Structures

Luiz Antonio Baccalá and Koichi Sameshima

Contents

9.1 Introduction

Several tools are available to describe the functional relationships between neurons by relating their activity to external sensory stimulus. Time domain analysis techniques, such as the correlation-based methods described in Chapter 8, have found widespread use in spike train analysis. Frequency domain techniques, though also applicable to spike analysis,[1] have had more of an impact in the analysis of EEG and local field potentials. Among the latter, the method of classical coherence analysis has ranked among the most popular. Both correlation and coherence analysis to a large extent relate the simultaneous analysis of pairs of neural elements.*

* The word element refers to a neural processing unit, either neurons or functional neural structures.

0-8493-3351-2/99/$0.00+$.50

For instance, in the classical coherence approach, the type of inference one can make relates to the explanatory power (in terms of variance) that the observed activity of one neural element has over the activity of another element through a linear interaction model in the frequency domain. Thus, high classical coherence between pairs of neural elements at a given frequency reflects the adequacy of the linear model in explaining the presence of mutual coactivation and the observation of a fair degree of synchronicity at that frequency.

The approach we discuss in this chapter is named the *directed coherence* method. Though also based on linear approximation models in the frequency domain, this method differs from classical coherence in two substantial ways. First, it is truly a multivariate technique, i.e., the simultaneous interaction of many neural elements can be jointly estimated — classical coherence studies, by contrast, must proceed through pair-wise neural element calculations. Second, and perhaps more important, this new method, unlike its classical counterpart, is capable of exposing the direction of information flow between structures. In fact, its use leads to the solution of an important problem: that of inference as to the existence of one possible driving element among an ensemble of observed neural elements.

Interestingly, despite being apparently restricted to relatively few practitioners, the method of directed coherence was actually introduced by researchers in connection with EEG analysis.[2-6] On a deeper level, however, its underlying theoretical basis stems from econometrics[7] in its attempts to search for empirical evidence of "cause" and "effect" between observed time series, such as inflation, investment, unemployment rates, and so on.

Central to the method, in neurobiology or elsewhere, is its ability to address and resolve the delicate issue of *cause* and *effect*[7,8] between observed time series. Thus, this method can prove useful in providing functional interaction clues in situations in which substantial neural activity persists in the absence of apparent external sensory stimuli, such as the REM sleep or epilepsy.

Another interesting and useful application of this method involves the systematic investigation for the presence of *feedback* loops between time series.[9] This issue, because it depends on more specialized mathematical details, is left to the Appendix at the end of this chapter. In the next section, to illustrate the use and interpretation of the method, we concentrate on examining some simple examples and compare the results derived from directed coherence to those obtained by using the more classical coherence approach.

9.2 Directed Coherence: Interpretation and Examples

To appreciate the method, we must "define" what causality between time series means. For simplicity, we restrict our attention to just two discrete time series, $x_1(t)$ and $x_2(t)$, obtained from the adequate sampling of measured neural activity. Conceptual extension to more time series is immediate. We say that $x_1(t)$ causes $x_2(t)$ if prediction of future values of $x_2(t)$ is significantly improved (statistically) by knowing the past of $x_1(t)$. The extent of this prediction improvement is measured by the *directed coherence* function (see Equation 12 for a precise computational definition). Thus, $\gamma_{21}(f)$ is related to the fraction of the power (variance) in the frequency domain of $x_2(t)$ (target) around frequency f whose prediction is accounted by the past of $x_1(t)$ (source) around that same frequency. The key, and perhaps surprising, ingredient is that $\gamma_{21}(f)$ need not equal and usually differs from $\gamma_{12}(f)$. This is so because, whereas $\gamma_{21}(f)$ measures the effect of the past of $x_1(t)$ on $x_2(t)$, $\gamma_{12}(f)$ reflects what happens when the past of $x_2(t)$ is considered in helping predict $x_1(t)$. This feature is absent for classical coherence $C_{12}(f)$ where exchange of $x_1(t)$ for $x_2(t)$ produces $C_{21}(f)$ which equals $C_{12}(f)$. This is illustrated in Figure 9.1A for the toy (linear) model (Figure 9.1B), that produced the observed time series depicted in Figure 9.1C. In the toy model, $x_1(t)$ and $x_2(t)$ are just delayed forms of one another plus noise. The coupling strengths between these series were respectively represented by α and β. For $\alpha = 0.9$ and $\beta = 0.1$, the effect of the past of $x_2(t)$ on $x_1(t)$ was large, whereas the past of $x_1(t)$ had little effect on $x_2(t)$. The directed coherences are summarized in Figure 9.1A, which contains four subplots arranged as though elements of a 2×2 matrix whose row index labels the signal *source* and column index labels the signal *target*. Along the main diagonal, rather than plot $\gamma_{ij}(f)$, the normalized influence of a series past on its own future, we depict the series spectra. The off-diagonal graphs contain $\gamma_{12}(f)$ and $\gamma_{21}(f)$ (solid lines), whose difference reflects the distinct values for α and β. For comparison, we also plotted $C_{12}(f)$ and $C_{21}(f)$ (dashed lines), whose equality is immediately apparent.

Note that, whereas both $\gamma_{ji}(f)$ and $C_{12}(f)$ have frequency domain interpretations and rely on spectral properties of $x_1(t)$ and $x_2(t)$, computation of $\gamma_{ji}(f)$ necessarily requires the use of *multivariate time series prediction* techniques (see Appendix), which one may dispense when estimating $C_{12}(f)$.

Moving to a physiologic setting, we first applied the directed coherence method to rats that received a chronic implant of a multielectrode array. The data employed in this analysis were derived from a series of local field potentials recorded in different regions of the rat hippocampus during desynchronized sleep containing strong oscillatory activity data (see Figure 9.2A). This is a typical example of an experimental scenario where causal relationships cannot be ascribed to any relevant external stimuli. The corresponding directed coherences between CA1 and CA3 for a two second-long data segment of the original record appears in Figure 9.2B in the form of the 2×2 matrix subplot display, similar to that of Figure 9.1A. This analysis revealed that hippocampal interaction from CA3 to CA1 is much stronger than that in the reverse direction.[10] This observation is consistent with the existence of well-known neuranatomical projections from CA3 to CA1.[11]

Because neural signals are widely recognized as nonstationary and because time series prediction techniques involved in the computation of $\gamma_{ij}(f)$ require a fair degree of stationarity, successful analysis of long records requires signal segmentation. A summary of the analysis for successive segments spanning the whole record of Figure 9.2A was constructed in Plate 8 by collating the results of the calculated $\gamma_{ij}(f)$ for two-second-long segments with a one-second-long overlap among adjacent segments. This produced the color-coded time-frequency plots of directed coherence evolution (see Plate 8A and 8B). All colors, except dark blue, represented significant values.[3] Thus, the same information flow pattern from CA3 to CA1 persisted during

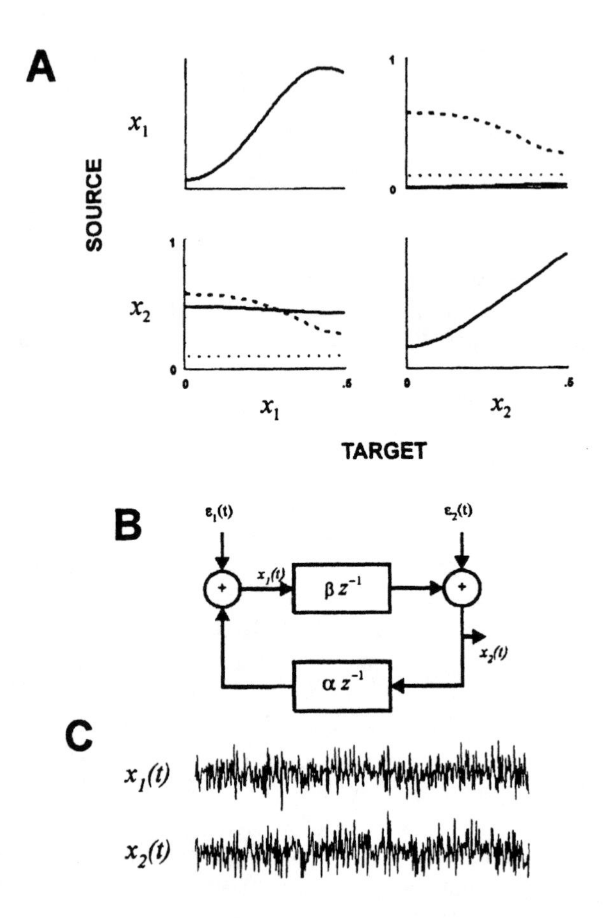

FIGURE 9.1
(A) Results of the estimated directed coherence (solid lines) are displayed along the off-diagonals of the 2 × 2 matrix subplot layout for the linear system (B) relating $x_1(t)$ and $x_2(t)$ simulated through

$$x_1(t) = \alpha x_2(t-1) + \varepsilon_1(t)$$

$$x_2(t) = \beta x_1(t-1) + \varepsilon_2(t)$$

where $\varepsilon_1(t)$ and $\varepsilon_2(t)$ is measurement noise with $\alpha = 0.9$ and $\beta = 0.1$, that resulted in the traces (C). The property $C_{12}(f) = C_{21}(f)$ of classical coherence, defined as

$$C_{ij}(f) = \frac{|S_{ij}(f)|^2}{S_{ii}(f)S_{jj}(f)},$$

where $S_{ii}(f)$ and $S_{ij}(f)$ are respectively auto- and cross-spectral densities of the series under analysis, is also illustrated by the dashed curves (A). The series power spectra are shown along the 2 × 2 matrix main diagonal. This display pattern is used throughout this type of analysis. Normalized frequency units were used in the x-axis. Dotted lines represent significance levels. Any values above these dotted lines are considered significant.

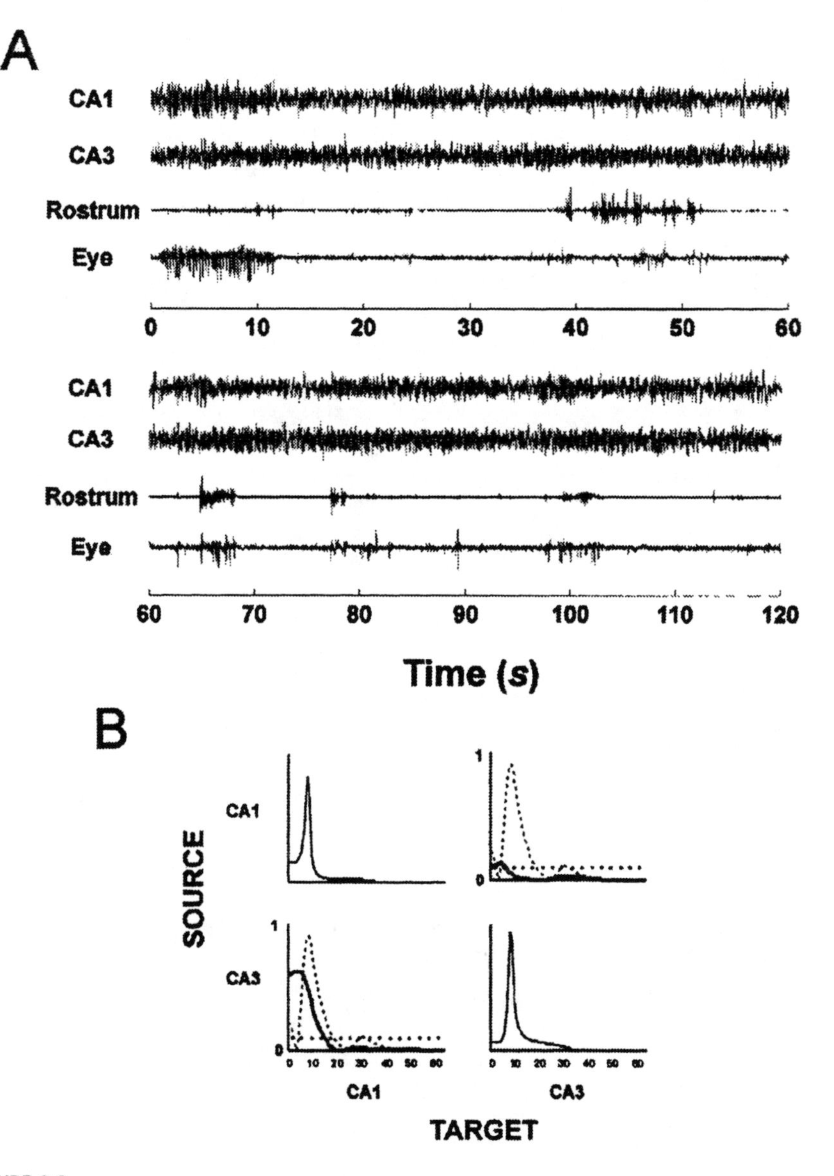

FIGURE 9.2
(A) Local field potentials lasting 120 *s* recorded from the hippocampal fields (CA1, CA3) of a sleeping rat using chronically implanted bipolar microelectrodes. It also shows the corresponding electromiographic traces recorded from whisker (Rostrum) and eye (Eye) muscles. The record contains a period of continuous desynchronized (REM) sleep in which several distinct electromiographically identifiable behavioral patterns are observed. Throughout this period, the hippocampus (CA1, CA3) shows oscillatory activity around 8 Hz, that is particularly clear during vigorous whisker or eye movements. For instance, during intervals 38-52, 65-68, 77-82, 88-92, and 98-102 seconds the oscillation becomes more regular and frequency increases to 10 Hz. (B) The dominance of the effect of CA3 over CA1 is examplified in the 2 × 2 matrix layout of directed coherence subplots (solid lines) for the signal segment between 88 and 90 *s*; classical coherence is also shown (dashed line). Values above $\gamma_{ij}(f) = 0.1$ are considered significant.[3] Solid lines represent power spectra along the main diagonal of the matrix in B.

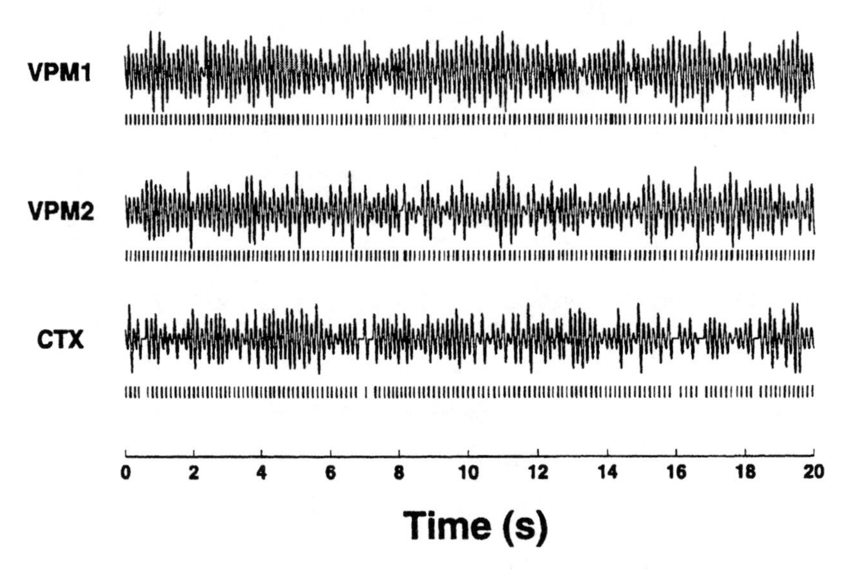

FIGURE 9.3
Illustrates the preprocessing used to generate signals suitable for directed coherence calculations of three 20-second-long series of spike trains recorded from the thalamus, VPM1 and VPM2, and the somatosensory cortex, CTX, of an awake rat during whisking behavior. For each element, the upper trace is the signal reconstructed based on spike time-stamp of the lower trace.

the whole record, except for short episodes when influence was bidirectional, suggesting the presence of occasional weak signal feedback (possibly through entorhinal cortex) from CA1 onto CA3. The simultaneous time-frequency plot of the corresponding classical coherence (Plate 8C), significant around 8 Hz throughout the whole record, indicated that both CA1 and CA3 were active and discharging rhythmically, but lead to no clue as to causal dependence.

Point processes can also be analyzed with this method with the help of signal preprocessing. The main reason for this requirement is that the ordinary time series prediction techniques involved in the computation of $\gamma_{ij}(f)$ have poor performance when spiky signals are employed. Given action potential time stamps denoted by t_i, one can obtain a discrete time series suitable for analysis via[1,3,12]

$$x(k) = \sum_i h(kT_s - t_i)$$

where $h(t)$ is the impulse response of a suitable filter and T_s is the sampling period. Filters of the forms

$$h(t) = \frac{\sin(W\pi t)}{W\pi t} \tag{1}$$

and

$$h(t) = Ke^{-\alpha t^2} \tag{2}$$

have been employed.[1,12] In our experience, because of its sharp frequency characteristics, Equation 1 constitutes a good choice for the analysis of oscillatory neural activity.

To use these filters properly, one must define the bandwidth of filter $h(t)$, given by W in the case of Equation 1. A natural constraint for W is Nyquist's criterion,[13] which governs accurate representation of sampled time data, namely

$$\frac{1}{T_s} \geq 2W. \tag{3}$$

Additionally, the filter's bandwidth must at least equal or exceed the frequency of the feature of interest, such as, for example, a typical burst firing rate of the original spike train. Also, to avoid the harmonics of the typical firing rate, W should be smaller than twice that rate. This is important to avoid the appearance of harmonics in the reconstructed series, as these impact time series prediction reliability in a deleterious way.

The construction of a signal in this way produces a time series with a nonzero mean value that must be extracted before use of time series prediction algorithms.[14] This is easy by writing $\Delta x(k) = x(k) - x(k-1)$, which is the final signal form used by the prediction algorithms.

As an illustration of this application, consider the simultaneously recorded multichannel action potential firing from the somatosensory cortex and thalamus of a behaving rat during exploratory activity,[15] shown with their reconstruction in Figure 9.3. This time, the analysis involved three series that produced the 3 × 3 displays of Figures 9.4A and 9.4B for two distinct data segments. These figures contain distinct $\gamma_{ij}(f)$ patterns; in Figure 9.4A information flows from cortex (CTX) to thalamus (VPM1), whereas a certain amount of information bidirectionality (feedback) is present in Figure 9.4B. A summary of these relationships is provided by the graphs of Figures 9.4C and 9.4D, respectively; the relative interaction strengths were represented by the width of the arrows connecting the elements.

As before, an analysis summary for successive segments is contained in Plate 9A and B, which shows how the relationships between VPM1 and CTX change dynamically in time. Though synchronously oscillating (Plate 9C), the nature of the interaction between VPM1 and CTX changed markedly over time. In other words, we conclude that $\gamma_{ij}(f)$ time-frequency plots permit us to "watch" how the pathways connecting neural elements switch "on" and "off" during neural processing.

9.3 Conclusions and Final Remarks

The evolution of neural recording instrumentation has led to the huge problem of analyzing an ever-growing number of simultaneous neural sensors,[16] bringing added

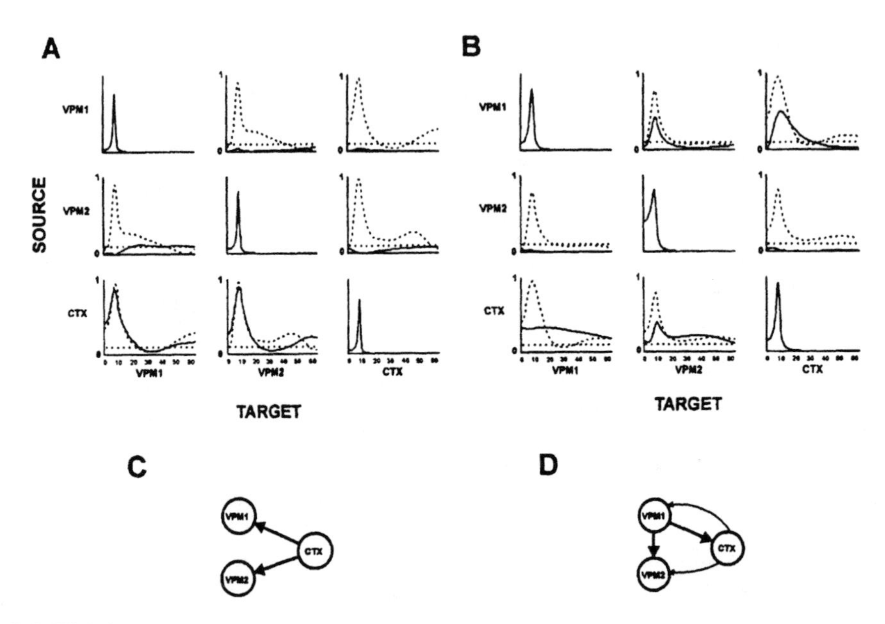

FIGURE 9.4
Directed coherence analysis for time segments between 4 to 6 *s* and 8 to 10 *s* are represented respectively in (A) and (B) as 3 × 3 subplot matrix layout displays, where distinct directed coherence patterns are distinguishable. Focusing on the analysis of VPM1 and CTX, one readily sees that there is cortical driving of the activity (A). The pattern of (B) is distinct, and the interaction becomes bidirectional. The joint couplings of these structures were summarized respectively in graphs (C) and (D), where coupling strengths are reflected in the thickness of the joining arrows.

challenge to traditional analysis techniques that cope efficiently with only a handful of neurons simultaneously. Reliance on multivariate time series procedures permits the analysis of many more neural elements simultaneously through directed coherence. It is difficult to state rigorously how many elements can be processed simultaneous, but analysis of 10 to 20 neural elements has been reported.[5] This shows the method can be an effective screening tool for sieving and sorting through the relationships between many processing neural elements.

A point beyond the present discussion is the important issue of statistical and procedural reliability, much of which is already available elsewhere.[3,8,9,17] One special area currently in need of more research concerns the optimization of these algorithms to handle instances of nonstationarity in a reliable and automatic way.

Despite its fundamental implication for neuroscience, systematic investigations of causal relations in the brain through directed coherence methods is not yet a standard and popular approach, even considering the original and pioneering contributions from neuroscientists to this field.[2,3,6,18] A possible cause for the technique's relatively low impact thus far lies in its demands on skilled practitioners who must not only understand the underlying neurobiological issues but also master a variety of multivariate time series analysis methods (see Appendix). Another reason may be that neuroscientists have other options in the study of causality. They can manipulate their subjects, e.g., surgically or via drugs. This gift is unavailable to researchers

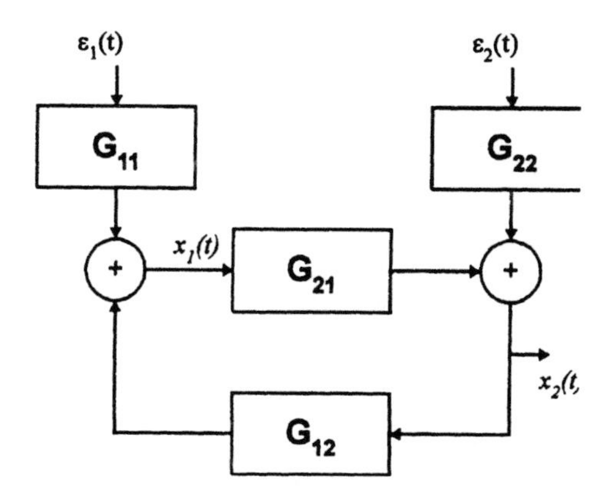

FIGURE 9.5
Block-diagram representation of the pair of variables $x_1(t)$ and $x_2(t)$ in a form suitable for feedback analysis. Feedback is absent whenever $G_{12} = 0$.

in other fields, who frequently lack the power to intervene, such as economists who have used these techniques rather frequently. Given the recent technological advances that grant an ever-increasing number of simultaneously recordable information from multiple channels (Chapter 3), neuroscientists need more powerful and speedy analysis tools. Directed coherence analysis offers an attractive option because it allows quick detection of events of interest that involve dynamically varying feedbacks which cannot be investigated very easily by traditional coherence and cross-correlation methods.

Classical coherence methods and cross-correlation function analysis* readily provide information about neural element coactivation and synchronicity, but neither, unlike the present method, is able to decompose time series influences in terms of how much mutual feedback they produce. Consequently, these methods are unable to pinpoint driving structural elements — the ones that receive no feedback from other elements. Here lies the chief difference and merit of directed coherence, which is capable of both through an analytical shift in emphasis. Rather than monitor element coactivation, directed coherence dynamically probes the activation of connection pathways. Thus, it is able to offer a picture of the instantaneous prevailing signal feedback structure responsible for the observed neural activity.

9.4 Appendix: Further Mathematical Details

Even though this method is, in principle, applicable to N series at a time, we restrict our discussion to just $x_1(t)$ and $x_2(t)$ presumed linearly related for simplicity.

* Note that both of these methods and directed coherence analysis constitute a first-order linear approximation method, and thus are unable to expose the nonlinear nature of the interaction between neural elements. Research is under way to extend these ideas to the case of nonlinear interactions.

When there is no effect of $x_1(t)$ onto $x_2(t)$, an accurate predictor model of the latter may be written as

$$x_2(t) = \sum_{i=0}^{M} g_{22}(i)\varepsilon_2(t-i), \tag{4}$$

where M is the model order, and $\varepsilon_2(t)$ represents a white noise process.

If upon an attempt to improve prediction of $x_2(t)$ via a new model taking into account the past of $x_1(t)$ as in

$$x_2(t) = \sum_{i=0}^{M_{22}} g_{22}(i)\varepsilon_2(t-i) + \sum_{i=1}^{M_{21}} g_{21}(i)x_1(t-i), \tag{5}$$

we come to realize that the prediction errors are significantly decreased, then we are entitled to conclude that the past of $x_1(t)$ *causes* $x_2(t)$.

In other words, if after suitable time series modeling procedures, it is impossible to reject $g_{21}(k) = 0$ for all k at some significance level, then causal influence from $x_1(t)$ onto $x_2(t)$ cannot be ruled out. The same argument applies equally well to

$$x_1(t) = \sum_{i=0}^{M_{11}} g_{11}(i)\varepsilon_1(t-i) + \sum_{i=1}^{M_{12}} g_{12}(i)x_2(t-i), \tag{6}$$

which we may rewrite together with (5) in matrix notation

$$\begin{bmatrix} x_1(t) \\ x_2(t) \end{bmatrix} = \begin{bmatrix} 0 & \mathbf{G}_{12} \\ \mathbf{G}_{21} & 0 \end{bmatrix} \begin{bmatrix} x_1(t) \\ x_2(t) \end{bmatrix} + \begin{bmatrix} \mathbf{G}_{11} & 0 \\ 0 & \mathbf{G}_{22} \end{bmatrix} \begin{bmatrix} \varepsilon_1(t) \\ \varepsilon_2(t) \end{bmatrix}, \tag{7}$$

where $\mathbf{G}_{ij}$ represent time delay operators

$$\mathbf{G}_{ij} = \sum_{k=\delta_{ij}}^{M_{ij}} g_{ij}(k)z^{-k}$$

with $z^{-k}\, x(t) = z(t - k)$ in which δ_{ij} is the Kronecker's delta function, equal to 1 whenever $i = j$ and zero otherwise (δ_{ij} guarantees exclusive use of the past of the other series).

Schematically Equation (7) is portrayed in the block diagram of Figure 9.4 where absence of causality from $x_1(t)$ onto $x_2(t)$ amounts to lack of feedback when $\mathbf{G}_{21} = 0$.[9]

After some algebra, it is possible to rewrite Equation (7) as

$$\begin{bmatrix} x_1(t) \\ x_2(t) \end{bmatrix} = \begin{bmatrix} \mathbf{H}_{11} & \mathbf{H}_{12} \\ \mathbf{H}_{21} & \mathbf{H}_{22} \end{bmatrix} \begin{bmatrix} \varepsilon_1(t) \\ \varepsilon_2(t) \end{bmatrix} \tag{8}$$

which isolates $x_i(t)$ on the right-hand side and $\varepsilon_i(t)$ on the left-hand side, and constitutes a standard representation of a multivariate random process analogous to Equation (4) in the univariate case.[19]

It is possible to show that nullity of $\mathbf{G}_{ij}$ $(i \neq j)$, which signals absence of causality, happens if and only if $\mathbf{H}_{ij} = \mathbf{0}$. This observation is key to causal inference using linear models.

Proposition.
The joint second-order statistics of stationary multivariate process may be described uniquely by its Fourier transform, S(f), which can be factored as

$$S(f) = \begin{bmatrix} S_{11}(f) & S_{12}(f) \\ S_{21}(f) & S_{22}(f) \end{bmatrix} = \begin{bmatrix} H_{11}(f) & H_{12}(f) \\ H_{21}(f) & H_{22}(f) \end{bmatrix} \sum \begin{bmatrix} H_{11}^{*}(f) & H_{21}^{*}(f) \\ H_{12}^{*}(f) & H_{22}^{*}(f) \end{bmatrix} \tag{9}$$

where $\sum = \begin{bmatrix} \sigma_{11}^2 & \sigma_{12} \\ \sigma_{21} & \sigma_{22}^2 \end{bmatrix}$ **is covariance matrix of** $\varepsilon_i(t)$, **and** $H_{ij}(f)$ **constitute the minimum phase frequency representation of the linear time delay operators** $\mathbf{H}_{ij} = \sum_k h_{ij}(k) z^{-k}$.[20,21]

In other words, for linear operators, multivariate spectral analysis can be used to estimate $\mathbf{H}_{ij}$ directly.

Expanding Equation (9), we may write the power spectral density for each series as:

$$S_1(f) = \sigma_{11}^2 \left| H_{11}(f) \right|^2 + \sigma_{22}^2 \left| H_{12}(f) \right|^2 \tag{10}$$

$$S_2(f) = \sigma_{11}^2 \left| H_{21}(f) \right|^2 + \sigma_{22}^2 \left| H_{22}(f) \right|^2 \tag{11}$$

This motivates the introduction of a normalization similar to that employed for classical coherence leading to the definition of **directed coherence**[2] as:

$$\gamma_{ij}(f) = \frac{\sigma_{ij} \left| H_{ij}(f) \right|}{\sqrt{S_i(f)}} \tag{12}$$

where $\left| \gamma_{ij}(f) \right|^2$ represents the fraction of the power in the ith series due to the past of the jth series. Hence, if $\left| \gamma_{ij}(f) \right| \approx 0$ influence from j *to I is absent.*

The simplest approach for obtaining $\mathbf{H}_{ij}$ is through Multivariate Auto Regressive (MAR) models of the form

$$\begin{bmatrix} x_1(t) \\ x_2(t) \end{bmatrix} = \sum_{k=1}^{L} \begin{bmatrix} a_{11}(k) & a_{12}(k) \\ a_{21}(k) & a_{22}(k) \end{bmatrix} \begin{bmatrix} x_1(t-k) \\ x_2(t-k) \end{bmatrix} + \begin{bmatrix} \varepsilon_1(t) \\ \varepsilon_2(t) \end{bmatrix} \tag{13}$$

whose solution leads directly to

$$\begin{bmatrix} H_{11}(f) & H_{12}(f) \\ H_{21}(f) & H_{22}(f) \end{bmatrix} = \begin{bmatrix} 1-\sum_{k=1}^{L} a_{11}(k)e^{-j2\pi fk} & -\sum_{k=1}^{L} a_{12}(k)e^{-j2\pi fk} \\ -\sum_{k=1}^{L} a_{21}(k)e^{-j2\pi fk} & 1-\sum_{k=1}^{L} a_{22}(k)e^{-j2\pi fk} \end{bmatrix}$$

Also, as a result of this type of analysis, one obtains an estimate of Σ. In practice, to grant causality, Σ must be statistically tested for diagonality as it may not result diagonal after estimation, i.e., one must check for $\sigma_{12} = \sigma_{21} = 0$, prior to rigorous application of the method. Some authors argue that MAR estimation algorithms aim at decorrelation between $\varepsilon_i(t)$, thereby allowing a variant of the method where the effect of Σ is completely ignored and Equation (12) is substituted by

$$\tilde{\gamma}_{ij}(f) = \frac{|H_{ij}(f)|}{\sqrt{|H_{11}(f)|^2 + |H_{12}(f)|^2}} \tag{14}$$

which is termed the Method of Directed Transfer Functions.[5] Inability to reject $\sigma_{12} = \sigma_{21} = 0$ can mean that $\varepsilon_1(t)$ and $\varepsilon_2(t)$ are due to the presence of an actual third common and undisclosed signal source ultimately driving both $x_1(t)$ and $x_2(t)$.

The main advantage of Equation (13) is the existence of efficient algorithms for its computation. Good numerical stability and low computational complexity characterizes the algorithms introduced in References 22 through 24 (used here). FORTRAN versions of both can be found in Reference 14.

Finally, one has to consider the delicate issue of model order determination. Among the several proposed criteria, the most popular is minimization of Akaike's

$$AIC(p) = n \ln[\det(\Sigma)] + 2N^2 p$$

function, where the best model order choice L is obtained by successively fitting models of increasing order p, with N standing for the number of multivariate time series and n for the number of time samples in each series. It has been suggested that model orders larger than $3\sqrt{n}/N$ are inadequate and should be avoided.[14] See Franaszczuk et al.[5] for other issues regarding model order choice.

Acknowledgments

We thank A.C. Valle and C. Timo-Iaria for the data in Figure 9.1 and E.E. Fanselow and M.A.L. Nicolelis for the data in Figure 9.3 and Plate 9. L.A.B. was supported by FAPESP 97/01690-6. and CNPq 301273/96-0, and K. S. by FAPESP 96/12118-9, PRONEX 41.96.0925.00, CNPq 301059/94-2 and FFM.

References

1. French, A. S. and Holden, A. V., Alias-free sampling of neuronal spike trains, *Kybernetik*, 8, 165, 1971.
2. Saito, Y. and Harashima, H., Tracking of information within multichannel EEG record — causal analysis in EEG, in *Recent Advances in EEG and EMG Data Processing*, Yamaguchi, N. and Fujisawa, K., Eds., Elsevier, Amsterdam, 1981, 133.
3. Schnider, S. M., Kwong, R. H., Lenz, F. A. and Kwan, H. C., Detection of feedback in the central nervous system using system identification techniques, *Biol. Cybern.*, 60, 203, 1989.
4. Kaminski, M. J. and Blinowska, K. J., A new method of the description of the information flow in the brain structures, *Biol. Cybern.*, 65, 203, 1991.
5. Franaszczuk, P. J., Bergey, G. K. and Kaminski, M. J., Analysis of mesial temporal seizure onset and propagation using the directed transfer function method, *Electroencephalogr. Clin. Neurophysiol.*, 91, 413, 1994.
6. Wang, G. and Takigawa, M., Directed coherence as a measure of interhemispheric correlation of EEG, *Int. J. Psychophysiol.*, 13, 119, 1992.
7. Granger, C. W. J., Investigating causal relations by econometric models and cross-spectral methods, *Econometrica*, 37, 1969.
8. Geweke, J., Inference and causality in economic time series, in *Handbook of Econometrics*, Griliches, Z. and Intrilligator, M. D., Eds., Elsevier, Amsterdam, 1984, 1101.
9. Caines, P. and Chan, C., Feedback between stationary stochastic processes, *IEEE Trans. Autom. Contr.*, AC 20, 498, 1975.
10. Sameshima, K., Baccalá, L. A., Ballester, G., Valle, A. C., and Timo-Iaria, C., Causality analysis of rhythmic activities of desynchronized sleep in the rat, *Soc. Neurosci. Abstr.*, 22, 27, 1996.
11. Brown, T. H. and Zador, A. M., Hippocampus, in *The Synaptic Organization of the Brain*, 3rd ed., Shepherd, G. M., Ed., Oxford University Press, New York, 1990, 346.
12. MacPherson, J. M. and Aldridge, J. W., A quantitative method of computer analysis of spike train data collected from behaving animals, *Brain Res.*, 175, 183, 1979.
13. Brigham, E. O., *Fast Fourier Transform and Its Applications*, Prentice-Hall, Englewood-Cliffs, 1988, 448.
14. Marple Jr, S. L., *Digital Spectral Analysis*, Prentice Hall, Englewood Cliffs, 1987.
15. Nicolelis, M. A. L., Baccalá, L. A., Lin, R. C. S. and Chapin, J. K., Sensorimotor encoding by synchronous neural ensemble activity at multiple levels of the somatosensory system, *Science*, 268, 1353, 1995.
16. Sameshima, K. and Baccalá, L. A., Trends in multichannel neural ensemble recording instrumentation, in *Methods in Neuronal Ensemble Recordings*, Nicolelis, M. A. L., Ed., CRC, Boca Raton, 1999.
17. Geweke, J., Meese, R. and Dent, W., Comparative tests of causality in temporal systems, *J. Econometrics*, 21, 161, 1983.

18. Franaszczuk, P. J., Blinowska, K. J., and Kowalczyk, M., The application of parametric multichannel spectral estimates in the study of electrical brain activity, *Biol. Cybern.*, 51, 239, 1985.
19. Priestley, M. B., *Spectral Analysis and Time Series*, Academic Press, London, 1981.
20. Gevers, M. R. and Anderson, B. D. O., Representation of jointly stationary stochastic feedback processes, *Int. J. Control*, 33, 777, 1981.
21. Gevers, M. R. and Anderson, B. D. O., On jointly stationary feedback-free stochastic processes, *IEEE Trans. Autom. Contr.*, AC27, 431, 1982.
22. Morf, M. A., Vieira, D. T., Lee, D. T., and Kailath, T., Recursive multichannel maximum entropy spectral estimation, *IEEE Trans. Geosci. Electron.*, 18, 85, 1978.
23. Nutall, A. H. "Multivariate linear predictive spectral analysis employing weighted forward and backward averaging. A generalization of Burg's algorithm," Naval Underwater Systems Center, 1976.
24. Strand, O. N., Multichannel complex maximum entropy (autoregressive) spectral analysis, *IEEE Trans. Autom. Contr.*, 22, 634, 1977.

Chapter

Population-Level Analysis of Multi-Single Neuron Recording Data: Multivariate Statistical Methods

John K. Chapin

Contents

0-8493-3351-2/99/$0.00+$.50

10.1 Introduction

Though it is clear that the brain processes information through interactions between very large populations of neurons, most neurophysiological analysis is still based on data from single neuron recording. In fact, most of what we "know" about the systems level of brain function is based on studies in which the spiking discharges of serially recorded single units are averaged around multiple repetitions of events, and then categorized according to their predominant responses to such events. Until recently, this approach was a matter of necessity, as little technology existed for achieving large-scale simultaneous recordings. Unfortunately, this practical necessity has spawned an investigative approach and philosophy which focuses too squarely on the neuron as the basis of brain functionality. For example, the "single unit" approach to evaluation of neural information is to average the unit's activity over multiple repetitions of each of several experimentally varied parameters. Though this is useful for approximating a "tuning function" of the neuron, it does not resolve neural coding functions in the context of normal information processing. Indeed, the brain processes information in real-time, i.e., it normally accomplishes its task without the need for multiple repetitions to provide statistical resolution of the data. Clearly the brain encodes information through the concurrent activity of neuronal populations, rather than from single neuron activity repeated over time. The essential problem to be discussed here is how to resolve the information encoded by such populations.

Interestingly, an earlier generation of neurophysiologists accustomed to EEG and field potential recordings, gave little philosophical importance to the individuality of neurons, and instead emphasized their cooperativity and mass action.[1,2] It seems assured, therefore, that the advent of technologies for even larger-scale neuronal ensemble recordings will once again nurture a philosophical emphasis on neuronal populations. Allied to this is the development in the information sciences of parallel distributed processing (PDP)[3,4] as an alternative to serial rules-based computation. From PDP we obtain a mathematical basis for distributed storage of information as patterns of synaptic weights across arrays of neuron-like processing elements. If the brain utilizes such distributed coding (of course in a vastly more complex form), what would be the appropriate method(s) for measuring that information in a functional ensemble of brain cells?

The analysis of multineuron spiking data generally requires an elaborate series of procedures, based largely upon the particular issue being studied. Often it is most appropriate to begin the analysis with a classical single unit approach. In sensory systems, for example, one may wish to analyze receptive fields by generating peri-event histograms. Receptive fields, however, are only one technique for characterizing the topographic mapping embedded among neurons. Whereas receptive fields depict the response of a single neuron to stimulation of several peripheral sites, population poststimulus time histograms (PPSTHs) depict the responses of several single neurons to stimulation of a single peripheral site. PPSTHs are advantageous in that they depict the actual responses of a neuronal network to stimulation, rather than the potential response of a single neuron (as an RF). Also unlike an RF, which requires a time-consuming sequence of multiple site stimulation, a PPSTH provides a "snapshot" of the state of the sensory map within a short time slice.

As such, it can be ideal for experiments in which sensory mappings may change over time. As an example of the advantages afforded by such snapshot representations, Plate 10 shows four PPSTHs depicting the averaged sensory responses of neuronal ensembles recorded simultaneously at four levels of the somatosensory system in a rat during repeated stimulation of a single whisker. (The four ensembles were recorded in the principal trigeminal [PrV] and spinal trigeminal [SpV] nuclei, the ventral posteromedial [VPM] thalamus, and the primary somatosensory [SI] cortex.) The neuronal response intensities are displayed through a 3D color map, while poststimulus time and the neurons comprising the population are represented by the x and y axes of the horizontal plane, and neuronal response magnitudes are represented vertically. In Plate 10 the poststimulus response patterns of all ensemble neurons are smoothed into a continuous surface whose topography is represented both in the peaks and valleys and color coding. Such PPSTHs provide a much greater visual impact than the alternative method of displaying multiple peri-event histograms, and in fact give the viewer an immediate sense of the concurrence of activity between the recorded neurons. In Plate 10, for example, the sensory responses to a single whisker stimulation at the lowest levels of the system (PrV and SpV) were very circumscribed in space and time, while at higher levels of the system (VPM and SI) these responses were successively more distributed over space and time. While the visual impression made by such PPSTHs is very compelling, they do little to reveal what coding patterns may be embedded in their complexity. To go further one must consider more quantitative approaches, such as the analysis techniques outlined below.

10.2 Population-Level Analysis of Simultaneously Recorded Neuronal Activity

10.2.1 Rationale for Population-Level Analysis: Distributed Coding

Techniques for neurophysiological analysis should be appropriate for addressing one's underlying hypothesis. If, for example, the hypothesis is that information processing in the system is limited to sharpening the contrast between neighboring labeled-line channels, then population-level analysis will not be necessary. If, on the other hand, one's hypothesis is that information processing in the system is distributed, then population-level analysis will be needed for the following reasons:[5-7] First, a distributed network is highly interconnected such that input information (e.g., from a sensory stimulus) tends to be spread over multiple processing elements (i.e., neurons). Second, each neuron is involved in the processing of multiple types of information. Third, ensemble output information is encoded as a pattern of neuronal activation, which itself is a function of the spread of input information as modified by intrinsic processing. Since a given neuron may be active during the output of several different types of information, the important measurement is not that neuron's individual response to a stimulus, but the pattern of other neurons that are also activated by that stimulus. In other words, the significance of each neuron's discharge is very largely determined by its contextual relationship with other neurons in the ensemble.

Our classical methods for analysis of single neuron recordings may be inadequate to the task of resolving the information content of such distributed systems. By comparison with analysis at the single unit level, use of multineuronal techniques for analysis can provide: (1) increased resolution of different informational factors distributed through neuronal networks, (2) appropriate mathematical or statistical analysis of these multivariate data sets, and (3) neuronal weightings suitable for construction of neuronal population functions (NPFs), which themselves can be used for subsequent statistical analysis.

10.2.2 Population Coding Techniques

In recent years, a number of excellent research papers and reviews have been published on the general issue of population coding,[8-10] and it is not our intention to reproduce those efforts. It is appropriate, however, to delineate the various categories of population coding schemes that have been proposed. Coding methods have been classified as either involving "probabilistic" or "basis function" methods.[11]

Probabilistic methods include various techniques (e.g., Bayesian reconstruction) in which the set of experimental conditions is set up as a probability distribution. Prediction of the experimental condition is based on neurons' measured probabilities of spiking during that condition. Basis function methods include:

1. *Population vectors.* These codes are constructed from measurements of neuronal activity, usually in serial order,[12-17] over a predefined set of experimental conditions. The measurements directly yield basis functions, defined by weighting coefficients for each neuron. The prediction of the experimental condition is then accomplished by using these coefficients to weight-sum (linearly recombine) the activity of the same neurons during a test run, yielding an output that directly predicts the experimental condition.
2. *Template matching.* These codes are similar except that each experimental condition has its own code (i.e., template). Prediction of the experimental condition requires that a given set of test data be compared against all templates, typically by searching for a match that yields the highest score after linear recombination using the different sets of coding coefficients.[18]
3. *Reciprocal basis functions.* These involve derivation of population codes by inversion of the covariance matrix between the neurons and the set of experimental conditions. Such derived functions can be used as templates and can be reinverted to yield the original data set.[19-22]

All of the above techniques involve generation of a neural population function by linear recombination of neuronal activity. In the most general sense, this is what occurs when single neurons integrate their synaptic inputs to create an output function. It is therefore of heuristic value to consider a hypothetical downstream neuron which receives all the output information from an ensemble of recorded neurons. Given what we know about neural electrophysiology, this neuron will be selectively responsive to temporally correlated axonal inputs which combine to raise their membrane potentials past threshold. In such a system, the information from the input neurons is not simply "transmitted" as a system of labeled lines, but transformed by the combination of multiple weighted synaptic inputs and further modified as a function of the various biophysical properties of the neuron. Ultimately, the neuronal output signal could be loosely defined as a "neural population function (NPF)," in that it represents a transformation based on the properties of the population of its neuronal inputs. Alternatively, one might prefer that the NPF reflects an analog function, such as the neuronal activation state (i.e., membrane potential, V_m). As has been shown in the field of PDP (artificial neural networks) such NPFs might be simplistically considered as deriving from a two-stage process, the first of which is a linear recombination of a matrix of synaptic inputs, followed by a nonlinear scaling or probability function. In some cases it may also be useful to rescale the inputs, such as with a square root transformation to convert a Poisson distribution to a more Gaussian distribution. Of course, in real neurons the initial recombination function can be highly nonlinear and may be highly dependent on dendritic architecture and intrinsic cellular properties. Nonetheless, many of these factors can be adequately approximated in the above two-stage linear–nonlinear process. For example, the hard nonlinear output (i.e., spike generation) function present in most mammalian neurons confers a spike-train coding that is more related to the derivative of V_m than

V_m itself. This, however, can be easily handled by a nonlinear postprocessor in one's population coding model. In general, we have observed that using linear functions as first approximators of population coding yield excellent general predictions of external experimental events, and that these can be improved by selective application of nonlinear postprocessors, ranging from simple thresholds to multilayer artificial neural networks. The discussions below, therefore, will focus on use of linear functions derived through use of various multivariate statistical techniques.

10.2.3 Why Use Multivariate Statistics for Analysis of Neuronal Ensembles?

We have found over the past several years that multivariate statistical procedures provide an excellent solution to the problem of neuronal population analysis.[23] The first argument for a multivariate statistical approach to neuronal population analysis is that there is an increasing need for advanced population-level analysis techniques that are also accessible to the typical neurobiologist accustomed to single unit methods. Second, virtually all of the general population analysis approaches discussed in the section above can also be carried out using an appropriate multivariate statistical procedure. The statistical approaches are well known and can be easily accessed in commercial statistics packages. They have been mathematically defined over many years, and provide a wealth of convenient measures (e.g., not just means and standard deviations and P-values, but variances, communalities, Mahalanobis distances, etc.). Though many other nonstatistical techniques are available, they do not provide the tools for evaluation of statistical significance, which is of ultimate importance in science. Of course, classical statistical techniques can also be a disadvantage in that they themselves can be dependent on certain assumptions. Some of these, such as the assumption of normality of the data, can usually be handled with data transformations or nonparametric models. Similarly, the assumption of underlying system linearity can pose problems, but these can often be solved using nonlinear transformations. In any case, once a complete statistical analysis is made, and the basic structure of the problem set is approximated, nonlinear analysis techniques, such as nonlinear regression and neural networks, can be applied. The presence of nonlinear complexities in the system should not diminish the utility of employing statistical methods for making first approximations, and then using more powerful nonlinear techniques, such as neural networks to further explore the problem. Thus we believe that the statistical approach is the best overall solution for the initial phase of analysis of multineuron data.

10.2.3.1 Choice of analysis technique

The choice of statistical technique is highly dependent on the goals of the investigation.[24] For example, some studies are initiated mainly to define the underlying functional structure in a group of neurons. Others are directed more toward using the neuronal population to predict external events. In most of these analyses, the simultaneously recorded neurons are treated as dependent variables, and experimental conditions (nested and crossed) as independent variables. Listed here are four typical analysis goals, and statistical techniques appropriate for achieving those goals (discussed in detail in subsequent sections):

1. Defining underlying factor structure, mapping, data compression, and dimensionality reduction: *Principal Components Analysis (PCA)* and *Factor Analysis (FA)* can be used to optimally compress the information content of neuronal populations into a few underlying factors or "components."
2. Hypothesis testing: *Multivariate Analysis of Variance (MANOVA)* is the best general purpose method for determining the statistical significance of various experimental manipulations on neuronal population activity.
3. Classification, discrimination, and prediction of a discrete dependent variable: *Canonical Correlation Analysis (CCA)* and *Discriminant Analysis (DA)* can be used, often in combination, to use neuronal population activity to map, classify, discriminate, and/or predict experimental variables.
4. Regression, i.e., prediction of a continuous dependent variable: *Multivariate Regression (MR)* can define optimal vectors for prediction of scalar quantities.

10.2.4 Using Multivariate Statistical Models to Understand Information Processing in Distributed Neuronal Networks

In virtually all the statistical techniques below, the measured "variables" constitute data obtained from single neurons within a simultaneously recorded multineuronal data set. In general, all experimentally measured variables, especially neurons, tend to exhibit a certain level of variability. In statistics this is quantified as "variance," the square of the standard deviation. To the extent that this variance is caused by knowable factors, such as a stimulus or experimental condition, it can be considered as information. All of the parametric statistical methods discussed here constitute different methods for determining how much of a variable's variance is "explained" by other factors, such as the experimental manipulation or other measured variables. The residual unexplained variance intrinsic to each variable is considered "noise" or "error." As such, the normal behavior of each measured variable may reflect a complex combination of information derived from different sources, intrinsic and extrinsic.

This statistical notion of variance allocation is not only consistent with our hypothesis of distributed multifunctional neuronal information processing, it also provides a direct means of measuring it. Thus, a single neuron may be found to contribute a portion of its variance to explaining each of several experimental factors, and this contribution may change as a function of context. Moreover, a certain portion of each neuron's variance may also be explained by other variables, demonstrating the existence of shared or "redundant" information across the data set. Such analyses of variance allocation can therefore be considered as important descriptive tools for understanding how multiple informational entities are mapped in space and time across a distributed neuronal network.

What issues constrain our ability to resolve such informational factors? First, it is not possible to record all neurons in a functional population or to control all of their inputs. Nonetheless, we have found that, as one increases the number of recorded elements, the proportion of unexplained variance decreases. Furthermore, if the network contains a distributed representation, one can assume that increasing the number of recorded neurons increases resolution of sampling of most or all informational factors, without necessarily increasing the number of factors to be resolved. Thus, if the sampling is homogeneous across the network, one can assume that a moderate number of neurons will yield a noisy resolution of most or all of the information in the network. This information should be revealed as recurring patterns of covariance between neurons, which represent output states of the network. How can these patterns be resolved? Though single neurons are quite noisy, the results below show that it is possible to statistically resolve recurring patterns of neuronal ensemble activity.

Another issue is the choice of temporal integration constants. Measures of neuronal activity are typically constituted as firing rate measurements integrated over time bins of arbitrary duration. We have observed that the proportion of unexplained variance generally decreases with increasing bin duration. This suggests that spike trains can take part in encoding informational factors over multiple time scales, and that the longer time scale information tends to be more redundant across the population. If so, we can expect to resolve a greater diversity of factors when using shorter temporal integration constants, but that these may require a larger number of neural recordings to resolve.

10.2.5 Pitfalls of Multivariate Statistical Analyses: Prevention of Spurious Results

Inappropriate use of multivariate techniques can allow chance occurrences to be misinterpreted as significant. This typically occurs when the number of samples is small relative to the number of variables. A rule of thumb is to obtain a samples/variables ratio of greater than 10 to 20. Fortunately, multineuron recording experiments typically generate very large numbers of samples, ranging from 10^2 to 10^6, depending on the experimental design. Nonetheless, one should routinely undertake additional measures to ensure significance of one's statistical findings. For example, data sets are often divided into two parts: the first to establish a finding or function, and the second to verify its validity. In MANOVA, one can also utilize *post-hoc* tests (e.g., Student–Newman–Keuls) to define significance of particular comparisons.

10.2.5.1 Normality assumptions of data distribution

All of the above statistical procedures are based on assumptions that the data have a "normal" (i.e., Gaussian) distribution. Unfortunately, single unit firing rate data tend to more closely approximate a Poisson (or "super-Poisson") distribution, in which the data are skewed against the zero boundary at low rates.[25] Such distributions can produce a classic problem wherein the data means are correlated with the variances. Fortunately, the statistical applications discussed here are known to be highly robust in terms of their ability to handle different distributions, especially when the number of data samples is high. Moreover, we have observed that neural spiking data tend to converge on normality as the number of spikes per sampling interval increases, or when multiple spike trains are summed. It appears, therefore, that neural spiking has an underlying normal distribution, and thus by the central limit theorem, sample deviations from that mean also converge on a normal distribution.

Of course, this does not exempt one from the necessity of being vigilant about the distribution problem, especially when using data with low spikes/sample, or small samples. If one's routine inspection of cluster plots and utilization of normality tests (e.g., the Kolmogorov–Smirnof test) reveals an excessively non-normal distribution, one can utilize standard procedures, such as the square-root transformation, for converting Poisson distributed data into Gaussian data. (It should be pointed out that we have repeatedly utilized such transformations on our data, invariably finding that the statistical results are little changed from those achieved using the untransformed data.) Finally, in cases where the data rates of different neurons are markedly different, one can choose to prestandardize the data for each variable by subtracting the mean and dividing by the standard deviation (see Chapter 7). (This is done automatically when one generates a correlation coefficient.) The overall conclusion, therefore, is that neuronal discharge data can be used with confidence in the statistical procedures described below, but that one must maintain a constant awareness and vigilance about the possible pitfalls.

10.2.6 Data Sets Used for Examples

Most of the examples described below utilize data sets recorded from multiple levels of the somatosensory system of awake behaving rats, in either the whisker or forepaw representations. In particular, several examples are shown using data from experiments in which 48 neurons were simultaneously recorded in awake rats during discrete stimulation of different facial whiskers. The recorded neurons were distributed through five levels of the somatosensory system, including the trigeminal ganglion (Vg), the principal (PrV) and spinal (SpV) trigeminal nuclei, the ventral posteromedial thalamus (VPM), and the SI cortex. A number of recent publications provide detailed explanations of these data, our techniques for surgical implantation of recording electrode arrays, and for the recording and discrimination of the neural spike-train data.[26-30] In our hands, such data sets provide the opportunity to investigate a variety of neural information, not only including the sensory properties of individual neurons, but also their mode of interaction over a range of behavioral conditions. Ultimately, we are interested in statistically defining how information is encoded by concurrent activity across the recorded ensemble of neurons. The following examples illustrate the rich diversity of different analyses that can be carried out utilizing the same multineuron data set. Most statistical analyses were carried out using STATISTICA (Stat-Soft, Tulsa, OK), though data manipulation requirements that were too large for this package utilized the author's own data analysis package (ANALYZE) or STRANGER (Biographics, Winston-Salem, NC).

10.3 Principal Components Analysis (PCA) and Factor Analysis (FA)

10.3.1 PCA Condenses the Significant Information in a Large Set of Variables into a Smaller Number of Principal Components

We initiate discussion of multivariate statistical techniques with PCA because it is a widely used technique for reduction and interpretation of multivariate data in many scientific areas, including neuroscience.[31-32] It is particularly significant for neuroscience because it is mathematically equivalent (in the linear sense) to unsupervised learning in a neural network.[33-36] In our approach we consider the variables in our analysis to be measurements obtained from the spiking discharge of single neurons over particular time periods in an experimental manipulation.[37] Thus, the activity of the neuronal population, measured in discrete time intervals over the experiment becomes the multivariate data set. If each neuron in a population can be considered as contributing a portion of its variance to "explain" different informational factors, then the location and magnitude of important information processing ought to be identifiable on the basis of patterns of "redundant" variance across the neuronal population. In contrast, information that is intrinsic to individual neurons (i.e., unexplained variance) might be considered as noise. PCA is essentially a method for reorganizing the information that is distributed across a set of partially correlated variables into an equal number of uncorrelated principal components. No information is lost, but most of the variance related to significant redundant information in the population is consolidated within the lower numbered components, while the noise (i.e., uncorrelated variance) is sequestered in higher numbered components. The first few components will be very useful not only because they concentrate most of the "signal" in the original population, while removing the noise, but also because each may identify a significant underlying factor important to information processing in the neuronal population. For our current purposes, this identification of underlying factors allows the further important step of reconstructing them into NPFs. In practical terms, we have found that PCA serves a valuable function as the first step in the analysis of multineuron data. By definition, it consolidates, but does not alter the information in a set of variables. As such, reconstructions of NPFs corresponding to the first three to six principal components (PCs) provide the investigator an excellent first approximation of the kinds of information contained in the recorded neuronal ensembles.

10.3.2 Covariance and Correlation Matrices

The first, and most time-consuming step in a PCA is to calculate a covariance or correlation matrix. (Correlation is equivalent to covariance between standardized variables, accomplished by subtracting the means and dividing by the standard deviations.) In our procedure, the parallel spike train data, which is recorded as discrete spikes, is first integrated over user-defined time intervals to obtain analog measures of firing intensity. To best preserve the timing information in neural spike trains we typically quantize the data from the whole experiment into equal time bins (typically, from 5 to 25 ms duration). Thus, a 1000-second experiment quantized into 10 ms bins yielded 10,000 data samples, each containing the spike count within that time interval. Correlation coefficients (r) are calculated for each neuron pair (x and y) using the formula:

$$r = \frac{\sum(xy)}{\sqrt{\left(\sum x^2 \sum y^2\right)}}$$

Statistical significance of the coefficient (r) is calculated by the formula,

$$r = tSr$$

where t = Students t, and

$$S_r = \sqrt{\frac{1 - r^2}{(1 - n)}}$$

where n is the number of data samples.

It is typical in multineuronal recording studies to obtain very large numbers of samples (10^3 to 10^5), but to obtain quite small r-values (<.1). If so, the above formula can be simplified as:

$$S_r \cong \sqrt{\frac{1}{n}}$$

This shows that the standard error of r becomes vanishingly small with increasing n. As such, the small r-values typically obtained when comparing correlations between neuronal firing over short time intervals can be shown (with large n's) to be significantly different from zero. As an example, Figure 10.1 shows a correlation matrix between 23 simultaneously recorded neurons in the VPM thalamus, calculated from data obtained in a 1821.49-sec segment of free behavior. It reveals a relatively homogeneous set of positive correlation coefficients (ranging from .0054 to .39). Despite the relatively low values for the correlation coefficients (r), they are highly statistically significant because of the large number of samples: since time integrals of 25 ms were used, a total of 78,840 sequential samples were obtained. From the above equation, this number of samples yielded a standard error of .00354, and thus $r =$ .011 was the confidence limit for a highly significant difference from zero (p<.001). In this matrix of 506 cross-correlation pairs, all but one r exceeded zero by at least this confidence limit. Moreover, the average r in this matrix was 0.137, which was 12.45 times the confidence limit.

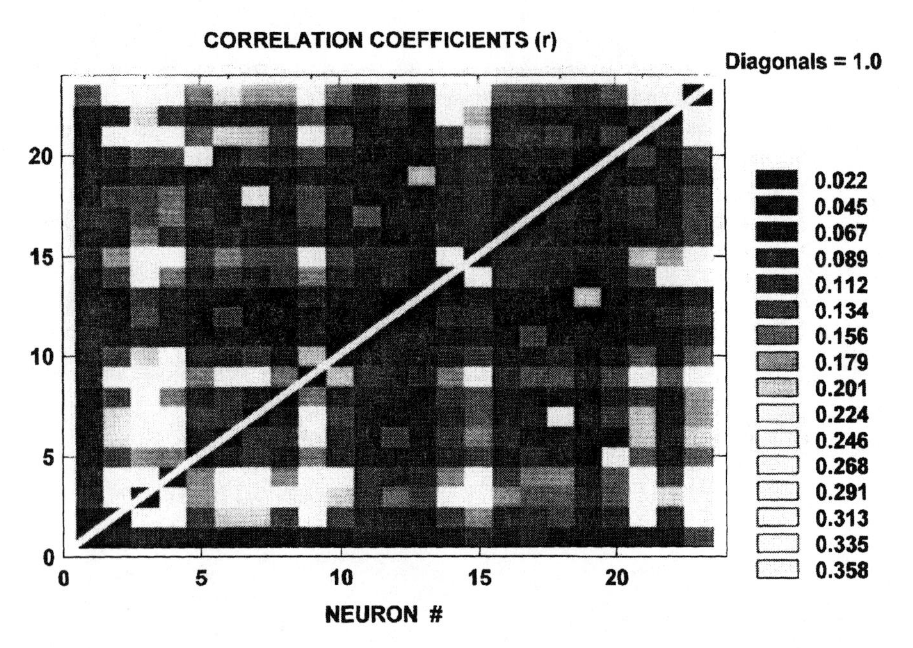

FIGURE 10.1
Correlation matrix showing functional relationships between 23 neurons in the VPM. Correlation coefficients (r) for each neuron–neuron pair (numbers on X and Y axes) are specified by shaded rectangles (scale shown at right). Correlation coefficients were calculated for 72,859 sequential 25 ms bins in this 1821.49-sec experiment. The r-values in this matrix are all positive, ranging from .006 to .386. All but two of these are significantly different from zero, using a $p \leqslant .001$ confidence level (see Methods).

The pattern of correlation between the neurons in this ensemble contains specific information on their functional interactions over a range of different behaviors. Attempts to directly analyze the patterns of simple correlations such as those in Figure 10.1 can be useful, but they are limited by the fact that they only involve pairwise comparisons. Unfortunately, attempts to directly analyze the pattern of correlations between large numbers of neurons invariably yield great complexity but very little meaning. A more appropriate method is to use a multivariate statistical technique such as PCA to reduce the complexity of a large $n \times n$ correlation matrix by deriving a much smaller series of factors or "components" which effectively summarize most of the information in the original matrix.

10.3.3 Theory of PCA

The theory of principal components states that every symmetrical covariance or correlation matrix relating p random variables $X_1, X_2, \ldots, X_n$ can be transformed into particular linear combinations by rotating the matrix into a new coordinate system, defined by matrices of weighting coefficients. This rotation is produced by multiplying the data from each of the original variables by their appropriate weighting coefficients. For each component, these weights comprise a vector called an *eigenvector*, and the variance "explained" by it is its *eigenvalue*. The original matrix is rotated such that the axis defined by the first principal component (PC1) is aligned in the direction of greatest variance, hence maximizing the eigenvalue. To obtain the second component (PC2), the matrix is rotated around the PC1 axis to obtain a second eigenvector, which again contains the greatest possible variance. This procedure is repeated until a set of n orthogonal (uncorrelated) components is obtained, arranged in descending order of variance. In this transformation, none of the information contained within the original variables is lost, and as such the derived components can be statistically manipulated in the same way as the original variables. Moreover, the transformation is useful because most of the significant total variance (i.e., correlated neuronal information) is concentrated within the first few uncorrelated PCs, while the remaining PCs mainly contain "noise" (i.e., uncorrelated neuronal information). The first few PCs not only provide a simpler and more parsimonious description of the covariance structure, they also concentrate the information which is normally spread across multiple variables (neurons) into a single, more statistically useful "factor."

After derivation of the PCs, they can be "reconstructed" into an NPF by using their weighting coefficients to weight-sum the time integrated data stream from each of the original neurons. Each NPF is therefore a continuous time series containing an analog value for each time bin within the experiment. Typically, these time bins are chosen to be the same as the time integrals used to calculate the original correlation matrix used for the PCA. As shown in Equation 1, each point Y_{pt} in a NPF is defined as the linear combination of the standardized, time integrated spike counts recorded during that time bin from the original neurons, X_{pt} multiplied by the appropriate eigenvector weights B_p.

$$Y_1 = e_i' \mathbf{X} = e_{11}X_1 + e_{21}X_2 + \ldots e_{p1}X_p$$

$$Y_2 = e_i' \mathbf{X} = e_{12}X_1 + e_{22}X_2 + \ldots e_{p2}X_p$$

•

•

$$Y_p = e_i' \mathbf{X} = e_{1p}X1 + e_{2p}X_2 + \ldots e_{pp}X_p$$

where e_i' are eigenvectors, and e_{ij} are eigenvector weightings.

In this case X_i is the spike count of neuron i within a particular time bin. It should be standardized by subtracting the mean and dividing by the standard deviation. Moreover, Y_p should be weighted by the eigenvalue λ_p, though this is unnecessary for our applications. The NPFs constructed using this technique constitute composite variables which, in statistical terms, can be treated much like the original variables (neurons). Since they are constructed from summations of weighted and standardized data, the NPF magnitudes can be thought of in terms of "population" firing rates, with units in Hz, but unrelated to actual firing rates of single neurons. These NPFs can be subject to traditional neurophysiological analyses, such as depiction in stripcharts and peri-event averages. They can also be used as variables in statistical analyses.

10.3.3.1 Example of eigenvalue rotation in PCA

The transformation of covariance data by PCA can be thought of as a rotation of the n-dimensional space which contains a scatterplot of the correlation between the n variables. As shown in Figure 10.2A, a scatterplot can be used to depict the correlation between the time integrated activities of simultaneously recorded neurons (unit 1 vs. unit 7). Figure 10.2A was constructed by quantizing the 1821-second experiment in Figure 10.1 into 3642 500 ms time intervals and plotting each interval as a point (vector) in an X–Y space defined by the two neurons, such that the point's position represents the functional state of the 2-neuron ensemble during that time interval. Using this long time integral, the correlation coefficient (r) between these two neurons was .61, which is highly significant ($p < .001$). This high positive correlation can be visualized by the preponderance of points lying along the 45° line in Figure 10.2A.

PCA can be performed using either correlation or covariance matrices. When correlation matrices are used, the data are standardized such that the mean discharge rates of all neurons are 0.0, and their variances are 1.0. Such a transformation is illustrated in the scatterplot in Figure 10.2B, which shows the data in Figure 10.2A after standardization. This transformation alleviates the problem of variance imbalance, which is often observed when studying functional relationships between neurons with markedly different firing rates. In a covariance matrix, the more rapidly firing neurons will tend to dominate. Standardizing these firing rates, as in a correlation matrix, will neutralize these imbalances. This investigation has routinely utilized both covariance and correlation matrices of the same data for PCA. In nearly all cases, the correlation matrices proved to be superior for deriving population vectors capable of resolving relevant features of sensorimotor information. For this reason correlation matrices were used exclusively in the analyses reported here. Nonetheless, the ultimate rationale for use of one or the other technique rests on a better understanding of the biological principals underlying information transmission in the nervous system.

FIGURE 10.2 (opposite)

(A) Scatterplot showing the distribution of joint activity states in an ensemble of two neurons (1 and 7) recorded simultaneously in the PrV nucleus. (Both of these were among 47 recorded across multiple levels of the somatosensory system). The position of each point depicts the firing rates of these 2 neurons over a single 500 ms time interval in a continuous experiment lasting 1800 s. The line through this point cluster shows their linear regression: ( ). Their correlation coefficient is .61 ($p < .01$). Since the coordinates of these points are integers, many overlap. (B) Scatterplot showing the same data after standardization by subtracting the mean and then dividing by the standard deviation. This removes excessive biasing of neurons with higher firing rates. (C) Half-normal probability plot showing that integrated spiking data converges on a normal distribution. The solid line shows the expected distribution of points in a perfectly normal (Gaussian) distribution. Points show the standardized scores from neuron 7 which were used to construct Figure 10.2B. (D) Scatterplot shows the same data from neurons 1 and 7 after axis rotation using PCA. The best linear fit through the original points shown by the regression line in Figure 10.2B is now oriented perfectly along (i.e., is "explained" by) the first principal component (PC1), with the second principal component explaining the "noise" in the perpendicular axis. The coordinates of the points in this new space defined by components 1 and 2 were calculated as weighted sums (i.e., the dot product) of the original standardized scores (from Figure 10.2B). The principal component weights used to achieve this rotation are shown in the text.

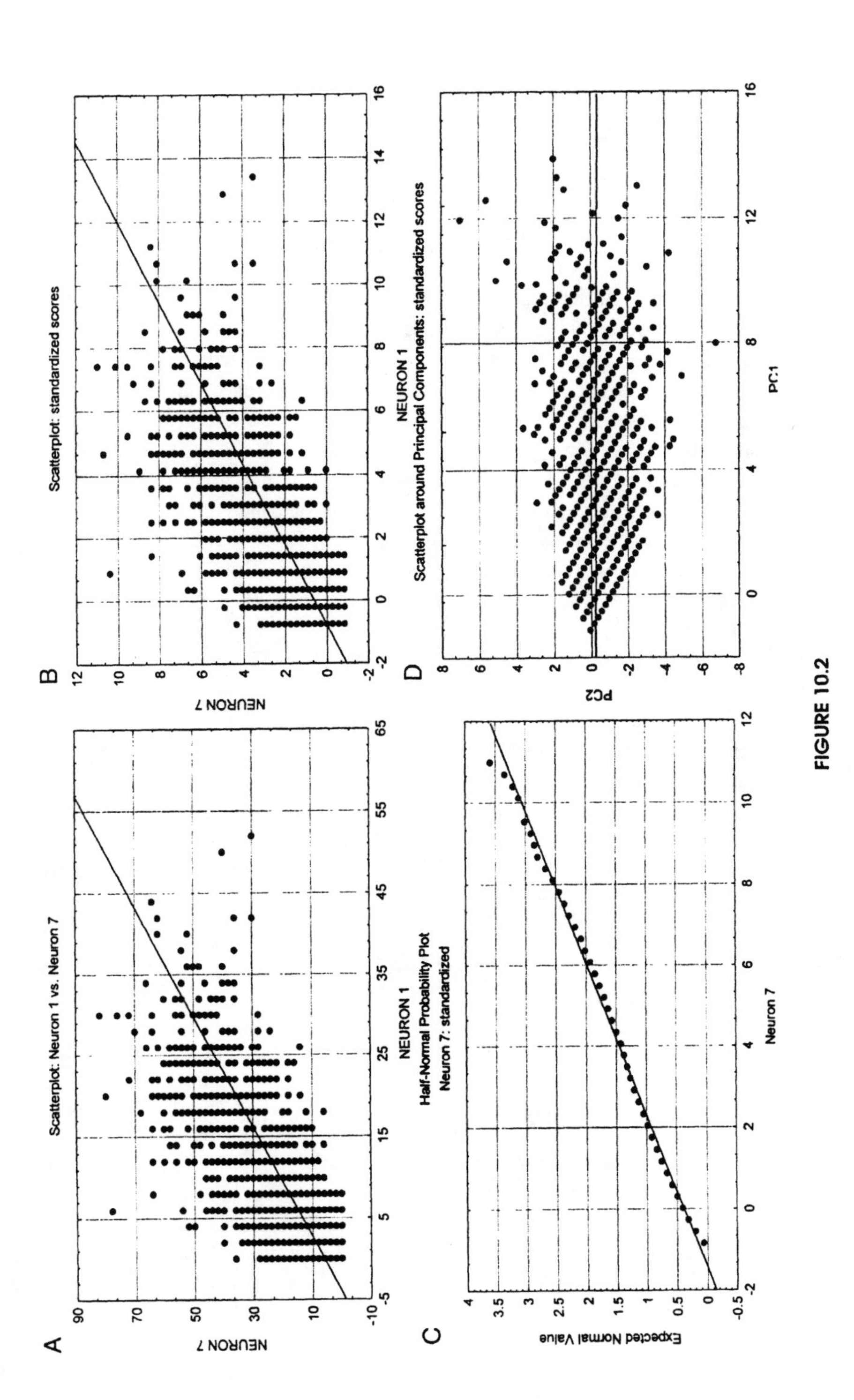
A
Scatterplot: Neuron 1 vs. Neuron 7
NEURON 7
NEURON 1
B
Scatterplot: standardized scores
NEURON 7
NEURON 1
C
Half-Normal Probability Plot
Neuron 7: standardized
Expected Normal Value
Neuron 7
D
Scatterplot around Principal Components: standardized scores
PC2
PC1

FIGURE 10.2

TABLE 10.1
Correlation Matrix between Neurons 1 and 2 and Eigenvectors (PC1 and PC2) after Rotation of the Correlation Matrix by PCA

	Neuron 1	Neuron 2	PC1	PC2
Neuron 1:	1.0	.61	.707	.707
Neuron 7:	.61	1.0	.707	–.707

Figure 10.2D illustrates the use of PCA to rotate the scatterplot in Figure 10.2B, which itself was constructed by standardizing the data in Figure 10.2A. The rotation of the 10.2B scatterplot was produced by using the following matrix of coefficients derived by the PCA for each component: The set of coefficients for each principal component (PC1 and PC2) is its eigenvector, as shown in Table 10.1. These coefficients produce the rotation in Figure 10.2D by multiplying each of the two neurons' standardized discharge rates by its appropriate coefficient, and then summing these two weighted scores for each component. This produces a counterclockwise rotation of the coordinate axis such that the X-axis (now called PC1) is aligned along the long axis of the scatterplot. The position of a point along this new axis now constitutes a useful measure of whatever factor(s) were originally responsible for producing the correlated discharge between the neurons.

The weighting coefficients in Table 10.1 represent the correlations of each variable (neuron) with the component. Since each component is apportioned some fraction of the total covariance of the neuron, the sum of squared coefficients across each row must equal 1.0. A further constraint is that the sums of squared coefficients of each component (column) must also equal 1.0. Thus the total variance contained in the ensemble can be represented either in terms of the combination of neurons (i.e., across rows in the eigenvector matrix), or the combination of principal components (i.e., across columns). The "eigenvalues," representing the percentage of the total variance explained by each of these principal components were calculated as 70.13 for PC1, and 29.87 for PC2. PC1 can be considered as a mathematical definition of a new coordinate system which parallels the major factor of interaction between these neurons, defined purely in terms of their correlation in this data set. PC2, which must be orthogonal to PC1, accounts for the remainder of the variance, which may represent noise, or a less significant type of interaction between the variables.

10.3.3.2 Statistical error of eigenvalues and eigenvector coefficients

Although PCA is a distribution free method it is sometimes useful to assess the statistical significance of the eigenvalues and principal component coefficients obtained. The following formulas can be used for calculation of standard error(s) of eigenvalues,

$$s(\lambda_h) = \lambda_h * \sqrt{2/n - 1}$$

and principal component coefficients:

$$s\left(b_{hj}\right)=\sqrt{\left[(1/n-1)\lambda_h\left(\sum_{k=1}^{p}\lambda_k\Big/\left(\lambda_{k-1}h\right)^2\left(b_{kj}^2\right)\right)\right]}$$

where n = number of samples, λ_h = eigenvalue of component h, b_{hj}> the jth coefficient of the hth component, and p = total number of components.

These apply to principal components analyses carried out on covariance matrices. However, since correlation matrices are equivalent to covariance matrices calculated from standardized data, the standard errors can be considered appropriate when standardized data are used to construct a covariance matrix. These formulas are technically correct except when used for data with small numbers of samples and for severely non-Gaussian distributed data. The typical situation for the simultaneous neural recordings here is to obtain a very large number (10^3 to 10^6) of samples, with near-Poisson distributed data. As discussed above, these statistics are quite robust when used with such large samples. Moreover, the data tend to converge on normality with increasing sample integration times and/or neuronal population size.

10.3.3.3 Influence of time integration interval on correlations and eigenvalues

The choice of the standard time interval over which to integrate spiking data is arbitrary, but must ultimately be based on assumptions about the time base(s) appropriate for processing in the system being studied. Since neurons are unlikely to interact significantly in only one time domain, any of several time integrals might be legitimately chosen for such an analysis, and taken into account when interpreting the results. In fact, the procedures outlined in this manuscript for mapping functional relationships between neurons are very useful for defining appropriate time bases. We have invariably found that one can obtain larger r-values, and also improve the normality of the distribution of scores, by increasing the time periods over which the spikes are integrated. This generally does little, however, to increase the statistical significance of the correlations, because it decreases the total number of samples. Furthermore, by increasing the time integral, one loses the fine temporal structure of data. Taking all of these considerations into account, we have found that time integrals between 10 and 25 ms are most appropriate for mapping sensory responses to passive and active whisker stimulation, and also provide the best balance between significance of r-values and temporal resolution of data.

The scree plots in Figure 10.3 support the notion that much of this information is lost when integrating over longer time integrals. This figure depicts the eigenvalues for the first 10 principal components using time intervals ranging from 5 ms to 2000 ms (from the data set used in Figure 10.1). Over this range the eigenvalues of PC1 consistently increase from 2.3 to 14.56. This phenomenon, which results from the increases in correlation coefficients, indicates that a successively higher proportion of fine timing information in the neural firing data is lost as one

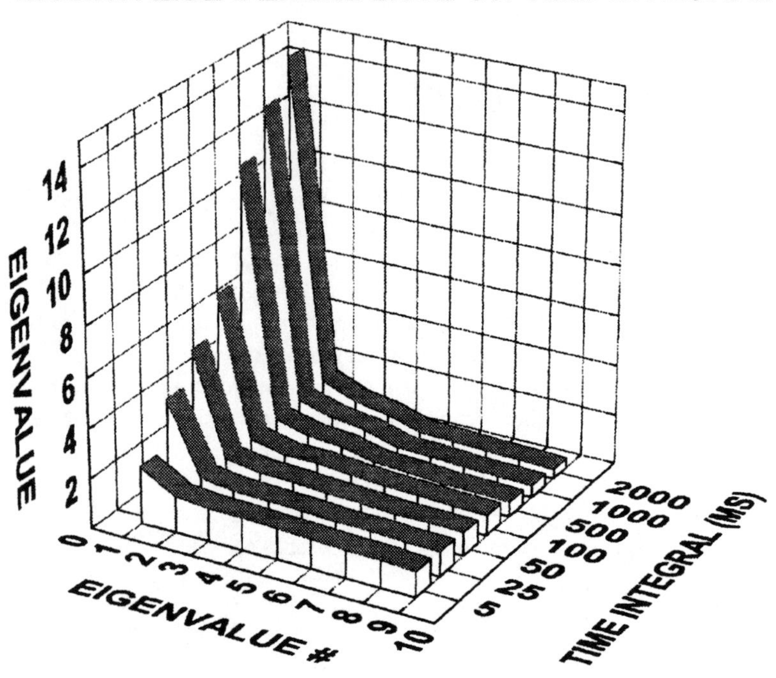

FIGURE 10.3
Scree plots show that the eigenvalues of the first principal component increase with greater time integrals. Eigenvalues were computed for seven different time integrals (from 5 to 2000 ms), which are the time segments of the experiment which constitute individual samples. Each strip in this 3D graph is a scree plot obtained when PCA was calculated for a different time integral, as shown on the X-axis (right). The number of the eigenvalue (i.e., the principal component number) is shown on the Y-axis (bottom), and the eigenvalue itself is shown on the Z-axis (vertical). Since these calculations were carried out using correlation coefficients, the variance of each neuron is standardized at 1.0 apiece, and therefore the total variance here was 23. Whereas eigenvalue #1 (EV1) for the 5-ms time integral was only 2.3 (10% of the total), that for the 2000-ms integral was 14.5 (63% of the total). This relationship was reversed with increasing eigenvalue numbers: EV10 for the 5-ms integral was 0.99, whereas that for 200-ms integral was 0.33.

increases the time integral. Since much information is lost, it can be effectively summarized with a smaller number of principal components. While this may be a useful means for reducing the complexity of analysis, it is not appropriate for studies such as in this example, which hopes to resolve covariance information with a time resolution sufficient to differentiate whisker movements occurring during active tactile exploration.

10.3.3.4 Interpretation of PCA results

Since PCA involves successive rotations of a covariance matrix, the configuration of each derived PC is a function of the previous (lower numbered) PCs. Thus, the PCs may not each represent a single factor underlying the population covariance, but instead may provide rules for separating between such factors. PCA is not, therefore, the method of choice for separating between completely independent factors, but is an excellent technique for defining a multidimensional mapping scheme for linear representation of a set of interdependent factors. As such, PCA should be ideal for analyzing processing of sensory information which normally involves interdependent factors.

10.3.4 Neural Population Functions (NPFs) Constructed Using PCA-Derived Weight Matrices

As an example, Figure 10.4 depicts use of PCA to construct NPFs using data obtained from 43 neurons recorded simultaneously from five levels of the somatosensory system (including trigeminal nuclei, VPM thalamus, and SI cortex) in an awake rat.[30] When reconstructed and integrated, these vectors produced continuous functions which depicted the activity of underlying global and local informational factors in the neuronal ensembles. Figure 10.4 (bottom) shows such a continuous reconstruction of the first principal component covering the time period from 64 to 69 seconds of an experiment in which the neurons were recorded. For comparison, the spiking activity of these neurons is shown as rasters at the top of Figure 10.4. Because of the stochastic nature of single neuronal firing, inspection of any one of these rasters reveals little about the overall pattern inherent in the ensemble. Furthermore, only a limited sense of this pattern can be obtained by inspecting the group of rasters: none of the single neurons produced spikes during every single oscillatory cycle. Nonetheless, the reconstructed first component (PC-1, bottom) shows a suddenly appearing episode of 8 to 12 Hz rhythmic discharge. On the other hand, the higher numbered PCs (not shown) did not reveal this clear oscillatory activity, but instead encoded features of neural activity more specific to the particular nuclei, or to specific features of sensory stimuli. As such, PCA separated global functions such as these oscillations from more local functions relating to various aspects of sensory processing. All of the first 4-5 PCs, however, encoded factors characteristic of the neuronal population, as opposed to single neurons.

These weighted-sum reconstructions of population vectors thus provide a means for increasing the statistical resolution of the multiple spike trains by combining its information into a single metric. The advantage of this is that such an NPF can be analyzed as a singular neurophysiological entity. In this case, careful video frame-by-frame analysis was carried out to determine how continuous time events might be encoded by the neural population. These showed that the oscillatory periods were predictive of the animal's behavior. Periods of spontaneous 8 to 12 Hz oscillation always began during periods of absolute resting behavior. About .2 to .7 s after the onset of this rhythmic discharge, however, the whiskers typically began to make small vibratory movements. The rhythmic oscillations were intermittently maintained throughout this period, and were phase synchronous with these small whisker movements. In contrast, large movements of the whiskers typically terminated the oscillations. Thus, global activity patterns of the neuronal population, as reconstructed using multivariate statistical techniques, provided a useful predictor of overall behavior that could not have been obtained using single neuron recording.

10.3.5 Factor Analysis (FA): Classification of Spatiotemporal Response Patterns

Factor analysis is an extension of PCA that allows one to increase the resolution of highly predictive sets of variables. In general, while PCA is often preferred as a method for data reduction, FA is often preferred when the goal of the analysis is to detect structure. It takes advantage of the fact that once PCA has been used to rotate the correlation matrix to create components aligned along maximal axes of correlation between the internal variables, further rotations can be performed to align the matrix along axes of maximal variance based on external (i.e., experimentally important) variables. As an example, Figure 10.5 shows use of factor analysis for classification of sensory response data from 23 simultaneously recorded VPM neurons. To classify the variety of spatiotemporal response patterns observed in the ensemble, a raw data matrix was first built in which each cell was treated as a variable. These matrices contained the measured magnitudes of each neuron's response to each whisker stimulated. Each of 8 consecutive poststimulus time epochs defined the cases (rows), and the cells themselves defined the columns. Thus, in an experiment in which 20 whiskers were stimulated and 23 cells were simultaneously recorded, the raw data contains 23 columns (neurons) and 160 cases (i.e., eight instantaneous firing rate samples for each of 20 whiskers stimulated). A correlation matrix was then derived from the raw data by calculating the correlation coefficients between equivalent measurements from pairs of variables (neurons). This matrix was then subject to PCA to yield a set of principal components (eigenvectors). These components were then subjected to a normalized varimax rotation to create a new set of factors that were better aligned for classification of spatiotemporal response patterns. In Figure 10.5, cellular data is clustered in an x–y plot whose axes are defined by the most significant factors derived. This provides a quantitative measurement of the functional association among the neurons during the stimulation of single whiskers. This procedure was next expanded using equivalent data derived from stimulation of several different whiskers. This quantifies the functional groupings among cells belonging to the simultaneously recorded ensemble, and their variation according to the whisker stimulated.

10.4 Multivariate Analysis of Variance (MANOVA)

MANOVA is the multivariate form of ANOVA, which is the most commonly used parametric technique for testing significant differences between means. It accomplishes this by partitioning the total variance (i.e., the sum of squared deviations from the overall mean). MANOVA calculates the ratio between the "between-groups" mean variance and the "within-groups" mean variance (or "mean square

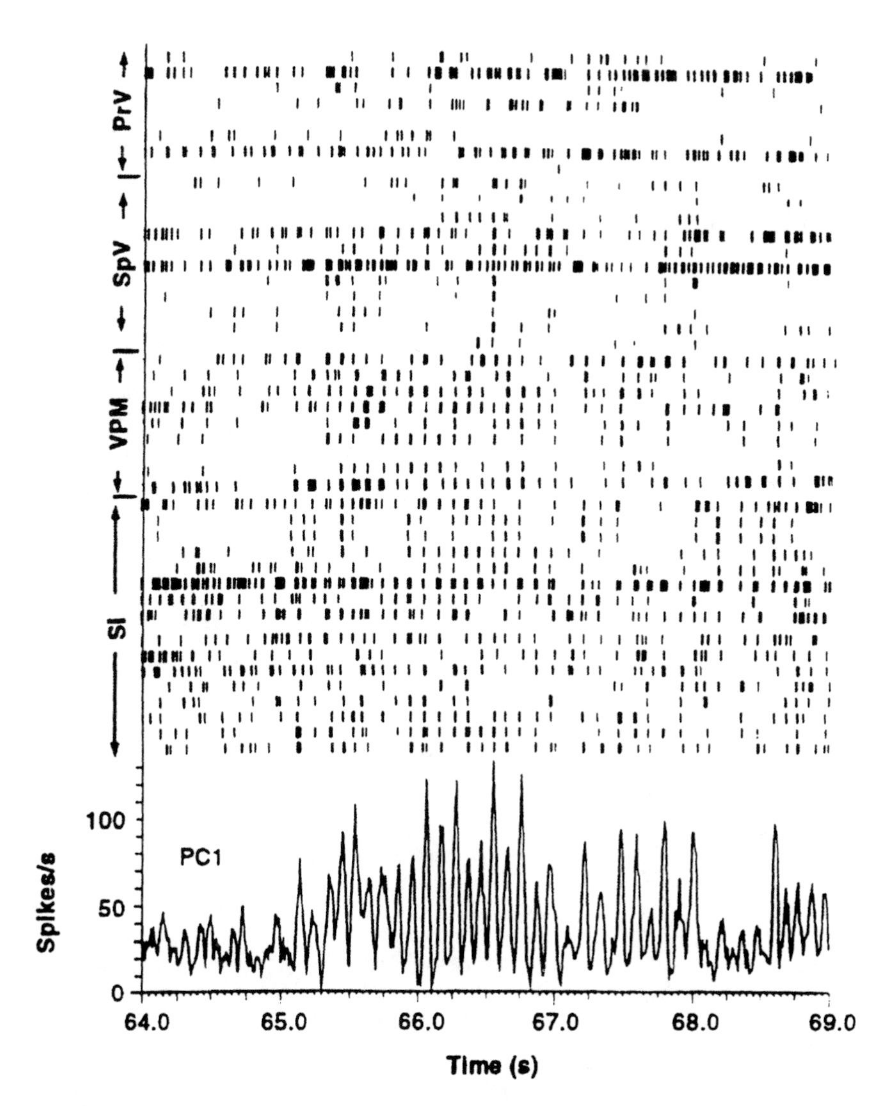

FIGURE 10.4
Reconstructed principal components provide resolution of global information in neuronal ensembles. This continuous stripchart shows 43 single-unit rasters (top) and reconstructed PC1 (bottom) over the same 5.0-s time period in an awake rat. The rasters show spiking activity of simultaneously recorded neurons in four system relays: PrV, SpV, VPM, and SI. Not shown are the five neurons recorded in Vg, which were inactive over this period. The vertical axis depicts the weighted neuronal population firing rate. (From Nicolelis, M.A.L., Baccalá, L.A., Lin, R.C.S., and Chapin, J.K., Synchronous neuronal ensemble activity at multiple levels of the rat somatosensory system anticipates onset and frequency of tactile exploratory movements, *Science*, 268, 1353, 1995. With permission.)

error"). This yields the "F" score, from which the significance (i.e., p-value) of the overall group differences can be determined. A major strength of MANOVA is its ability to handle rather complex multifactor experimental designs. For example, the effects of multiple nestings or crossings of different experimental manipulations can

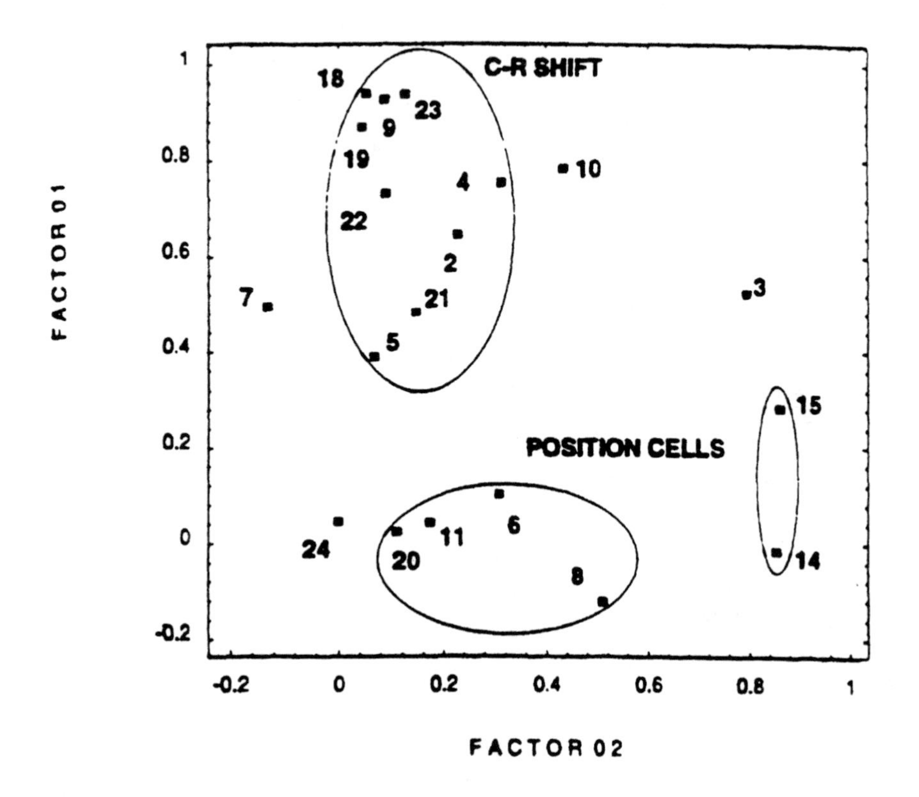

FIGURE 10.5
Functional classification of VPM neurons is shown in a scatterplot depicting the correlation of the sensory responses of 23 VPM neurons, obtained by independent stimulation of 20 whiskers, to the two most significant factors derived from the application of factor analysis to the same data. Cells exhibiting high correlations with Factor 1 (vertical axes) have the largest overall RFs and the largest RF shifts. Cells highly correlated with Factor 2 had their short-latency principal whiskers located in the most rostral whiskers, while neurons with lower correlation with Factor 2 had their short latency principal whiskers located in progressively more caudal locations. The combination of these two factors in this scatterplot clearly differentiated two heterogeneous populations of VPM neurons: "caudal-to-rostral-shifting (C-R SHIFT)" cells and "POSITION" cells. Clusters of these cells are indicated by ellipses, one for the C-R cells and two for the POSITION cells (From Nicolelis, M.A.L. and Chapin, J.K. (1994) The spatiotemporal structure of somatosensory responses of many-neuron ensembles in the rat ventral posterior medial nucleus of the thalamus. *J. Neurosci.*, 14(6):3511-3532. With permission.)

be assessed within the same overall calculation. As such, MANOVA cannot only define the significance of group-related differences between the means of individual variables, but it can also detect group-related *interactions* between those variables. Finally, MANOVA allows one to go beyond the simple one-hypothesis-one-test type of experimental design, and instead test many hypotheses within the same overall experiment. This kind of *a posteriori* hypothesis testing is made possible by different *"post-hoc"* tests that can be administered after the initial MANOVA calculation.

If the same set of dependent variables is measured during application of different experimental conditions (i.e., groups or "independent variables"), a "repeated measures" MANOVA design can be performed. This generally is more powerful than

TABLE 10.2
MANOVA: Summary of All Effects (factors)

Effect	df: Effect	MS: Effect	df: Error	MS: Error	F	p-level
1-All	1	368.21	198	502.89	.73	.3932
2-Neurons	18	2.65	3564	.11	24.14	0.0000
3-Latencies	10	4.70	1980	.11	43.27	0.0000

"between-groups" designs in which different variables are measured during different experiments. Fortunately, multineuron recording experiments allow one to use the repeated measures designs in situations where single serial unit recording would force the use of between-groups designs.

Table 10.2 provides a brief illustration of how one can use MANOVA to analyze the complexities inherent in multineuronal recording data. For this demonstration we used a subset of 10 neurons located in the trigeminal nuclei, from the 48 neuron data set described above in which various facial whiskers were repeatedly deflected. A data file was made consisting of per-trial spiking rates of these neurons within two different latency response periods (1 to 15 and 16 to 30 ms poststimulus) to 100 discrete deflections of each of two different whiskers (E3 and B4). A repeated measures MANOVA with three nested factors was used to determine the main sources of statistical differences between these 10 neurons' responses to stimulation of the two whiskers. Three different factors were tested in terms of their contribution to the neural encoding of the two whiskers: (1) the mean responses of all 10 neurons, (2) the mean poststimulus responses of each individual neuron, and (3) each neuron's responses with the short (1 to 15 ms) vs. long (16 to 30 ms) latency epoch.

The results show that the mean discharge of the 10 neurons did not significantly distinguish between the two whiskers (Effect 1: $F = .73$; $p = .39$), but that there were highly significant differences in the patterns of responses across the neuronal ensemble (Effect 2: $F = 24.14$; $p<10^{-8}$). Moreover, there were even more highly significant differences between these response patterns when one separately considered the short and long response latency epochs (Effect 3: $F = 43.27$; $p<10^{-8}$). Since factors 1, 2, and 3 are nested, it was inappropriate to test interactions between the factors. In a non-nested design, however, calculations of the F-values of such interactions would be an important addition to the analysis. Finally, a number of *post-hoc* tests are provided in most statistical packages that allow one to evaluate the significance of each individual variable (i.e., neuron or latency response epoch) in differentiating between the experimental groups.

10.5 Canonical Correlation (CCA) and Discriminant Analysis (DA)

CCA and DA are often employed together to determine whether a set of measured variables (i.e., recorded neurons) can classify and discriminate between experimental groups. CCA is used first to define a space in which to view the separation of experimental trials into these groups. DA is then used construct a discriminant function to directly discriminate between these groups.

10.5.1 Canonical Correlation Analysis (CCA)

Though CCA and DA may be used in tandem, they were originally configured for very different purposes. CCA is similar to PCA in that it is essentially a tool for analysis of covariance structure. While PCA maps patterns of correlation within one set of variables, CCA maps patterns of correlation between two sets of variables. Like PCA's eigenvalue-maximizing rotation of an autocorrelation matrix to yield a series of mutually orthogonal components, CCA makes a series of eigenvalue-maximizing rotations of a cross-correlation matrix to yield a series of mutually orthogonal "roots." When two sets of variables (e.g., A and B) are correlated, two sets of roots are extracted, one for A and one for B. The roots are eigenvector matrices used to make linear combinations of the original variables to create "canonical variables." For each successive root, these canonical variables of A and B are chosen such that their correlations with each other ("canonical correlations") are maximized. Like PCA, therefore, CCA concentrates high-dimensional relationships between large numbers of variables into a simplifed few canonical variables. The advantage of CCA is that one can directly map the patterns of covariance between the neurons and experimental parameters. It is therefore an obvious choice when one wishes to use multineuronal recordings to classify experimental groups. Here its utility will be demonstrated for the simplified case of correlations between a set of neuronal variables and a single grouping variable.

As an example, we use a data set in which ensembles of neurons were recorded in the VPM thalamus of the somatosensory system during stimulation of different whiskers.[38] We wished to determine whether these neuronal ensembles were capable of predicting the identities of the stimulated whiskers. For this we utilized CCA to identify and quantify the associations between the set of dependent variables (i.e., the recorded neurons) and the independent variable (i.e., which whisker was stimulated). CCA constructs a correlation matrix between the dependent and independent variables and then performs an eigenvalue transformation on this matrix, similar to that described for PCA above. As such, CCA defines the sets of linear combinations of the two sets of variables which are most highly correlated, but mutually orthogonal. These are the "canonical variates" or "roots," which define an optimal space into which neuronal ensemble recordings can be defined. For example, Figure 10.6A shows a scatterplot in which each dot represents the responses to a single whisker stimulus of ensembles of neurons simultaneously recorded in the VPM thalamus. The x and y axes of this plot are defined by canonical roots 1 and 2. The coordinates for each dot were calculated by summing together the poststimulus spiking responses of all neurons, each response weighted by its neuron's appropriate coefficient in roots 1 and 2. Three partially differentiable clusters of dots can be observed in this plot, corresponding to the three stimulated whiskers. Though the two roots (1 and 2) defined by CCA are defined as the two best factors for classifying these trials;

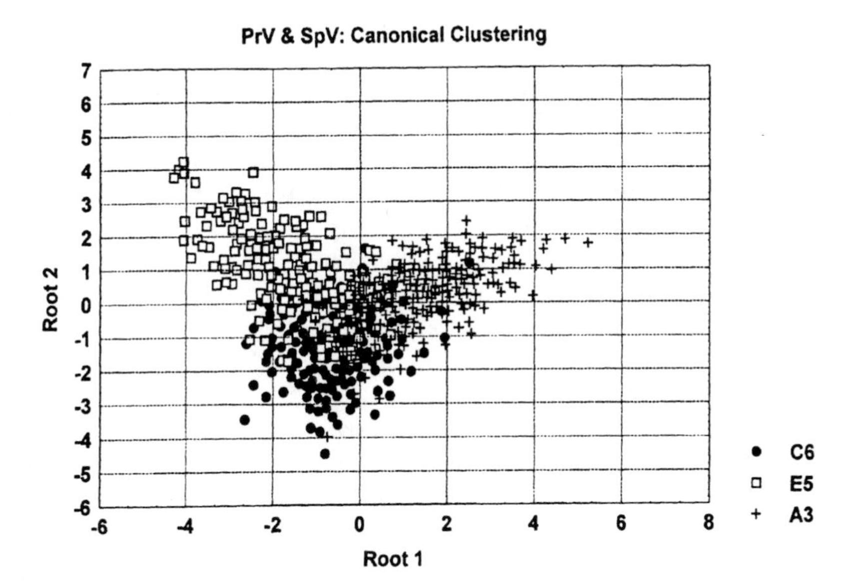

FIGURE 10.6
Use of CCA and DA for classification and discrimination: multineuron ensembles predict stimulus location. (A) Encoding of stimulus location by VPM neuronal ensembles as depicted in a space defined by CCA. Scatterplots depict distribution of canonical scores, computed from orthogonal canonical roots, for a series of single trials obtained during stimulation of three different whiskers on the face (respectively depicted by squares (E5), circles (C6), and pluses (A3)).

adding information from additional roots might further differentiate the clusters, especially between particular groups. However, these would add further dimensions, complicating the visualization of the problem. Thus, after one defines a space for visualization of such clusters, it is more appropriate to utilize discriminant analysis (DA) to optimize classification between the set of experimental groups.

10.5.2 Discriminant Analysis (DA)

Discriminant analysis is used to define a "predictive model" in which a set of measured variables can, in linear combination, optimally discriminate between experimental groups. Initially this involves carrying out a series of MANOVA-like analyses to determine whether the combination of a beginning set of variables can statistically distinguish between a set of groups (i.e., by calculating an F and p-value). Next the model can be refined by adding or deleting variables in a stepwise fashion, recalculating the F after each step. The changes in F caused by adding or removing a variable reveal its value to the overall discrimination, and whether it should be included or excluded from the final model. This procedure is computationally and conceptually equivalent to stepwise multiple regression, except that Wilk's lambda values are calculated, allowing multivariate F-scores to also be calculated. After the model is finished, a set of discriminant functions is derived, one for each experimental group, such that:

$$\text{Prediction of experimental group "A"} = a + b_1*x_1 + b_2*x_2 \ldots + b_n*x_n$$

where a is a constant offset and b_{1-n} are regression coefficients which quantitate the relative contribution of each neuron (x_n) to prediction of experimental group "A." For greatest reliability, such discriminations should be made on equivalent data sets which were not used to generate the discriminant functions. In addition to pure discrimination, DA provides a host of results that are potentially useful to the neurobiologist, many of which are illustrated in examples below.

10.5.2.1 Use of DA to define population coding

In our utilization of DA, the "independent variables" are typically one or more measures obtained from each of many simultaneously recorded neurons. The dependent variables are experimental groupings, such as different stimulus sites. This use of DA to define discriminant functions in multineuron data might therefore be considered as quantitatively defining the codes embedded within the distributed network of recorded cells.[39,40] The following example shows typical procedures in which discriminant analysis was used to define codings of sensory information embedded in the simultaneously recorded somatosensory system neuronal ensembles. This example utilizes a subset of the data from the experiments described above in which 47 neurons were recorded during controlled stimulation of each of eight points on the face. Here, the short latency (5 to 15 ms) and long latency (16 to 30 ms) responses of each of nine of these neurons (#'s 31–39; 31–32 in VPM; 33–39 in SI) were measured for each of 300 stimulations of each of three whiskers: B4, A3, and B2. Using the CSS-Statistica discriminant analysis program, all of these measures were evaluated for their individual and joint statistical relation to the three whiskers stimulated. (The analysis includes 18 "variables": both short and long latency responses for each of the nine neurons, 3 "groups": 3 whiskers, and 900 "cases": 300 stimulations of each whisker. An additional group of equivalent cases was set aside for later use to test the derived classification functions.)

The statistical significance of the overall comparison between groups is indicated by the Wilks' lambda, F, and p values at the top of Table 10.3: the lower the Wilks' lambda and p-level, the higher the level of statistical discriminability provided by the combination of variables. The "F-to-remove" and associated "p" values for each variable indicate their statistical ability to discriminate between the three groups. This technique can also be used to analyze the predictive value of subsets of neuron ensembles separately. When DA was carried out using just short or long latency responses (9 variables × 3 groups × 900 cases), the "short latency" Wilks' lambda was .70348 (F: 18.971), while the "long latency" Wilks' lambda was .44472 (F: 49.231). The lower Wilks' lambda and higher F values show that the long latency responses predicted the stimulated whisker much better than the short latency responses.

Classification functions: To assess the ability of the recorded neuronal ensembles to discriminate the stimulated whisker, three classification functions were constructed. As shown in Table 10.4, these consist of a matrix of coefficients, relating each short or long latency neuronal response to each whisker. Each of these defines a group centroid which can be used to predict the stimulated whisker from the pattern

TABLE 10.3
Discriminant Analysis Using both Short and Long Latency Neuronal Responses to Whisker Stimulation

Cells (SL)	F-remove (2,878)	p-level	Cells (LL)	F-remove (2,878)	p-level
31S	5.6093	.003796	31L	3.3389	.035925
32S	5.8944	.002865	32L	10.1323	.000045
33S	12.4328	.000005	33L	2.8593	.057842
34S	12.4792	.000005	34L	1.9377	.144646
35S	.8759	.416853	35L	87.6189	.000000
36S	17.7564	.000000	36L	5.6658	.003590
37S	1.8322	.160668	37L	16.8868	.000000
38S	10.1628	.000043	38L	20.0180	.000000
39S	3.7280	.024423	39L	106.1493	.000000

The variables consist of the short latency (SL) and long latency (LL) responses of each of the nine neurons (31–39).

Wilks' lambda: .34720 F: 34.003 p<.000000

of neural responses, on a trial-by-trial basis. For this, the program calculates the squared Mahalanobis distances from the neural response patterns observed in each trial (case) to each of the group centroids. (Mahalanobis distance is Euclidean distance normalized according to the statistical reliability of the measurements.) Each case is classified as belonging to the group (whisker) whose centroid is closest.

Table 10.4B shows examples of classifications of single stimulus trials using the classification functions. Posterior probabilities were calculated from the Mahalanobis distances used for classification. As shown here, the incorrect classifications may differ markedly from the correct ones, indicating that these stimuli somehow produced entirely different neural response patterns. The circumstances surrounding these stimulations can be checked using *post-hoc* analysis, for example by analysis of videotaped records of the experiment.

Table 10.5 shows the overall results of this classification. The use of all 18 variables allowed an average 70.6% prediction of the stimulated whisker. The cumulative benefit provided by all of the variables is demonstrated by comparing Table 10.5A with Tables 10.5B and 10.5C, which show the results of using only short or long latency responses for the classification. Both of these subsets yielded much smaller overall percentages of correctly classified trials. It should be noted that the long latency subset (nine variables) yielded better predictions for the B2 whisker, but much worse predictions for the B4 whisker. This demonstrates that smaller subsets of variables may provide very good specific discriminations between two groups, but cannot discriminate well between larger numbers of groups.

10.5.2.2 Use of DA to map sensory coding

The matrices of classification functions, such as those in Table 10.4 will not only be used for classification (as in Table 10.6), but more generally to define the sensory

TABLE 10.4
Classification

A. Functions for classification of neuronal population responses to stimulus of three different whiskers

Cells (SL)	Whiskers B4	A3	B2	Cells (LL)	Whiskers B4	A3	B2
31S	1.84686	.81576	.89963	**31L**	.81146	.71313	.17360
32S	1.57253	.87454	.40644 3	**32L**	.70471	.88243	.04658
33S	1.08537	1.01166	.51406	**33L**	.75087	.34002	.79700
34S	1.97317	.35103	.59860	**34L**	.08981	.68171	.41949
35S	.08821	–.52601	–.1922	**35L**	.13073	3.00059	.17451
36S	1.49409	.41220	.84497	**36L**	.26959	.61415	.93945
37S	1.08269	.52295	.85085	**37L**	.51417	1.47305	.55941
38S	.68139	1.23278	.72454	**38L**	1.20305	.45664	1.13993
39S	1.16373	1.25671	.86946	**39L**	.20060	.73868	.27348

Constants: B4: –4.55288, A3: –6.48649, B2: –3.02127

B. Case-by-case Classifications: Posterior Probabilities

Observed: Case	Classif.	Whisker-B4 p = .33408	Whisker-A3 p = .33296	Whisker-B2 p = .33296
1	**Whisker-A3**	**.000930**	**.994704**	**.004366**
2	Whisker-A3	.207446	.330115	.462439
3	**Whisker-A3**	**.000466**	**.997653**	**.001881**
4	**Whisker-A3**	**.020994**	**.973271**	**.005735**
5	Whisker-A3	.564713	.254659	.180628

(Correct classifications in bold)

codes embedded within ensembles of recorded neurons. These will allow quantitative measurement of the changes in sensory coding produced by behavioral set and discrimination training. Furthermore, they will be used to quantitatively define how peripheral sensory inputs are mapped in the "neural space" defined by the recorded neurons. A simple example of this is shown in Table 10.7, which shows the relative "distances" between the three whisker stimuli. When used with multiple types of stimuli, this will be used to define the coding for each type of sensory input in terms of position within an N-dimensional neural space.

10.5.3 Use of DA to Predict Behavioral Responses

An example of the use of discriminant analysis is shown using data from the experiment in Figure 10.7, in which a partially trained rat responded with a CR to the CS+ in only about half the trials. Seven neurons were simultaneously recorded, all of which exhibited

TABLE 10.5
Classification Matrix: Short + Long Latency Responses

(Rows: Observed classifications, Columns: Predicted classifications)

A: All responses

Group	Percent Correct	Whisker-B4 p = .33333	Whisker-A3 p = .33333	Whisker-B2 p = .33333	Total
Whisker-B4	64.66666	194	15	91	300
Whisker-A3	74.74747	14	225	61	300
Whisker-B2	73.15437	59	21	220	300
Total	70.60133	267	261	372	

B: Only Short-latency Responses

Group	Percent Correct	Whisker-B4 p = .33333	Whisker-A3 p = .33333	Whisker-B2 p = .33333	Total
Whisker-B4	56.66667	170	55	75	300
Whisker-A3	51.17057	45	153	102	300
Whisker-B2	53.51171	55	84	161	300
Total	53.72636	270	292	338	

C: Only Long-latency Responses

Group	Percent Correct	Whisker-B4 p = .33333	Whisker-A3 p = .33333	Whisker-B2 p = .33333	Total
Whisker-B4	35.66667	107	11	182	300
Whisker-A3	74.83221	15	223	62	300
Whisker-B2	81.54362	34	21	243	300
Total	63.80846	156	255	489	

TABLE 10.6
Squared Mahalanobis Distances

Groups:	Whisker-B4	Whisker-A3	Whisker-B2
Whisker-B4	0.000000	6.991520	1.540656
Whisker-A3	6.991520	0.000000	5.545127
Whisker-B2	1.540656	5.545127	0.000000

TABLE 10.7

Discriminant Analysis		Classification Functions		
Neuron	F-remove	p-level	CR	No-CR
GZ-1	85.85	.000	1.085	.225
GZ-2	8.67	.003	1.134	.843
GZ-3	31.04	.000	1.477	.873
DZ-1	2.69	.101	.314	.102
DZ-2	3.64	.056	−.161	.104
DZ-3	2.54	.111	.078	.320
MI-1	.00	.991	−.130	−.129

some response to the CS+: Three neurons were located in the SI cortical granular zone (SI-GZ), three in the SI dysgranular zone (SI-DZ), and one in the motor (M1) cortex. Here the discriminant analysis was used to determine how well the ensemble of recorded neurons could predict whether the rat would respond to the CS+. Data for both groups were measured during the 20 ms time epoch following CS+ stimuli. They were classified according to whether or not the rat responded properly to the CS+, resulting in a reward. Using the CSS-Statistica discriminant analysis program, all variables were evaluated for their statistical relation to the two conditions. Table 10.7 shows that the three neurons in the SI-GZ had high F-to-remove values and very low associated p-levels. No other neurons were significant by themselves.

This analysis demonstrates several points: (1) This technique is able to quantitatively and statistically define the contribution of each neuron to the discrimination of the CS+. (2) The anatomical location(s) of neurons which contribute to the discrimination can be easily determined. (3) The discrimination generally improves by adding more neurons to the population code.

To assess the ability of the recorded neuronal ensembles to discriminate the presence or absence of a CS+, classification functions were constructed (Right: Table 10.7). Using these, the behavioral response to each of 520 CS+ presentations was predicted, as shown in Table 10.8. Though the three SI–GZ cells alone were able to classify 61.5% of the CS+ trials, this improved to 67.5% when all seven neurons were used. Thus, the prediction improved even though four neurons with nonsignificant F-to-remove values were added to the prediction matrix. This demonstrates the value of subliminal neuronal responses when they are utilized in a population coded scheme.

10.6 Multiple Regression Analysis (MRA)

10.6.1 Linear MRA Models

MRA is the statistical technique of choice for defining the mathematical relationship between a set of independent predictor variables (e.g., a population of neurons) and

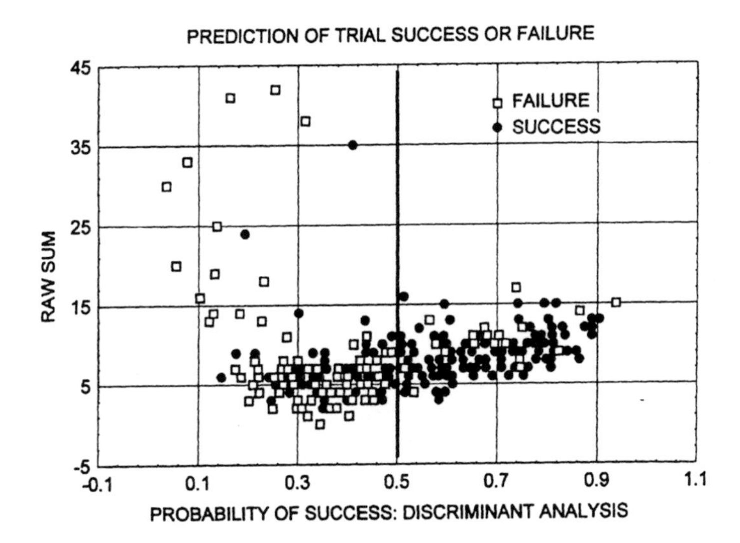

FIGURE 10.7
DA functions predict a rat's success in an operantly conditioned task based on cue-related activity of cortical neuronal ensembles. See text for details of conditioning paradigm and neuronal recordings. DA utilized the measured responses to the sensory cue of seven cortical neurons to calculate probabilities that the animal would move in response to the sensory cue. This experiment consisted of 521 trials, of which the first 200 were used to calculate the discriminant functions. Of the 321 remaining trials, the animal successfully obtained its water reward in 186, and failed in 135. For each of these trials, DA calculated a probability of success, based on the weighted responses of the seven cortical neurons to the sensory cue. Overall, successful trials were correctly predicted in 83.32% of cases, and unsuccessful trials in 61.5% of cases. Whereas the mean probability for successful trials was 0.578, for failed trials it was 0.418. This scatterplot compares the efficacy of DA functions vs. raw averages in defining a space for predicting success or failure in this task. Each of 321 single trials is shown as solid circles (success) or open squares (failure). The horizontal axis depicts probabilities, computed by DA, that the sensory cue would successfully trigger a correct behavioral response. The threshold for predicting success was $p = .5$, depicted by a vertical line. The vertical axis depicts the raw sum of spiking by the seven neuron ensemble during the same 20 ms post-cue periods. These provide a direct demonstration of the superiority of the DA method for prediction.

a dependent variable (e.g., a sensory input or motor output function). In its simplest form, MRA will yield a linear equation of the form:

$$Y = a + b_1X_1 + b_2X_2 + \ldots + b_pX_p$$

in which Y = the dependent variable (e.g., manipulandum angle), X_n = the independent variables (e.g., neuronal population activity), a = an offset, and b_n = coefficients for weighting the independent variables. This relation is easily calculated from a sample data set by linear least squares estimation, which computes a line through the observed data points such that the squared deviations of the points through that line are minimized. This calculation also yields the coefficient of determination ($\mathbf{R^2}$), which quantitates the predictive reliability of the model (in a range from 0 to 1). Beyond this, it is very important to test the model through application of a number of "residual analysis" procedures, which can identify outliers, and gross violations of the assumptions of MRA.

TABLE 10.8
Classification of Single Trials
(All neurons used for classification.)

Predicted classifications:

Actual classifications	% Correct	p = .484 CS+	p = .515 Non-CS+
CS+	67.46	170	82
Non-CS+	80.97	51	217
TOTAL	74.42	221	299
When only GZ-1,2,3 used:		p = .484	p = .515
CS+	61.50	155	97
Non-CS+	83.69	45	223
TOTAL	72.69	200	320

TABLE 10.9
Multiple Regression of 10 (of 92) Neuron Response Variables Against Rostrocaudal Whisker Position

Neuron	BETA	St. Err. of BETA	B	St. Err. of B	t(2807)	p-level
Intercept	4.926634	.119574	41.20161	0.000000		
N1	−.022911	.016162	−.131811	.092983	−1.41759	.156420
N2	−.007541	.016682	−.051242	.113359	−.45203	.651284
N3	−.028123	.016533	−.104276	.061303	−1.70100	.089053
N4	.026900	.016678	.111905	.069380	1.61292	.106874
N5	.012307	.015956	.161611	.209530	.77130	.440592
N6	.019994	.015933	.322876	.257290	1.25491	.209616
N7	.006936	.016051	.039457	.091308	.43213	.665684
N8	.048888	.016074	.310880	.102214	3.04146	.002376
N9	.020242	.017561	.108172	.093845	1.15267	.249145
N10	.001786	.017326	.015948	.154685	.10310	.917889

Regression Summary:
Multiple R = .5836, R2 = .3406, Adjusted R2 = .3190
$F(92,2807) = 15.766$, $p < 10^{-13}$, Std. Error of estimate: 1.8538

To illustrate use of MRA for analysis of neural population data, the following example utilized the measured short and long latency responses of the total 46 neurons recorded in the trigeminal system, comprising 92 different independent variables (the first 10 of which are summarized in Table 10.9). The dependent variable was a measure of rostrocaudal position of each of the eight different facial whiskers which were stimulated in this experiment. The database consists of measurements of 2900 separate stimulus trials, over which each of the eight whiskers was deflected a mean 362.5 times.

In Table 10.9, "BETA" represents standardized regression coefficients, while "B" represents nonstandardized regression coefficients. Stardard errors of both measure the dispersion of the observed values about the regression line. Regressions are shown for all the variables (neurons; N1 to N10) and the intercept. The significance level for each neuron is shown by Student's t (degrees of freedom in parentheses) and calculated p values. These test the null hypothesis that beta = 0. In the regression summary, "multiple R" is the coefficient of multiple correlation; "R2" measures the reduction in the total variation of the dependent variable due to the multiple independent variables – R2 = 1 – (residual SS/total SS), and "Adjusted R2" is adjusted by dividing the error sum of squares and total sums of square by their respective degrees of freedom. Finally, the multivariate "F" is also shown with its standard error and p-value.

The overall results indicated a very highly significant ($p < 10^{-13}$) multivariate encoding of rostrocaudal position, even though the overall correlation (multiple R) was only .58. Moreover, only 31% of the total population variance (adjusted R2) was explained by relative rostrocaudal position. This exercise demonstrates, therefore, how the large amounts of data that can be obtained from simultaneous neuronal recordings can be used to resolve relatively subtle experimental effects. For the same reason, of course, there is an increased possibility of attributing excessive significance to unimportant, and possibly spurious, outliers. For this reason, it is important to approach these analyses with a properly skeptical perspective.

10.7 Nonlinear Analysis of Neuronal Population Activity

10.7.1 Nonlinear Estimation in Regression Analysis

As with all of the parametric statistical techniques described above, traditional MRA depends on the assumption that predictor variables have a linear relation to the dependent variable. Often, however, relationships between variables have distinct nonlinearities. Fortunately, linear MRA, such as that described above, can be used as a starting point for a variety of nonlinear estimation procedures which can define virtually any significant relation between independent and dependent variables. The first step in identifying such nonlinearities is to examine scatterplots between the independent and dependent variables, and then utilize various nonlinear estimation techniques to best identify any nonlinearities involved. These techniques often use a weighted least squares method to estimate a nonlinear "loss function" (i.e., the error). This procedure generally includes a linear least squares component and also a nonlinear component. Depending on the types of nonlinearities revealed by these procedures, any of the following nonlinear regression models could be implemented:

1. *Linearizing mathematical transformation:* Often, the observed nonlinearities can be corrected by simple mathematical transformations. For example, if the relationship between variables fits an exponential or trigonometric function, the values of the independent variables can be transformed by the inverse functions, yielding a linear model.

2. *Polynomial regression* is used when relations between independent and dependent variables are curvilinear and can be optimally fitted to particular polynomial expressions.
3. *Piecewise linear regression:* Individual neurons may have different linear relationships with motor output over different segments of the output function's range. This can be handled by specifying different regression formulas for the different segments.
4. *Breakpoint regression:* Similarly, the expressions can include additional parameters to reflect sudden steps in the regression line.

10.7.2 Artificial Neural Networks (ANNs)

The rapidly expanding area of ANNs provides a much more elaborate means of applying nonlinear analyses to neuronal population recordings. While it is clear the ANNs will play an increasingly important role in neurobiology, the subject cannot be covered properly here, and therefore the reader is referred to Chapter 7 and other references.[41]

Briefly, there is a certain logic to using ANNs for neurophysiological analysis, since they are themselves based (somewhat loosely) on the structure of real neuronal networks. Indeed, the neuron model provides the unique computational approach used by ANNs. For example, whereas traditional regression analysis, even nonlinear regression, tends to depend on the analytical solution of linear least squares, neural networks approach the same problem through iterative approximation using a synaptic adaptation ("learning") algorithm. The former can be computationally intensive, because it requires matrix inversion, and thus it scales with the square of the number of variables, ANNs can be more computationally efficient in that they scale with the number of variables, not their square. On the other hand, ANNs cannot be guaranteed to find the optimal solution, even in the linear sense. Moreover, ANNs do not provide outputs that can be readily used for hypothesis testing and statistical inference.

On the other hand, distinct parallels exist between the functional problems that can be addressed by ANNs and multivariate statistics. In fact, a number of ANNs have been specifically designed to provide a nonlinear means of solving traditional statistical problems, such as principal components.[33-36]

References

1. Lashley, K.S., Mass action in cerebral function, *Science,* 73, 245, 1931.
2. Pribram, K.H., A review of theory in physiological psychology, *Ann. Rev. Psychol.,* 11, 1, 1960.
3. Rumelhardt, D., McClelland, J., and the PDP Research Group, *Parallel Distributed Processing,* Cambridge, MA, MIT Press, 1986.
4. Anderson, J.A. and Rosenfeld, E., Eds., *Neurocomputing: Foundations of Research,* Cambridge, MA, MIT Press, 1988.
5. Erickson, R.P., Stimulus coding in topographic and nontopographic afferent modalities: on the significance of the activity of individual sensory neurons, *Psychol. Rev.,* 75(6), 447, 1968.

6. Churchland, P.M., *A Neurocomputational Perspective,* MIT Press, Cambridge, MA, 1989.
7. Anderson, C.H. and Van Essen, D.C., Neurobiological computational systems, in *Computational Intelligence: Imitating Life,* Eds. J.M. Zurada, R.J. Marks II, and C.J. Robinson. New York: IEEE Press, 1994.
8. Abbot, L.F., Decoding neuronal firing and modelling neural networks, *Q. Rev. Biophysics,* 27, 291, 1994.
9. Bialek, W., Rieke, R., de Ruyter van Steveninick, R.R., and Warland, D., Reading a neural code, *Science,* 252, 1854, 1991.
10. Rieke, R., Warland, D., de Ruyter van Steveninick, R.R., and Warland, D., *Spikes: Exploring the Neural Code,* MIT Press, Cambridge, MA, 1997.
11. Zhang, K., Ginzburg, I., McNaughton, B.L., and Sejnowski, T.J., Interpreting neuronal population activity by reconstruction: unified framework with application to hippocampal place cells, *J. Neurophysiol.,* 79, 1017, 1998.
12. Georgopoulos, A.P., Kettner, R.E., Schwartz, A.B., Neuronal population coding of movement direction, *Science,* 233, 1416, 1986.
13. Georgopoulos, A.P., Kettner, R.E., Schwartz, A.B., Primate motor cortex and free arm movement to visual targets in three-dimensional space. II. Coding of the direction of movement by a neuronal population, *J. Neurosci.,* 8, 2928, 1988.
14. Georgopoulos, A.P., Lurito, J.T., Petrides M., Schwartz, A.B., Massey, J.T., Mental rotation of the neuronal poulation vector, *Science,* 213, 231, 1989.
15. Lee, C., Rohrer, W.H., Sparks, D.L., Population coding of saccadic eye movements by neurons in the superior colliculus, *Nature,* 332, 357, 1988.
16. Schwartz, A.B., Direct cortical representation of drawing, *Science,* 265, 540, 1994.
17. Di Lorenzo, P.M., Across unit patterns in the neural response to taste: vector space analysis, *J. Neurophysiol.,* 62, 823, 1994.
18. Lehky, S.R. and Sejnowski, T.J., Neural model of stereoacuity and depth interpolation based on a distributed representation of stereo disparity, *J. Neurosci.,* 10, 2281, 1990.
19. Gaal, G., Calculation of movement direction from firing activities of neurons in intrinsic co-ordinate systems defined by their preferred directions, *J. Theor. Biol.,* 162, 1, 103, 1993.
20. Gaal, G., Relationship of calculating the jacobian matrices of nonlinear systems and population coding algorithms in neurobiology, *Physica D,* 84, 3, 582, 1995.
21. Gaal, G., Population coding by simultaneous activities of neurons in Intrinsic coordinate systems defined by their receptive field weighting functions, *Neural Networks,* 6, 4, 499, 1993.
22. Sanger, T.D. (1994) Theoretical considerations for the analysis of population coding in motor cortex, *Neural Comput.,* 6, 12, 1994.
23. Chapin, J., Nicolelis, M., Yu, C.-H., and Sollot, S., Characterization of ensemble properties of simultaneously recorded neurons in somatosensory (SI) cortex, *Abstr. Soc. for Neurosci. Ann Mtg.,* 1989.
24. Johnson, R.A. and Wichern, D.W., *Applied Multivariate Statistical Analysis*, Prentice-Hall Inc., Englewood Cliffs, NJ, 1988.
25. Moore, G.P., Perkel, D.H., and Segundo, J.P., Statistical analysis and functional interpretation of neuronal spike data, *Ann. Rev. Physiol.,* 28, 493, 1966.
26. Chapin, J.K. and Lin, C-.S., Mapping the body representation in the SI cortex of anesthetized and awake rats, *J. Comp. Neurol.,* 229, 199, 1984.

27. Chapin, J.K. and Lin, C.-S., The somatosensory cortex of rat, in R. Tees and B. Kolb Eds. *The Neocortex of Rat,* Academic Press, New York, 1990.
28. Nicolelis, M.A.L., Lin, C.-S., Woodward, D.J., and Chapin, J.K., Distributed processing of somatic information by networks of thalamic cells induces time-dependent shifts of their receptive fields, *Proc. Natl. Acad. Sci.,* 90, 2212, 1993.
29. Nicolelis, M.A.L. and Chapin, J.K. The spatiotemporal structure of somatosensory responses of many-neuron ensembles in the rat ventral posterior medial nucleus of the thalamus, *J. Neurosci.,* 14(6), 3511, 1994.
30. Nicolelis, M.A.L., Baccalá, L.A., Lin, R.C.S., and Chapin, J.K., Synchronous neuronal ensemble activity at multiple levels of the rat somatosensory system anticipates onset and frequency of tactile exploratory movements, *Science,* 268, 1353, 1995.
31. Richmond, B.J., Optican, L.M., and Spitzer, H., Temporal encoding of two-dimensional patterns by single units in primate primary visual cortex, I. Stimulus-response relations, *J. Neurophys.,* 64, 351, 1990.
32. Richmond, B.J. and Optican, L.M., Temporal encoding of two-dimensional patterns by single units in primate primary visual cortex, II. Information transmission, *J. Neurophys.,* 64, 351, 1990.
33. Oja, E., A simplified neuron model as a principal component analyzer, *J. Math. & Biol.,* 15, 267, 1982.
34. Oja, E., Neural networks, principal components, and subspaces, *Intern. J. Neural Systems,* 1, 61, 1989.
35. Oja, E., Principal components, minor components, and linear neural networks, *Neural Networks,* 5, 927, 1992.
36. Sanger, T.D., Optimal unsupervised learning in a single-layer linear feedforward neural network, *Neural Networks,* 2, 459, 1989.
37. Chapin, J.K. and Nicolelis, M.A.L., Neural network mechanisms of oscillatory brain states: characterization using simultaneous multi-single neuron recordings, *Electroenceph. Clin. Neurophysiol. Suppl.,* 45, 113, 1996.
38. Nicolelis, M.A.L., Lin, C.-S., and Chapin, J.K., Neonatal whisker removal reduces the discrimination of tactile stimuli by thalamic ensembles in adult rats, *J. Neurophys.,* 78(3), 1691, 1997.
39. Deadwyler, S.A. and Hampson, R.E., The significance of neural ensemble codes during behavior cognition, *Ann. Rev. Neuroscience,* 20, 217, 1997.
40. Hampson, R.E. and Deadwyler, S.A., Ensemble codes involving hippocampal neurons are at risk during delayed performance tests, *PNAS,* 93(24), 13487, 1996.
41. Maren, A.J., Harston, C.T., and Pap, R.M., *Handbook of Neural Computing Applications,* Academic Press, 1990.

Chapter 11

Pitfalls and Problems in the Analysis of Neuronal Ensemble Recordings During Behavioral Tasks

Robert E. Hampson and Sam A. Deadwyler

Contents

11.1 Introduction

A major advantage of recording from neuron ensembles in the hippocampus, or any brain area, is the potential for deciphering some of its well-described behavioral and cognitive correlates. The ultimate goal of ensemble recording and analysis is to determine how activity within the ensemble relates to behavioral and/or cognitive events, i.e., "What does the ensemble *encode*?" In the hippocampus, for example, the anatomy has been studied extensively such that connections between the principal

0-8493-3351-2/99/$0.00+$.50

cell groups are well characterized and the local "functional" circuitry is currently under intense investigation.[1] Neurons have been recorded in all major subfields in the hippocampus, and cell identification via firing signature is not a problem in most cases. In the same manner, anatomical connections between subfields are also known; therefore, it is possible to position recording electrodes along specific anatomic projections in an attempt to record ensembles of neurons with suspected anatomic connectivity. Given these factors, it should be possible using multineuron recording techniques to determine how neural activity within hippocampal circuits is integrated with behavioral and cognitive events. However, as in many systems, the techniques used to analyze the multineuron data from such "ensembles" are critical, and not all types of ensemble analyses are appropriate. Misapplication of an ensemble technique can at best misrepresent the data, or at worst lead the investigator seriously astray. In the following sections, we will examine several common uses of ensemble analyses, and determine "pitfalls" or problems often associated with such applications.

11.2 Simultaneous vs. Serial Ensemble Recordings

There are several means of "constructing" ensembles of neurons. Early attempts employed ensembles from serially (i.e., not simultaneously) recorded single neuron data. Employing sophisticated data integration and sorting techniques, researchers have demonstrated examples of ensemble encoding within behavioral contexts from such serially recorded data.[2-6] A major issue therefore, is what facets of ensemble activity are overlooked by constructing ensembles from serially recorded data vs. recording simultaneously?

In a recent study, ensembles of hippocampal neurons were recorded from rats performing a spatial delayed-nonmatch-to-sample (DNMS) task.[7-10] Between 10 and 16 single neurons were identified within each ensemble (by electrode position, waveform characteristics, and correlation to behavioral events), and recorded over a number of behavioral sessions. Data from 13 different animals (ensembles) collected over 100 DNMS trials per day, for a total of 500 to 1000 DNMS trials per ensemble, were characterized and cataloged.[10] Multivariate statistical analyses of the firing characteristics of each ensemble revealed that specific and repeatable firing patterns could be derived from the ensemble activity that corresponded to task phase, lever position, and correct or error performance on individual DNMS trials.[8,10,11] It was therefore possible to determine the "information content" of the ensemble on the basis of how well single trial ensemble activity could be used to predict which combination of factors was present on that trial.[6,12,13] The results showed that the statistically derived patterns of activity could be used to accurately predict the occurrence of these factors on about 87% of all trials for each of the seven ensembles. Based on the success rate at predicting the three categories of behavioral events (phase, position, and performance) it was determined that the simultaneously recorded ensemble information content (I_{ens}) was 2.73 bits (out of a possible 3.0 bits).[10]

Was this successful determination of an ensemble code for behavior due in large part to the simultaneous nature of the ensemble recordings? If the neuron ensemble was *reconstructed* from single neuron recordings at different times or from different animals, would the information content of the ensemble have been the same? Reports by many researchers have assumed that the information content of serially reconstructed ensembles is the same as for simultaneously recorded ensembles.[3,14-16] In fact, Gochin and co-workers state in their study that simultaneous vs. reconstructed ensembles were not different.[6] However, such an assertion is based on two critical assumptions: (1) that information content can be estimated accurately from serial data, and (2) that the original analysis is sensitive enough to isolate all of the ensemble's firing features. In the hippocampal ensemble study described above,[10] serially reconstructed ensembles were shown to encode less than half of the information of simultaneously recorded ensembles (Figure 11.1). The reason for this failure was the encoding of the spatial position of the DNMS response which was disrupted by the reconstruction process.[10] Figure 11.1 demonstrates how the type of ensemble construction interacts with information content, and hence the number of neurons required to accurately encode that information. In this circumstance, an ensemble of 10 simultaneously recorded neurons could encode behavioral events as accurately as ensembles reconstructed from 100 serially recorded neurons.

Given this result, it is important to determine *how* an ensemble of only 10 simultaneously recorded neurons could perform so much better than ensembles reconstructed from the same neurons, but in different combinations. Analysis of individual neuron firing characteristics of 168 hippocampal CA1 and CA3 complex spike cells [7,8,17] revealed that each neuron in the ensembles could encode DNMS task information in one of three ways: (1) as single event encoding in which the cell increased firing only during isolated events on each trial; (2) multievent "conjunctive" firing, in which the cell increased firing according to two coexisting dimensions of the event, such as left and sample, or right and nonmatch within the trial; and (3) multievent "disjunctive," in which there is a reciprocity of cell firing for different events across trials (e.g., left sample on one trial, left nonmatch on another). Of the 168 hippocampal neurons analyzed, roughly half (51%) showed multievent conjunctive firing, one third (33%) showed single-event firing characteristics, and the remaining proportion (16%) was of the multievent disjunctive type. It is the presence of the multievent cells which allows the hippocampal ensembles to encode all three dimensions (bits) of DNMS information with small numbers of neurons. In the DNMS task, the three bits that were encoded actually represent three *orthogonal* dimensions of information: task phase, lever position, and trial performance. In other words, each bit of information encodes a feature of the task that is independent from the other two bits, i.e., phase vs. position, phase vs. performance, position vs. performance. In a recent study,[18] it was shown that ensembles constructed exclusively of multievent disjunctive neurons successfully encoded the same three bits of DNMS information with only three neurons, *but only if simultaneously recorded.* This derives from the fact that multievent firing characteristics are dependent on covariance between neurons within an ensemble to encode multiple dimensions. When the covariance across all neurons is incomplete (from not recording all neurons simultaneously),

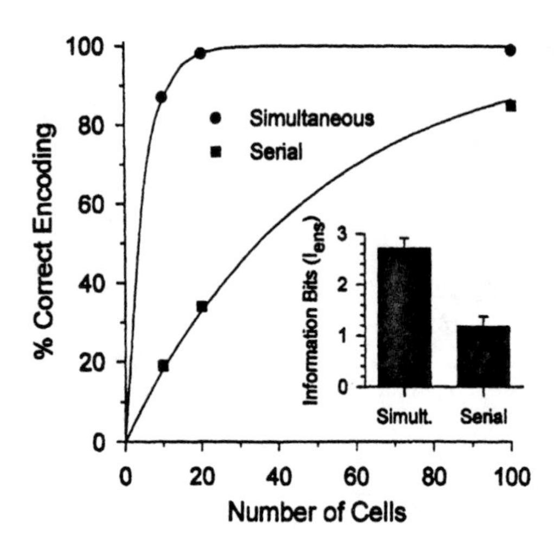

FIGURE 11.1
Number of neurons required to encode the same delayed nonmatch to sample task information (phase, position, and performance) on a given trial for parallel recorded ("simultaneous") and reconstructed ("serial") ensembles. Behavioral event classification accuracy for each type of ensemble was calculated based on ensembles of 10, 20, and 100 neurons (solid symbols) and fitted to a power function relating ensemble size to accuracy. It can be seen that the encoding obtained from the simultaneous ensembles would allow > 90% correct encoding of the three bits of information with fewer than 20 neurons. By contrast, the encoding by the serial ensembles would require > 100 neurons to achieve the same accuracy. Inset: Ensemble information content I_{ens}, calculated for simultaneous vs. serial ensembles. Mean (± SEM) I_{ens} (in bits) was calculated for simultaneous ensembles (10 neurons recorded simultaneously) according to success in predicting each behavioral event classification (i.e., sample vs. nonmatch, correct vs. error, left vs. right) on the basis of derived measures of ensemble firing. I_{ens} calculated for serial ensembles (constructed by recombining neurons from the total dataset across ensembles, preserving relative trial sequence) was less than half that calculated for simultaneous ensembles due to disruption of information encoded as covariance between neurons across trials.

one or more of the encoded dimensions is lost as "noise" due to lack of coherence, and the neurons can only encode single events. Thus, reconstructed ensembles are limited because even if the neuron was multievent, the likelihood that it would be synchronized in "encoding" with other neurons to represent all task features is decreased when neurons are serially recorded in a random manner. Thus, the most relevant dimension or "bit" in the task encoded by the ensemble is usually all that is apparent. In tasks that contain more than two discrete stimuli, such as multiple video images, the analysis will identify more than one "bit" of information; however, since the information is all of the same type (i.e., all visual stimuli with the same behavioral context), there is still only one relevant dimension required to be discretely encoded by the ensemble on any trial.[6] Hence, "reconstructed" ensembles in this case would exhibit the same information content as simultaneously recorded ensembles because all the task required was single-event neurons. Whenever more than one dimension must be encoded, serial reconstruction of ensembles with single neuron data is likely to underestimate the information content encoded by neural ensembles.

11.3 Ensemble Population and "Intention" Vectors

Another popular type of ensemble analysis is a direct examination of ensemble firing rate patterns corresponding to specific motor movements or behavioral events. This analysis has been termed the "population vector" and has been extensively applied by Georgopoulos, Schwartz, and others.[2,14,15,19,20] The analysis is potentially quite powerful, especially in terms of encoding directional and velocity parameters. However, there are two necessary precautions that must be taken when applying population vector analyses. The first precaution is to use separate datasets for computing vs. testing the vectors. The second is to select an appropriate sampling time for accumulating ensemble activity.

Population vectors consist of a set of measurements of the mean firing rates for each neuron in the ensemble corresponding to discrete times when unique events, or stimuli occur.[21] The utility of a population vector analysis depends on identifying discrete conditions or "states" (such as tracking visual images, or degrees of rotation of a monkey forearm) which correspond to a unique set of mean firing rates across the ensemble for each state. In the application by Georgopoulos et al.,[22,23] the mean firing of selected neurons recorded in motor cortex was computed corresponding to forearm movements performed in monkeys and an ensemble was constructed by combining neurons which were selectively "tuned" to different movements. A set of population vectors was obtained which corresponded to specific angles of arm rotation, then successfully compared to new recordings to test the accuracy of the population vector which "predicted" the forearm angle prior to movement.[14] These investigators were among the first to show that ensemble population vectors are more effective at predicting *intended* arm movements (hence the label "intention vectors"[24]) than single neuron recordings. Other reports have applied this technique to ensembles of neurons encoding spatial,[25,26] visual,[27] and auditory [28,29] as well. Recent refinements of the technique allow for calculation of population vectors from small numbers of neurons that need not be constructed only from "selected neurons."[30,31]

Since population vectors are constructed by first selecting neurons, and then averaging ensemble activity associated with discrete stimuli, events or locations, it is expected that the prediction error between the vectors and the dataset from which they were constructed would be quite low. However, this is not a valid test of the predictive properties. In most studies the computed vectors are tested against a novel dataset,[15,32] or by demonstrating that the population vectors *predicted* limb movement before that movement was performed.[19] In a slightly different application, [26] population vectors were calculated from one half of a spatial exploration session to predict locations in the second half of the same session. However, those same vectors were not as accurate in a new or expanded environment. This principle is illustrated in Figure 11.2, which compares predicted (solid line) vs. actual (dashed line) trajectories of a rat exploring a 60 cm-by-60 cm square enclosure in a similar type study. In Figure 11.2 (left), population vectors were computed and tested against data from the same session, and the mean of the differences between each predicted and actual location (i.e., the prediction error) was 3.88 ± 0.42 cm. When the population vectors were tested against data from the next behavioral session (24 hours later), the mean

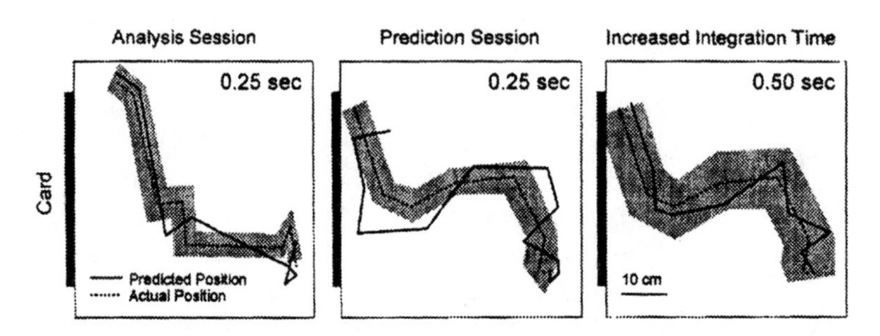

FIGURE 11.2
Prediction of location by ensemble population vectors. Actual vs. predicted positions are illustrated for a 5 s traversal across a 60 cm × 60 cm enclosure. Hippocampal CA1 and CA3 "place cells" (n = 10) were recorded from a male Long-Evans rat during performance of a food pellet-chasing exploration task.[52] Neural ensemble activity was collected for 30 minutes while the rat explored the chamber, and consumed .5 gm flavored food pellets randomly introduced into the chamber. The rat's position was determined by video tracking of a red LED mounted on the recording cable, and registered with 1.5 cm × 1.5 cm "pixel" resolution. Predictive population ("intention") vectors were constructed by averaging all ensemble activity that occurred while the animal occupied each "pixel." Left: The animals' positions during a 5 sec trajectory across the chamber is shown by the dashed line. The positions predicted by comparison of the population vectors with sliding 0.25 s averages of ensemble activity during the same traversal is shown by the solid line. When tested against the data from which the population vectors were computed, mean difference between actual and predicted location (prediction error) was 3.88 ± 0.42 cm. For comparison with the population vectors, ensemble spike train data was averaged (integrated) in 0.25 s intervals. The shaded region defines the area of uncertainty in the animal's location given the fact that the average traversal speed of 12 cm/s would allow the animal to travel up to 3 cm in the same 0.25 s interval. Center: Population vectors computed from the session shown at left were compared with ensemble spike train data recorded 24 hours later. Mean prediction error using a 0.25 s averaging time was 6.15 ± 0.74 cm. The shaded area shows a similar region of uncertainty in animal position as shown at left. Right: The same data as in the Center pattern was compared using an averaging interval of *0.50* s. Mean prediction error dropped to 3.16 ± 0.37 cm with longer averaging of the ensemble spike trains. However, at the longer averaging time, uncertainty in actual animal position was increased to 6 cm (enlarged shaded region). The fact that the predicted traversal is entirely within the gray band indicates that greater location uncertainty negated the improvement in prediction error produced by increased averaging time.

prediction error increased to 6.15 ± 0.37 cm (Figure 11.2, center panel). The accuracy of the prediction could be improved if the amount of time over which the ensemble activity was averaged was increased (i.e., to obtain a more representative vector).[26] However, since the animals moved at a variable rate of speed throughout the session, the averaging time "adjustment" resulted in an increase in variability of the predicted location of the rat. Figure 11.2 illustrates the increased variability in position prediction as an increase in the width of the shaded area around the trajectories. For the shorter averaging times (left and center), this variability was approximately 3 cm; thus the limiting factor in accuracy of the prediction was the mean error between predicted and actual position. At longer averaging times, even though the mean prediction error was reduced to 3.16 ± 0.37 cm, the variability was increased to 6 cm (Figure 11.2, right). Since the mean error at shorter averaging times (6.15 cm) and the variability at longer averaging times (6 cm) were not significantly different, increasing the averaging time in order to reduce errors in predicting spatial locations did not necessarily improve the accuracy of the prediction. Thus, since neural ensemble and animal behavior are never completely static, there is a trade-off between the accuracy of the prediction that can be obtained using population vectors, and the inherent variability introduced by the averaging neural ensemble data over time to obtain those vectors.

Population vectors provide a potentially powerful tool for analysis of neural ensembles. The key constraints in applying the analysis are: appropriate testing of the vectors against data that was *not* used for computation of the vectors, and an ensemble sampling or integration time that does not introduce large errors in prediction. The latest techniques in linear estimation [30-32] allow calculation of the vectors with (relatively) small ensembles (i.e., <40 neurons). Within the stipulated constraints, population vectors provide a powerful analysis technique of ensemble data based on the mean firing rate of all neurons within well-defined experimental contexts.[15,33]

11.4 Cross-Correlation: Connectivity vs. Coactivity

At its most basic level, ensemble codes can be analyzed simply by looking for repeating patterns of firing across populations of recorded neurons.[34-37] These patterns may consist of spatial combinations of two or more neurons within the ensemble, a temporal pattern within single neurons, a particular sequence of firing between several neurons, or combinations of spatial and temporal oscillations. Ensemble spike trains can therefore be scrutinized using cross-correlation techniques with the construction of cross-correlation histograms (CCHs) for this purpose. Detection of specific spatiotemporal firing patterns across many neurons may suggest ways in which those neurons may be connected, or may simply result from common "driving" influences on several different cells. Calculation of pairwise cross-correlation histograms (CCHs) between neurons provides an indication of the firing relationships of neurons in the ensemble. Such "spike-triggered averages" (CCHs) provide rapid indications of the possible "coupling" (or temporal relationships) between pairs of neurons[35,36,38] Although it is appealing to use CCHs to map temporal connectivity and patterned firing between neurons within an ensemble,[39] such procedures are not well suited to circumstances where nonstationary processes (i.e., behavior) change the firing rate during data acquisition.[40] In addition, as discussed elsewhere,[41] it can be difficult to distinguish true connectivity from mere coactivity between neurons using CCHs.[42,43]

Figure 11.3 models how both monosynaptic connectivity and coactivity of neurons within an ensemble can produce apparent increases in correlated firing. The random firing of any given neuron within an ensemble is typically a Poisson process.[44] When comparing the firing of two neurons, that function becomes two overlapping Gaussian distributions of spike occurrences (see Figure 11.3D). The firing of two neurons that are tightly monosynaptically coupled reveals the same Gaussian functions offset by the synaptic delay (Figure 11.3A). The firing of two neurons that are "coactively" driven via monosynaptic connections to a third neuron also show similar Gaussian functions that are more closely overlapped in time; however, the peak of the distribution reflects the highest density of overlap, and spike occurrences are symmetric around that point (Figure 11.3B). When the common driving neuron is not directly connected to both neurons, the overlapping Gaussian functions can vary considerably (Figure 11.3C). The CCHs and Gaussian functions in Figure 11.3 illustrate the calculation of cross-correlation under these three conditions. The CCHs corresponding to monosynaptically connected neurons (Figure 11.3A) demonstrate that true connectivity between neurons occurs within short temporal intervals (e.g., ±10 ms) offset from the 0 timepoint. For neurons that are monosynaptically *coactive* (Figure 11.3B), on the other hand, the firing peak in the CCH occurs near 0, but at a slightly longer timebase (±20 ms), indicating less rigorous temporal coupling. For both instances, however, the overall mean firing rates do not change, regardless of the actual correlation, because the spike counts for both neurons are constrained by tight coupling within very short temporal intervals (±10 or ±20 ms). Figure 11.3C illustrates CCHs obtained from two neurons that are coactive and driven by a common source, but not directly coupled to that source. In this case, the peak in the CCH results from the overlapping Gaussian probability over a much longer timebase (±100 ms) of each neuron firing independent of a common influence. Figure 11.3D compares the temporal span of the relationships depicted in A through C. The type of polysynaptic coactivity typified by Figure 11.3C occurs over a much broader time scale than the monosynaptic interactions shown in Figure 11.3A and 11.3B. The peak in this CCH results from

FIGURE 11.3 (opposite)

Illustration of the effects of connectivity vs. coactivity on cross-correlation histograms (CCHs). Cross-correlation histograms are shown for neuronal firing modeled to correspond to one of three conditions: Fixed monosynaptic connectivity (A), in which the recorded cell fires within a fixed time interval after the triggering cell; monosynaptic coactivity (B), in which the cells used to construct the CCHs (filled circles) are both driven by a common source; and polysynaptic coactivity (C), in which the recorded cells are driven by a common, yet remote source. Each CCH was constructed as a spike-triggered histogram (with a ±100 ms analysis epoch) from a pair of simulated spike trains (10,000 spikes each, random 2 to 10 Hz firing). The relationship of the CCHs to the overlapping Gaussian firing distributions of the neurons is shown below each histogram. (A) CCHs were constructed using Pearson-product moment cross-correlations (*rho*) of 0.9 to represent tight coupling of spike trains, in which 80% of the spikes from the target neurons fired within +3.5 to +6.5 ms (i.e., one synaptic delay) of triggering spikes (dashed vertical lines) from the projection neuron. Mean firing rate across the entire CCH was normalized to the firing rates of both spike trains, and was constant at 0.005 spikes/s/trigger for any *rho* value. A single Gaussian distribution of neural firing is depicted, offset by 5.0 ms, illustrating that monosynaptically coupled target neurons will fire with the same distribution of the trigger neuron, offset by the synaptic delay. Scale represents the mean firing rate used for the simulation. (B) CCHs depict the cross-correlation between the two *coactive* (but unconnected) neurons, the peak firing rate in the CCH was shifted to the time of occurrence of the trigger spike (time 0), because both neurons in the simulation fired their initial spikes in the train at the same latency from the driver neuron spike. The overlapping Gaussian distributions are identical, indicating that each neuron is tightly coupled to the common driving neuron, but they are only slightly offset from each other. This produces the widening in the CCH, yet does not alter mean or peak firing rate. (C) For the CCHs depicting polysynaptic interactions (right), the peak *and* mean rate across the CCH increase as a direct function of the degree of overlap of the two Gaussian firing distributions. Due to the remote nature of the common driving neuron, the distributions differ in size, shape, and degree of overlap, directly influencing the mean and peak firing rates of the CCH. Note that only when firing distributions overlap in this manner, does mean or peak rate change in the CCH; therefore, cross-correlation coefficients derived from the CCH can only reflect changes in *coactivity* resulting from overlapping firing distributions. (D) Comparison of the three Gaussian distributions on the same timebase (±100 ms) illustrates the contrast in the temporal resolution of monosynaptic connectivity and coactivity vs. polysynaptic coactivity.

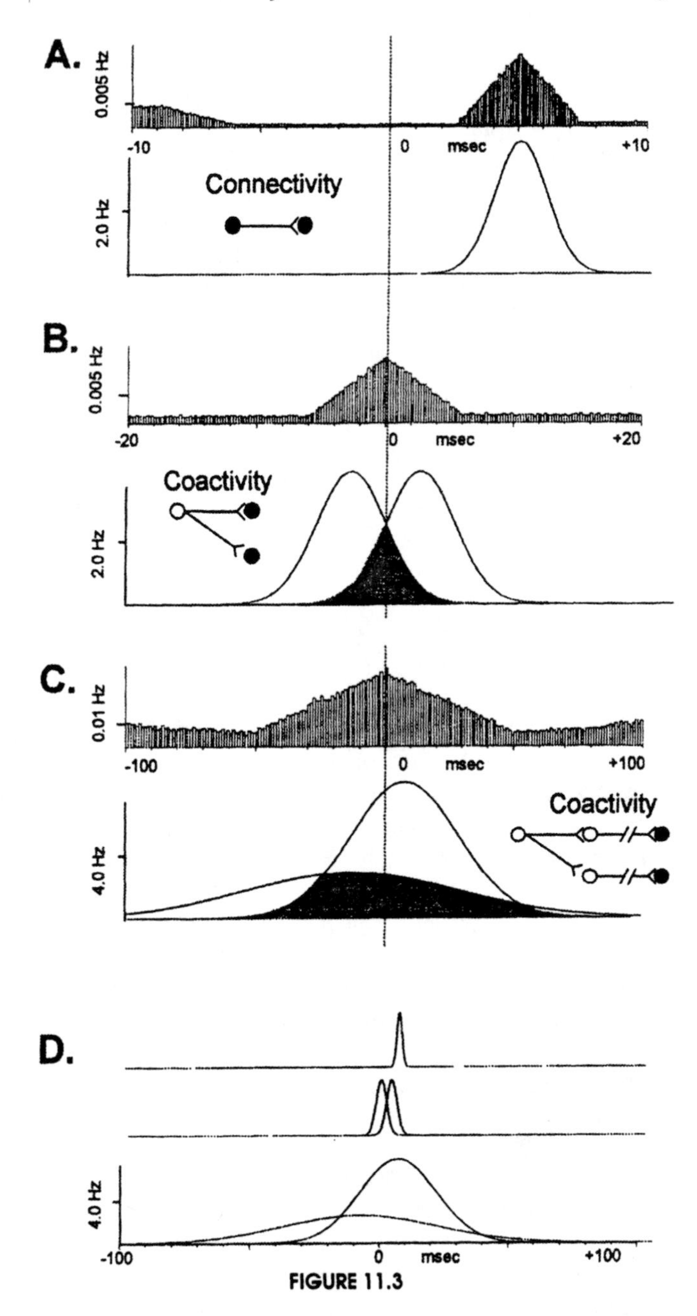

FIGURE 11.3

coincidental, i.e., contiguous, firing of the two neurons (reflecting coactivity), but can be totally unrelated to any physical *connectivity* between the neurons. Both background and overall mean firing rate in the CCH vary as a function of the degree of overlap in the Gaussian firing distributions. Given these demonstrations, if only the peak firing rates were measured, it would be difficult to determine from the CCHs whether the increased rate resulted from tightly coupled "synaptic" connectivity, or from a convergence of firing probabilities in cells that respond in unison within the same temporal interval to the same triggering stimulus.

Figure 11.3 points out another problem in cross-correlation analysis, namely the relationship between the CCH and the value assigned to indicate the degree of the cross-correlation. As originally defined by Perkel and Gerstein,[34] the CCH was used *without* calculating an overall correlation coefficient.[36] It can be seen in Figure 11.3 that neither peak nor mean firing rates reflect the actual Pearson-product cross-correlations (*rho*) used to generate the CCHs. Unlike simply measuring peak or mean firing, one measure from the CCH which *does* correspond to the applied *rhos* is the *difference* between the peak and the baseline firing. This measure is commonly used in reporting cross-correlation "coefficients," however, it does not distinguish differences in coactive CCHs because peaks tend to be "flattened" and baseline firing varied because of a disproportionate number of spikes. Thus the derivation of a cross-correlation "coefficient" from a CCH is imperfect, and can either under- or overestimate temporal relationships between ensemble neurons.

11.5 Use of Cross-Correlation in Ensemble Analysis

Given the above caveats, it should nonetheless be emphasized that cross-correlation analysis can be important for determining functional connectivity between neurons within ensembles. While CCHs can provide valuable insight in this context, application of this analysis to behavioral contexts with long time-courses (i.e., seconds instead of milliseconds) should be carefully considered. The following reports utilizing cross-correlations in the analysis of neural ensembles illustrate not only the problems inherent in the application, but also how the *interpretation* of cross-correlations is changing with use.

When cross-correlation coefficients are calculated for pairwise combinations of neurons, some investigators have suggested that increased correlation could be the result of strengthened synaptic connection formed during an intervening manipulation (i.e., behavioral or physiological).[45] There are several problems with this assumption: (1) calculating cross-correlation *coefficients* from the CCHs, requires an explanation of what CCH parameter is represented (see above); (2) such derived CCH measures do not necessarily indicate increased synaptic connectivity between cells, as much as they reflect coactivity possibly from a single (perhaps irrelevant) generator; and (3) in this context, changes in CCH-derived cross-correlations are much more likely to reflect a change in coactivity rather than a change in connectivity.

Even though the above factors require careful attention, the technique of computing pairwise cross-correlations on hippocampal place cell firing has been extensively applied in many different instances[39,46-50] without such caution. In early instances,[45] it was indeed concluded that the changed CCH parameters represented altered connectivity between neurons. Later reports modified the interpretation of CCH parameters to a presumption of coactivity between neurons.[39,46] O'Keefe and Recce[44] provided support for this when they demonstrated that the within-field firing of hippocampal place fields was controlled largely by phase-locking to the underlying theta rhythm.[47,48,51,52] Hence, unless two place cells have completely congruent place fields, the theta phase would be slightly different for each neuron, and the two spike trains of suspected correlated neurons could not therefore be phase-locked, but would exhibit a strong tendency to fire within the same (relatively large) 200 ms temporal interval (see References 38 and 43) defined by the CCH. In one of the few attempts to derive anatomic relationships from known electrode placements using cross-correlational techniques, Hampson et al.[53] showed that *any* "suspected" connectivity assessed using the CCH measures was likely derived via common underlying coactive generators (i.e., anatomic connections) which were not altered by manipulations of behavioral circumstances.

Unfortunately, the evolution of the use of CCHs in ensemble analyses illustrates that it is all too easy to interpret cross-correlations in whatever manner the author chooses. One example is the suggested derivation of "temporal bias" in conjoint cell firing from CCHs.[39,52] The derivation of the temporal bias measure was prompted by the evidence that theta phase differences in place cell firing,[44,51,52] reflected sequential activation as the respective place fields were encountered during the rat's traversal on a linear track. As described, a temporal bias could result from as little as one spike difference in the two temporal halves of the CCH.[39,42,43] However, in order to conclusively demonstrate that temporal bias indicated a substrate for directional synaptic connectivity underlying behavioral traversal, the same pairwise CCHs *must show* a reversal of temporal bias when animals run *in the opposite direction* through the same two place fields (see Figure 11.4A and 11.4B).

The above assumption,[39,52] that temporal bias as reflected in the CCH reflects the order in which place fields are encountered, was tested in animals that traversed the same two place fields in opposing directions. In Figure 11.4C, all place fields were traversed from left-to-right, and the temporal bias between pairs of cells with overlapping place fields was calculated (upper histogram inset). Temporal biases for left-to-right traversals were plotted against the overall bias calculated irrespective of the traversal direction (lower histogram inset). The same measures are shown in Figure 11.4D for right-to-left traversals, with the CCH for right-to-left traversals shown as the upper histogram inset. The same overall temporal bias measure was used in both Figures 11.4C and 11.4D. If, in fact, the temporal bias reflected the order of traversal through the fields, one might expect biases for the same cell pairs to be opposite for the two traversal directions (see Figures 11.4A, 11.4B). The inset CCHs in Figures 11.4C and 11.4D all showed similar temporal biases, suggesting that at least the current direction of traversal was not responsible for the temporal bias in pairwise cell firing. The only factor that could produce the *same* temporal bias irrespective of the direction of traversal was if both cells in the pair were driven by a common source. In other words, overlap between place fields, direction of traversal, and temporal bias were irrelevant, and the "correlation" between place cells is, as described above, reduced strictly to coactivity between cells that are driven in common with other cells within the hippocampal circuit.

The problems associated with use of CCHs as a technique for assessing the dynamics of neural ensembles are significant, especially with respect to what is encoded. In the first place, it is difficult to assess, and there is no consensus as to whether CCHs represent connectivity, coactivity, or other arbitrarily derived relationships between neurons, because as shown above, significantly changed coefficients could

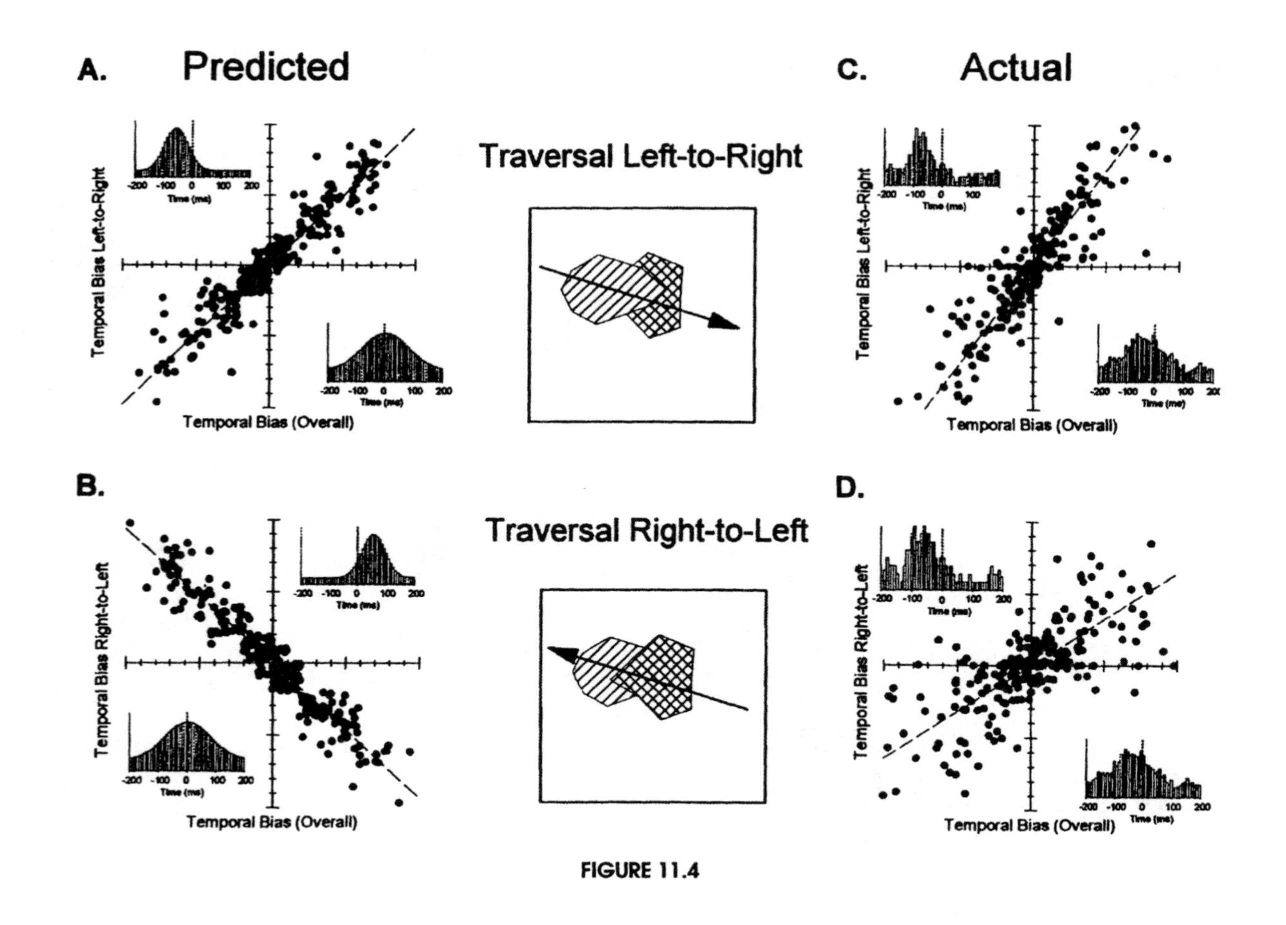
A.
Predicted
C.
Actual
Traversal Left-to-Right
Traversal Right-to-Left
B.
D.
Temporal Bias Left-to-Right
Temporal Bias Right-to-Left
Temporal Bias (Overall)
Time (ms)

FIGURE 11.4

represent several different things. As indicated, changing interpretations of the CCH contribute to this confusion, and often totally different "measures" of cross-correlation appear depending on the point that is advanced in the report.[9,36,42,45] Still, other "derived" measures (temporal bias),[39] are expounded with only enough empirical data to support a single conclusion, rather than a rigorous test of the "true" meaning of the derived measure across different behavioral and experimental contexts. Finally, despite the potential for reflecting functional connectivity, cross-correlation remains a *pairwise* analysis, and as such is not constrained by population statistical models. In fact, the ensemble is essentially ignored in order to serially analyze pairs of neurons. Thus, not only will averaging of cross-correlational measures provide spurious information about connectivity between neurons, it will necessarily mask, distort, or exaggerate interactions with other neurons in the ensemble.

11.6 Multivariate Analyses of Neural Ensembles

Analyses of this type which have previously been applied to neural ensembles include linear discriminant analysis [6,54,55] and principal components analysis (see also Chapters 7 and 10).[3,5,56] The differences between these multivariate analyses and the analyses described above are many. For instance, unlike cross-correlations, multivariate analyses simultaneously take account of the activity of all neurons in

FIGURE 11.4 (opposite)

Temporal bias vs. direction of traversal. (A) Temporal bias was calculated as the difference between the sum of all activity in the CCH from –200 ms to 0 ms and 0 ms to +200 ms,[38] for 300 pairs of simulated place cells with overlapping place fields. Cross-correlations were simulated such that where place fields overlapped, pairs of cells would fire in temporal order based on order of entry into the place fields. The CCH for a representative modeled cell during left-to-right traversals is inset at upper left; the CCH incorporating traversals *in all possible directions* is inset at lower right. The correspondence of temporal bias for left-to-right traversals compared to the temporal bias irrespective of traversal direction is shown by the scatter diagram. (B) Temporal bias for the same simulated cell pairs calculated for right-to-left traversals. Note that the lower inset CCH remained the same as in A, however, the CCH depicting right-to-left traversals was reversed. The reversal in slope of the scatter diagram for right-to-left traversals indicates the pattern predicted if temporal bias reflects the order of activation of place cells, (and hence, order of entry into place fields). (C) Temporal bias was computed from pairwise CCHs for 96 hippocampal place cells from 10 rats (>600 CCHs) performing a spatial delayed nonmatch-to-sample task.[53] Animals traversed the same area of the apparatus in two opposite directions during different phases of the task (center illustrations). Representative CCH for left-to-right traversals is inset at upper left. The overall "in-field" CCH calculated from all traversals throughout the session regardless of direction is inset at lower right. Temporal bias for left-to-right traversals vs. overall is plotted in the scatter diagram. The data were best fit by a linear function (dashed line) with slope = 1.36 and intercept = 0 ($r^2 = 0.75$; $F_{(1,644)} = 2300.1$, $p<0.0001$). The distribution of points in the upper right and lower left quadrants was significantly higher ($X^2_{(4)} = 23.76$, $p<0.001$) than in the other two quadrants. (D) Temporal bias for right-to-left traversals for the same cell pairs shown in C. Best fit linear function (dashed line): slope = 0.64 and intercept = 0 ($r^2 = 0.56$; $F_{(1,644)} = 514.0$, $p<0.0001$). The distribution of points in the upper right and lower left quadrants, although showing more scatter than in A, showed the same significant left-to-right temporal bias ($X^2_{(4)} = 15.32$, $p<0.001$). Note that although the CCH for all traversals (lower right inset) remained the same as expected, there was also no change in the location in the peak of the CCH for right-to-left traversals. The slopes for the linear regressions in C and D were both significantly greater than 0 ($p<0.01$), but were not significantly different from each other ($p = 0.22$).

the ensemble. In addition, although population vectors do incorporate the firing of all neurons within the ensemble, the vector is an "instantaneous" average of ensemble activity at a fixed point in time, which must be recalculated over many time points. Multivariate analyses incorporate both spatial (i.e., relationship between neurons within the ensemble) and temporal (time before and after events) information in the analyses. Finally, instead of producing one measure of variability for a given neuron (i.e., firing rate), the components derived can "discriminate" firing rate changes produced by independent factors acting on the ensemble.[41]

Multivariate analyses are an attractive alternative to the use of cross-correlations to assess ensemble activity. Recent studies in this and other laboratories have applied canonical discriminant analysis techniques to the assessment of neural ensembles.[8-10,17,57] This type of analysis subdivides the sources of variance within the ensemble firing into components which can be used to discriminate the behavioral events encoded within the ensemble activity.[8,58,59] The canonical discriminant analysis (CDA) used extensively in our laboratory is a form of linear discriminant analysis which has allowed the identification of variance sources corresponding to three orthogonal dimensions of a behavioral (DNMS) task (task phase, response position, and trial performance; see References 8 and 10). The analysis allowed the identification of two separate sources of behavioral errors encoded within hippocampal neural ensembles.[8,9] Recent studies have utilized these measures to "predict" performance[18] by correlating the strength of ensemble encoding with susceptibility to behavioral errors and proactive interference.[8,9,17,18,57] In those investigations, the CDA partitioned the variance in ensemble firing rate according to identified fixed temporal intervals corresponding to key behavioral and/or cognitive events within the trial, but was not capable of analyzing activity associated with the time between those events.

A recent refinement of the CDA has been the implementation of a "sliding analysis window," which allows examination of ensemble encoding at experimenter-determined temporal segments throughout all phases of the task.[18,57] In this version, the CDA is iteratively recalculated throughout all segments of the trial by sliding the entire "window" of calculations across successive temporal segments, using the same coefficients of the discriminant functions derived from key events within the trial. As shown in Figure 11.5A, the CDA was applied successively to the intertrial interval (ITI), sample response (SR), last nosepoke (LNP) in the delay interval, and nonmatch response (NR) events. Five sets of canonical root functions were derived, with the first three corresponding to task phase and type of response, and the fourth and fifth corresponding to levers pressed in the sample and nonmatch phases. By recalculating discriminant functions derived from one aspect of the task at all other times within the trial, it was shown that patterns of ensemble firing with data collected predictably varied in accordance with the behavioral contingencies of the task. The plot shown in Figure 11.5B shows discriminant scores for the CAN4 discriminant function, which tracked ensemble activity associated with response position, throughout the entire trial. The function shows reciprocal firing changes during the trial which decreased across the delay interval and increased in opposite sign to the sample phase, indicating the anticipation and registration of the nonmatch response. The fluctuation in this component was the mirror image on the opposite

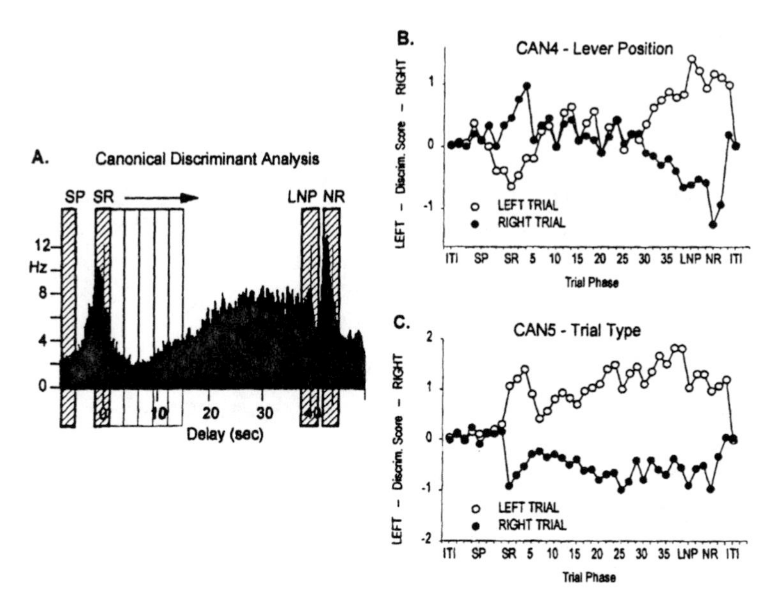

FIGURE 11.5
Calculation of discriminant scores across the timecourse of a DNMS trial. Upper Left: (A) Ensemble firing rates during the intertrial interval (ITI – 3 s prior to sample presentation, SP), ±1.5 s each at the sample response (SR), last nosepoke (LNP), and nonmatch response (NR) (shaded bars) for 10 simultaneously recorded neurons were sorted according to combinations of behavioral events (phase, lever position, and performance). Cross-covariances between neurons, time points, and events were calculated, and eigenvectors were extracted which corresponded to the coefficients of canonical root discriminant functions. Discriminant scores were calculated by weighting and summing the firing rate of each neuron at each point in time by the respective root coefficient.[7,17] Continuous discriminant scores were recalculated using the same coefficients as above, at 3-s intervals throughout the DNMS trial (open bars), successively throughout the timecourse of the DNMS trial (arrow). (B) Continuous discriminant scores plotted to show the presence or absence of ensemble encoding (represented by the canonical root) at various times throughout the DNMS trial. Scores were calculated for the discriminant function (canonical root 4 – CAN4) that always discriminated left from right lever response position regardless of task phase. Note that encoding for CAN4 peaked during sample and "crossed over" during the delay and subsequent nonmatch in accordance with the response position contingency of the DNMS task. (C) Scores derived from canonical root 5 (CAN5) that encoded the same value in both sample and nonmatch for a given type of trial (i.e., left sample-right nonmatch vs. right sample-left nonmatch). Encoding for CAN5 peaked during the sample, was diminished during the early part of the delay, then increased again at longer delays and during the nonmatch phase. Note that encoding was negligible for both roots during the ITI.

type trial. Other CDA components also behaved in a similarly predictable manner in other portions of the trial. The "sliding CDA" thus provided insight into the firing of the ensemble with respect to how key variance sources fluctuate over an entire trial. It is difficult to resist the notion that these factors are in some manner responsible for behavioral performance as regulated by the "ensemble code." Analysis of error trials revealed that these patterns were disrupted in predictable ways on error trials.[18,57]

The "sliding CDA" technique shows how it is possible to modify a multivariate analysis, which already provides directly relevant information regarding ensemble encoding of behavior, to provide even greater insight into how the ensemble operates with respect to the discrete behavioral events at times when those events are not occurring. By assessing the intervals between fixed events, the manner in which the ensemble code is developed or fluctuates with respect to other factors within the trial becomes apparent. Ordinarily, to find patterns of ensemble activity between events (i.e., not time-locked) would require either a massive search algorithm, as originally proposed by Gerstein et al.,[37,38,60] or nonlinear analysis techniques.[41,59,61] The sliding CDA approach provides a means of approximating the relevant outcomes of both of these without the labor-intensive calculations.

11.7 Analyzing Neural Ensemble Activity During Behavior: Which Analyses?

The above discussion of serially reconstructed ensembles, population vector analysis, and cross-correlational analyses should not be taken as a condemnation of these techniques. Rather, each of these analyses can provide valuable insight into ensemble activity if the respective limitations are known and accounted for. The same applies, of course, for the multivariate approach. Investigations of ensemble activity provides critical information at the level of circuit and network properties underlying information encoding, representation, transformation, and retrieval. Other analysis, such as assessing coherent oscillations between neurons and structures,[62] population statistics,[6,8,21,27,54,56] mapping of temporal changes in ensemble activity across different phases of a behavioral trial,[63] and derivation of population-based neural codes,[8,11,17,18,41,57] etc., will provide more detailed assessments of ensemble correlates of behavior in the future.[59] We have only begun to understand the significance of this approach for understanding the neural basis of behavior. Future studies will provide several important links between multiple neuronal representations at the single cell level and the significance of this combined representation for cognitive and behavioral processes.

Acknowledgments

The preparation of this manuscript and the work discussed in it were supported by grants to R. Hampson and S. Deadwyler from NIDA (DA08549 REH, DA03502, DA00119 SAD). The assistance and technical support of D.R. Byrd, J.K. Konstantopoulos, J. Brooks, and T. Bunn is greatly appreciated.

References

1. Freund, T. F. and Buzsaki, G. Interneurons of the hippocampus. *Hippocampus,* 6, 347, 1996.
2. Schultz, W., Apicella, P., and Ljungberg, T. Responses of monkey dopamine neurons to reward and conditioned stimuli during successive steps of learning a delayed response task. *J. Neurosci.*, 13, 900, 1993.

3. Miller, E. K., Li, L., and Desimone, R. Activity of neurons in anterior inferior temporal cortex during a short-term memory task. *J. Neurosci.,* 13, 1460, 1993.
4. Miller, E. K., Li, L., and Desimone, R. A neural mechanism for working and recognition memory in inferior temporal cortex. *Science*, 254, 1377, 1991.
5. Eskandar, E. N., Richmond, B. J., and Optican, L. M. Role of inferior temporal neurons in visual memory: I. Temporal encoding of information about visual images, recalled images, and behavioral content. *J. Neurophysiol.,* 68, 1277, 1992.
6. Gochin, P. M., Colombo, M., Dorfman, G. A., Gerstein, G. L., and Gross, C. G. Neural ensemble coding in inferior temporal cortex. *J. Neurophysiol.,* 71, 2325, 1994.
7. Hampson, R. E., Heyser, C. J., and Deadwyler, S. A. Hippocampal cell firing correlates of delayed-match-to-sample performance in the rat. *Behav. Neurosci.,* 107, 715, 1993.
8. Deadwyler, S. A., Bunn, T., and Hampson, R. E. Hippocampal ensemble activity during spatial delayed-nonmatch-to-sample performance in rats. *J. Neurosci.,* 16, 354, 1996.
9. Hampson, R. E. and Deadwyler, S. A. Ensemble codes involving hippocampal neurons are at risk during delayed performance tests. *Proc. Natl. Acad. Sci. U.S.A.,* 93, 13487, 1996.
10. Hampson, R. E. and Deadwyler, S. A. LTP and LTD and the encoding of memory in small ensembles of hippocampal neurons, in *Long-term Potentiation, Volume 3*, Ed. M. Baudry, J. Davis, MIT Press, Cambridge, MA, 1996, 199.
11. Deadwyler, S. A. and Hampson, R. E. Ensemble activity and behavior: What's the code? *Science*, 270, 1316, 1995.
12. Shannon, C. E. Monograph B-1598, v.27, in *Bell System Technical Journal*, 1948.
13. Hamming, R. W. *Coding and Information Theory*, Prentice-Hall, Englewood Cliffs, NJ, 1986.
14. Georgopoulos, A. P., Lurito, J. T., Petrides, M., Schwartz, A. B., and Massey, J. T. Mental rotation of the neuronal population vector. *Science*, 243, 234, 1989.
15. Georgopoulos, A. P. Current issues in directional motor control. *TINS*, 18, 506, 1995.
16. Gawne, T. J. and Richmond, B. J. How independent are the messages carried by adjacent inferior temporal cortical neurons? *J. Neurosci.,* 13, 2758, 1993.
17. Hampson, R. E., Rogers, G., Lynch, G., and Deadwyler, S. A. Facilitative effects of the ampakine CX516 on short-term memory in rats: Correlations with hippocampal ensemble activity. *J. Neurosci.,* 18, 248, 1998.
18. Hampson, R. E., Rawley, J., Simeral, J. D., Byrd, D. R., Brooks, J. K., and Deadwyler, S. A. Disruption of encoding but not recognition by cannabinoids via differential action on hippocampal memory circuits. *Soc. Neurosci. Abstr.*, 23, 509, 1997.
19. Pellizzer, G., Sargent, P., and Georgopoulos, A. P. Motor cortical activity in a context-recall task. *Science*, 269, 702, 1995.
20. Schwartz, A. B. Motor cortical activity during drawing movements: population representation during sinusoid tracing. *J. Neurophysiol.,* 70, 28, 1993.
21. DiLorenzo, P. M. Across unit patterns in the neural response to taste: Vector space analysis. *J. Neurophysiol.,* 62, 823, 1989.
22. Georgopoulos, A. P., Kalaska, J. F., Caminiti, R., and Massey, J. T. On the relations between the direction of two-dimensional arm movements and cell discharge in primate motor cortex. *J. Neurosci.,* 2, 1982.
23. Georgopoulos, A. P., Schwartz, A. B., and Kettner, R. E. Neuronal population encoding of movement direction. *Science*, 233, 1416, 1986.

24. Alexander, G. E. and Crutcher, M. D. Preparation for movement: neural representations of intended direction in three motor areas of the monkey. *J. Neurophysiol.,* 64, 133, 1990.
25. Funahashi, S., Bruce, C. J., and Goldman-Rakic, P. S. Mnemonic coding of visual space in the monkey's dorsolateral prefrontal cortex. *J. Neurophysiol.,* 61, 331, 1989.
26. Wilson, M. A. and McNaughton, B. L. Dynamics of the hippocampal ensemble code for space. *Science,* 261, 1055, 1993.
27. Young, M. P. and Yamane, S. Sparse population coding of faces in the inferotemporal cortex. *Science,* 256, 1327, 1992.
28. Kuwada, S., Batra, R., Yin, T. C., Oliver, D. L., Haberly, L. B., and Stanford, T. R. Intracellular recordings in response to monaural and binaural stimulation of neurons in the inferior colliculus of the cat. *J. Neurosci.,* 17, 7565, 1997.
29. Fitzpatrick, D. C., Batra, R., Stanford, T. R., and Kuwada, S. A neuronal population code for sound localization. *Nature,* 388, 871, 1997.
30. Salinas, E. and Abbott, L. F. Vector reconstruction from firing rates. *J. Comput. Neurosci.,* 1, 89, 1994.
31. Lee, D., Port, N. P., Kruse, W., and Georgopoulos, A. P. Neuronal population coding: Multielectrode recording in primate cerebral cortex, in *Neuronal Ensembles: Strategies for Recording and Decoding,* Eds. H. Eichenbaum, J. Davis, Wiley, New York, 1998, 117.
32. Georgopoulos, A. P. Population activity in the control of movement. *Int. Rev. Neurobiol.,* 37, 103, 1994.
33. Eichenbaum, H. Thinking about brain assemblies. *Science,* 261, 993, 1993.
34. Perkel, D. H., Gerstein, G. L. and Moore, G. P. Neuronal spike trains and stochastic point processes: II Simultaneous spike trains. *Biophys. J.,* 7, 419, 1967.
35. Gerstein, G. L. and Perkel, D. H. Simultaneously recorded trains of action potentials: analysis and functional interpretation. *Science,* 164, 828, 1969.
36. Gerstein, G. L. and Perkel, D. H. Mutual temporal relationships among neuronal spike trains. Statistical techniques for display and analysis. *Biophys. J.,* 12, 453, 1972.
37. Abeles, M. and Gerstein, G. L. Detecting spatiotemporal firing patterns among simultaneously recorded single neurons. *J. Neurosci.,* 60, 909, 1988.
38. Gerstein, G. L., Perkel, D. H., and Subramanian, K. N. Identification of functionally related neural assemblies. *Brain Res.,* 140, 43, 1978.
39. Skaggs, W. E. and McNaughton, B. L. Replay of neuronal firing sequences in rat hippocampus during sleep following spatial exposure. *Science,* 271, 1870, 1996.
40. Schwartz, A. B. and Adams, J. L. A method for detecting the time course of correlation between single-unit activity and EMG during a behavioral task. *J. Neurosci. Meth.,* 58, 127, 1995.
41. Deadwyler, S. A. and Hampson, R. E. The significance of neural ensemble codes during behavior and cognition, in *Annual Review of Neuroscience,* Vol. 20, Eds. W. M. Cowan, E. M. Shooter, C. F. Stevens, R. F. Thompson, Annual Reviews, Inc., Palo Alto, CA, 1997, 217.
42. Aertsen, H. J. and Gerstein, G. L. Evaluation of neuronal connectivity: Sensitivity of cross-correlation. *Brain Res.,* 340, 341, 1985.
43. Aertsen, H. J., Gerstein, G. L., Habib, M. K., and Palm, G. Dynamics of neuronal firing correlation: Modulation of "effective connectivity." *J. Neurophysiol.,* 61, 900, 1989.

44. O'Keefe, J. and Recce, M. L. Phase relationship between hippocampal place units and the EEG theta rhythm. *Hippocampus*, 3, 317, 1993.
45. Wilson, M. A. and McNaughton, B. L. Reactivation of hippocampal ensemble memories during sleep. *Science*, 265, 676, 1994.
46. Gothard, K. M., Skaggs, W. E., Moore, K. M., and McNaughton, B. L. Binding of hippocampal CA1 neural activity to multiple reference frames in a landmark-based navigation task. *J. Neurosci.*, 16, 823, 1996.
47. O'Keefe, J. and Burgess, N. Geometric determinants of the place fields of hippocampal neurons [see comments]. *Nature*, 381, 425, 1996.
48. Burgess, N. and O'Keefe, J. Neuronal computations underlying the firing of place cells and their role in navigation. *Hippocampus*, 6, 749, 1996.
49. Tonegawa, S., Tsien, J. Z., McHugh, T. J., Huerta, P., Blum, K. I., and Wilson, M. A. Hippocampal CA1-region-restricted knockout of NMDAR1 gene disrupts synaptic plasticity, place fields, and spatial learning. *Cold. Spring. Harb. Symp. Quant. Biol.*, 61, 225, 1996.
50. McHugh, T. J., Blum, K. I., Tsien, J. Z., Tonegawa, S., and Wilson, M. A. Impaired hippocampal representation of space in CA1-specific NMDAR1 knockout mice [see comments]. *Cell*, 87, 1339, 1996.
51. Tsodyks, M. V., Skaggs, W. E., Sejnowski, T. J., and McNaughton, B. L. Population dynamics and theta rhythm phase precession of hippocampal place cell firing: a spiking neuron model. *Hippocampus*, 6, 271, 1996.
52. Skaggs, W. E., McNaughton, B. L., Wilson, M. A., and Barnes, C. A. Theta phase precession in hippocampal neuronal populations and the compression of temporal sequences. *Hippocampus*, 6, 149, 1996.
53. Hampson, R. E., Byrd, D. R., Konstantopoulos, J. K., Bunn, T., and Deadwyler, S. A. Hippocampal place fields and spike-train correlation: Relationship between degree of place field overlap and cross-correlation. *Hippocampus*, 6, 281, 1996.
54. Schoenbaum, G. and Eichenbaum, H. Information coding in the rodent prefrontal cortex. II. Ensemble activity in orbitofrontal cortex. *J. Neurophysiol.*, 74, 751, 1995.
55. Nicolelis, M. A., Lin, R. C., and Chapin, J.K. Neonatal whisker removal reduces the discrimination of tactile stimuli by thalamic ensembles in adult rats. *J. Neurophysiol.*, 78, 1691, 1997.
56. Nicolelis, M. A. and Chapin, J. K. Spatiotemporal structure of somatosensory responses of many-neuron ensembles in the rat ventral posterior medial nucleus of the thalamus. *J. Neurosci.*, 14, 3511, 1994.
57. Deadwyler, S. A., Rogers, G., Lynch, G., and Hampson, R. E. Facilitated encoding of task-relevant events in ensembles of hippocampal neurons by the ampakine CX516 (Cortex Pharmaceuticals). *Soc. Neurosci. Abstr.*, 23, 509, 1997.
58. Stevens, J. *Applied Multivariate Statistics for the Social Sciences*, Lawrence Erlbaum Associates, Hillsdale, 1992, 273.
59. Hampson, R. E. and Deadwyler, S. A. Methods, results and issues related to recording neural ensembles, in *Neuronal Ensembles: Strategies for Recording and Encoding*, Eds. H. Eichenbaum, J. Davis, Wiley, New York, 1998, 207.
60. Dayhoff, J. E. and Gerstein, G. L. Favored patterns in spike trains. I. Detection. *J. Neurophysiol.*, 49, 1334, 1983.

61. Marmarelis, V. Z. Identification of nonlinear biological systems using Laguerre expansions of kernels. *Ann. Biomed. Eng.*, 21, 573, 1993.
62. Nicolelis, M. A., Baccalá, L. A., Lin, R. C., and Chapin, J. K. Sensorimotor encoding by synchronous neural ensemble activity at multiple levels of the somatosensory system. *Science*, 268, 1353, 1995.
63. Schultz, W. Behavior-related activity of primate dopamine neurons. *Rev. Neurol. (Paris)*, 150, 634, 1994.

Index

F

G

H

I

J

K

L

M

N

Q

R

S

T